Food Gels: Fabrication, Characterization, and Application

Food Gels: Fabrication, Characterization, and Application

Guest Editor

Hao Cheng

MDPI

Basel • Beijing • Wuhan • Barcelona • Belgrade • Novi Sad • Cluj • Manchester

Guest Editor
Hao Cheng
School of Food Science and
Technology
Jiangnan University
Wuxi
China

Editorial Office
MDPI AG
Grosspeteranlage 5
4052 Basel, Switzerland

This is a reprint of the Special Issue, published open access by the journal *Gels* (ISSN 2310-2861), freely accessible at: https://www.mdpi.com/journal/gels/special_issues/XYYQ8U372D.

For citation purposes, cite each article independently as indicated on the article page online and as indicated below:

Lastname, A.A.; Lastname, B.B. Article Title. *Journal Name* **Year**, *Volume Number*, Page Range.

ISBN 978-3-7258-5897-2 (Hbk)
ISBN 978-3-7258-5898-9 (PDF)
https://doi.org/10.3390/books978-3-7258-5898-9

Contents

About the Editor

Hao Cheng

Hao Cheng (Associate Professor): Hao Cheng is an associate professor in the State Key Laboratory of Food Science and Technology at Jiangnan University. He obtained his master's degree in food engineering from the School of Food Science and Technology, Jiangnan University, in 2016. He studied in the Faculty of Agriculture, Life and Environmental Sciences at the University of Alberta as a visiting scholar from September 2018 to October 2019. He obtained his Ph.D. in food science and technology from the School of Food Science and Technology, Jiangnan University, in 2020. Since 2020, he has worked as an associate professor in the State Key Laboratory of Food Science and Technology at Jiangnan University. His research mainly focuses on protein structure–function relationships, biopolymer-based nutraceutical delivery systems and the development of functional foods. During his work, individually and with colleagues, he has undertaken several national and industrial projects, including with the National Natural Science Foundation of China. He has published 50 research articles during the last 5 years, among which 2 research articles were selected as front covers. He has obtained six patents and co-edited two book chapters.

gels

Editorial

Food Gels: Fabrication, Characterization, and Application

Wanwen Chen [1,2,3] and Hao Cheng [4,*]

1 Wuxi Fisheries College, Nanjing Agricultural University, Wuxi 214081, China; chenwanwen@ffrc.cn
2 Key Laboratory of Integrated Rice-Fish Farming Ecology, Ministry of Agriculture and Rural Affairs, Freshwater Fisheries Research Center, Chinese Academy of Fishery Sciences, Wuxi 214081, China
3 Sino-US Cooperative International Laboratory for Germplasm Conservation and Utilization of Freshwater Mollusks, Freshwater Fisheries Research Center, Chinese Academy of Fishery Sciences, Wuxi 214081, China
4 State Key Laboratory of Food Science and Resources, School of Food Science and Technology, Jiangnan University, Wuxi 214122, China
* Correspondence: haocheng@jiangnan.edu.cn

1. Introduction

Food gels, typically formulated from proteins, polysaccharides, and lipids, are viscoelastic systems capable of entrapping water (hydrogels), oil (oleogels), and air (aero-gels) within their three-dimensional networks [1]. These gels play a crucial role in modern food formulation and production due to their versatile functional properties, including tailoring the structure of food to achieve a desired appearance, sensory attributes, and textures; incorporating bioactive compounds with high physicochemical stability and bioavailability; enabling the creation of customized food shapes via 3D printing; serving as fat replacers to reduce excessive saturated fatty acid, cholesterol, and calorie intake; and stabilizing metastable food structures and extending their shelf life [1–3]. The rational design of food gels can therefore enhance not only food quality and modification, but also its nutritional and health benefits. However, due to their inherent complexity compared to synthetic polymer gels, the relationships between material selection, fabrication strategies, microstructure, and mechanical and functional properties remain insufficiently understood in specific food systems. Moreover, as scientific research often overlooks practical considerations, bridging the gap between theory and application represents a critical challenge in scaling up the production of innovative gel-based foods with enhanced characteristics and functionalities.

This Special Issue comprises recent advancements in food gel technology and their applications across multiple domains, encompassing surimi-based seafood analogs that demand mechanically robust networks and dairy products requiring high emulsification stability and efficient nutrient integration. The innovation of composite gels, which incorporate bioactive components such as enzymatic crosslinkers (e.g., transglutaminase), polysaccharides (e.g., cassia bean gum and soy protein isolate), and nutritional fortificants (e.g., omega-3 fatty acids and iron microcapsules), has further expanded their functional capabilities and health-promoting potential. The research landscape encompasses critical advancements, including low-salt surimi gelation strategies utilizing amino acid supplements and high-pressure processing, emulsion restructuring in milk via homogenization and ultrasonic treatment, and the design of double emulsions ($W_1/O/W_2$) for enhanced nutrient encapsulation. Moreover, it addresses emerging sustainable resource utilization approaches, such as gelatin extraction from underutilized raw materials (e.g., Yanbian cattle bones, and crayfish muscle), biotherapeutic delivery through protein–polysaccharide complexes, and advanced processing technologies like microfluidization and membrane emulsification to improve texture and stability.

 https://doi.org/10.3390/gels11110886

The multidisciplinary nature of food gel research requires the integration of protein science, colloidal chemistry, process engineering, and nutritional biology. This interdisciplinary synergy has resulted in significant advances in understanding inter-component interactions, optimizing mechanical properties for targeted food matrices, and developing innovative crosslinking strategies that improve functionality without compromising nutritional quality. Through these developments, food gels continue to pioneer the creation of next-generation food products that support global health and environmental sustainability goals.

2. Overview of Papers Published in This Special Issue

This Special Issue, "Food Gels: Fabrication, Characterization, and Application", brings together nine research articles and two review papers highlighting recent advancements in food gels in terms of their fabrication, characterization, and application. These contributions explore innovative synthesis methods, novel material properties, and diverse applications in products including milk, margarines, Surimi, meat, and bioactive delivery systems.

The study "A Comparative Study of Soy Protein Isolate-κ-Carrageenan Emulsion Gels and Bigels for the Encapsulation, Protection, and Delivery of Curcumin" by Gray et al. examines curcumin delivery performance across soy protein isolate-κ-carrageenan emulsion gels and bigels incorporating glycerol monostearate. The research demonstrates that bigels with optimized glycerol monostearate content significantly enhance curcumin stability and controlled release profiles compared to emulsion gels. Bigels exhibited superior curcumin retention during storage and delayed gastric release while promoting intestinal liberation through improved lipolysis. The findings highlight bigels' structural advantages for hydrophobic polyphenol delivery in functional food applications.

The study "Improvement of Gel Properties of *Nemipterus virgatus* Myofibrillar Protein Emulsion Gels by Curdlan" by Wu et al. examines the enhancement of emulsion gel properties through the incorporation of curdlan (Cur) into myofibrillar protein (MP) matrices and its application in emulsified surimi. The research demonstrates that Cur, primarily interacting with MP via hydrogen bonding, promotes the formation of a uniform and dense composite network structure, significantly improving adsorption capacity at the oil/water interface. At an optimal concentration of 6% (w/v), Cur substantially enhances the gel's hardness, strength, water-holding capacity, and viscoelastic properties. Furthermore, the application of Cur/MP emulsion gels in surimi products outperforms direct oil addition, yielding superior texture, gel strength, and water retention. These findings advance the development of protein–polysaccharide emulsion gels and highlight their potential as functional fat substitutes in seafood products.

The study "Encapsulation of *Lactobacillus reuteri* in Chia-Alginate Hydrogels for Whey-Based Functional Powders" by Cid-Córdoba et al. examines the efficacy of electrohydrodynamic atomization (EHDA) and drip mode (DM) encapsulation techniques for protecting *Lactobacillus reuteri* DSM 17938 within chia mucilage–sodium alginate hydrogels, and their integration into whey-based functional beverage powders. Both EHDA and DM achieved a high encapsulation efficiency (99.0%) and maintained bacterial viability above 9.9 $\log_{10}$ CFU/mL post-lyophilization, demonstrating the hydrogel's exceptional protective capacity. Microstructural analysis revealed well-preserved cell morphology and homogeneous distribution within the polymer matrix, with SEM images confirming spherical, porous microcapsules with surface characteristics influenced by the encapsulation method.

The study "Development, Characterization, and Stability of Margarine Containing Oleogels Based on Olive Oil, Coconut Oil, Starch, and Beeswax" by Naves et al. examines the formulation of margarines using oleogels structured with extra virgin olive oil, coconut oil, corn starch, and beeswax as alternatives to conventional saturated and *trans*-fat-rich spreads. Their key findings reveal that oleogel-based margarines exhibit higher melting temperatures (46–49 °C) and broader melting ranges due to beeswax's structuring abilities, alongside lower enthalpy values (1.9–2.8 mW) attributed to the higher unsaturated oil content. A color analysis showed a distinct greenish hue from the olive oil, contrasting with commercial samples. Crucially, the oleogel margarines displayed superior thermal stability, resisting oil exudation during thermal cycling, while their commercial counterparts suffered phase separation. The microstructural differences included larger, irregular water droplets in the oleogels, yet their oxidative stability remained acceptable (peroxide values $\leq$ 9.73 mEq O_2/kg) owing to olive oil's natural antioxidants. This research confirms the viability of oleogel-based margarines as functional, healthier options with enhanced thermal performance, oxidative resistance, and reduced saturated fat content.

The paper "Development and Application of Anthocyanin-Based Complex Polysaccharide Gels Based on Blueberry Pomace for Monitoring Beef Freshness" by Zhi et al. examines the fabrication of pH-responsive gels using anthocyanins extracted from blueberry pomace incorporated into chitosan/polyvinyl alcohol (CS/PVA) and starch/PVA (S/PVA) matrices for real-time beef freshness monitoring. The study demonstrated that CS/PVA-BA gels achieved optimal elongation at break (high flexibility), low hydration (8.33% water content), and potent antioxidant activity, while S/PVA-BA gels exhibited superior tensile strength and enhanced pH-sensitive colorimetric responses. Structural analyses confirmed molecular compatibility through hydrogen bonding between anthocyanins and polymer networks. Applied to chilled beef storage at 4 °C, the gels displayed visible color transitions from magenta-red (initial spoilage at 48 h) to bluish-gray (advanced spoilage by day 6), correlating with biochemical spoilage markers (TVB-N > 15 mg/100 g, TVC > 4.0 lg CFU/g). These findings establish a multifunctional platform for intelligent packaging that integrates real-time freshness indication with antioxidant protection, highlighting the potential of waste-derived anthocyanins in sustainable food safety solutions.

The paper "A Mathematical Model of Myosin Heavy Chain Dynamics in the Disintegration of Golden Threadfin Bream *Nemipterus virgatus* Surimi Gel" by Nakamizo et al. investigates the disintegration mechanism of surimi gel in golden threadfin bream, a species characterized by low transglutaminase activity and high protease activity at elevated temperatures. The study focuses on the competition between non-enzymatic polymerization and proteolytic degradation of myosin heavy chain (MHC), a key protein governing gel network formation. Using a kinetic model based on SDS-PAGE analysis of MHC dynamics during heating at 60 °C, the research demonstrates that the model accurately captures MHC depletion, revealing significant degradation of both unpolymerized and polymerized MHC. This degradation directly correlates with the reduced mechanical strength of the gel, highlighting the role of proteolytic activity in gel disintegration. The mathematical framework provides a predictive tool for optimizing heating conditions and controlling the surimi gel's properties, facilitating the application of underutilized fish species in surimi processing. This work advances the understanding of protein dynamics in food gels and offers practical strategies for improving the quality of seafood products.

The paper "High-Quality Application of Crayfish Muscle in Surimi Gels: Fortification of Blended Gels by Transglutaminase" by Wang et al. investigates the integration of crayfish muscle (0–10%) into silver carp surimi gels and the compensatory role of transglutaminase (TGase) in mitigating structural deterioration. The study demonstrates that crayfish muscle incorporation without TGase progressively reduced the gel's strength, water-holding capacity (WHC), and structural homogeneity due to disrupted hydrogen bonding, hydrophobic interactions, and disulfide bond formation, as evidenced by decreased whiteness, increased porosity, and free water migration. Conversely, 0.6% TGase addition counteracted these effects by catalyzing ε-(γ-Glu)-Lys crosslinks, enhancing protein aggregation, and promoting shifts in secondary structures from α-helices to β-turns, resulting in denser gel networks, improved WHC, and restored mechanical properties, even at 7.5% crayfish inclusion. TGase-mediated restructuring optimized water distribution and reduced electrostatic repulsion, facilitating disulfide bond formation despite the activity of crayfish-derived proteases. This synergy enables utilization of high-value crayfish byproducts in surimi products, offering a strategy to enhance texture, moisture retention, and sustainability in seafood processing.

The paper "Physicochemical and Functional Properties of Yanbian Cattle Bone Gelatin Extracted Using Acid, Alkaline, and Enzymatic Hydrolysis Methods" by Zhang et al. examines the extraction efficiency, structural integrity, and functional performance of gelatin derived from Yanbian cattle bones using three distinct methods: acid hydrolysis (with hydrochloric acid), alkaline hydrolysis (with sodium hydroxide), and enzymatic hydrolysis (with papain). The study demonstrated that enzymatic hydrolysis with papain achieved the highest yield (25.25%) and optimally preserved collagen's native structure, resulting in superior hydroxyproline content (19.13 g/100 g), gel strength (259 g), viscosity (521.67 cP), and thermal stability compared to acid and alkaline methods. Structural analyses confirmed that papain extraction minimized protein degradation and maintained the triple-helical conformation, while the amino acid composition revealed enhanced levels of functional residues (e.g., glycine, proline, and hydroxyproline). These findings highlight enzymatic hydrolysis as a mild and efficient approach for producing high-quality, halal-compliant gelatin from underutilized bone byproducts, offering a sustainable alternative for food, pharmaceutical, and biomedical applications.

The paper "Preparation of Cassia Bean Gum/Soy Protein Isolate Composite Matrix Emulsion Gel and Its Effect on the Stability of Meat Sausage" by Zou et al. examines the development and application of emulsion gels formulated with soy protein isolate (SPI) and varying concentrations of cassia bean gum (CG) (0–2%) as fat substitutes in meat sausages. The study demonstrated that emulsion gels with 1.75% CG concentration exhibited optimal structural properties, including the highest gel strength (586.91 g), elasticity (0.94 mm), chewiness (452.94 mJ), water-holding capacity (98.45%), and thermal stability, attributed to enhanced hydrogen bonding and a compact, homogeneous microstructure. When applied as fat replacers in meat sausages, the 1.75% CG/SPI emulsion gel at a 50% substitution level maintained cooking loss, emulsification stability, color, texture, and antioxidant activity comparable to full-fat sausages, while also improving freeze–thaw stability. These findings highlight the potential of CG/SPI emulsion gels as effective fat alternatives for producing low-fat meat products with preserved quality and enhanced functional properties.

The review "Surimi and Low-Salt Surimi Gelation: Key Components to Enhance the Physicochemical Properties of Gels" by Walayat et al. examines strategies to mitigate the textural deterioration of surimi gels in low-salt (reduced NaCl) formulations, focusing on the role of gelation enhancers such as microbial transglutaminase, polyphenols, phosphates, hydrocolloids, amino acids (e.g., *L*-arginine, *L*-lysine), and plant/animal proteins. The study demonstrates that these additives compensate for reduced salt by promoting protein crosslinking, hydrogen bonding, and hydrophobic interactions, thereby improving gel strength, water-holding capacity, and structural integrity. Notably, specific amino acids and protein combinations effectively maintained the properties of the gel comparably to those of full-salt controls, highlighting their potential for producing healthier surimi products without compromising on techno-functional quality. These findings provide a practical framework for the industry to address sodium reduction challenges while meeting consumer demands for nutritious, low-salt seafood alternatives.

The review "Emulsion Structural Remodeling in Milk and Its Gelling Products" by Yao et al. examines the restructuring of milk's native oil-in-water (O/W) emulsion into three distinct types—restructured single emulsion, mixed emulsion, and double emulsion ($W_1/O/W_2$)—through the incorporation of alternative lipids (e.g., fish oil and flaxseed oil) or pre-formed emulsions, alongside the impact of processing technologies (including heat treatment, high-pressure processing, homogenization, ultrasonic treatment, microfluidization, freezing, and membrane emulsification) on microstructure and functional properties. The study demonstrated that these processing methods significantly reduce the dispersed phase size, modify the interfacial layer composition (e.g., enhancing protein adsorption or milk fat globule membrane integrity), and alter the aqueous phase morphology, thereby improving emulsion stability, shelf-life, and sensory characteristics while enabling nutrient encapsulation in double emulsions. This structural remodeling approach facilitates the development of low-fat, nutritionally fortified dairy products with tailored textural and functional attributes, addressing consumer demands for health-focused and sustainable food options.

3. Conclusions and Future Perspectives

I would like to acknowledge all the authors who have contributed to this Special Issue, which underscores the significant advancements made in modifying food gels, highlighting their versatility in food applications such as surimi products, meat sausages, dairy items, and gelatin-based materials. The investigations included delve into fundamental gelation mechanisms, advanced processing techniques, and innovative additive strategies to enhance physicochemical properties, reduce salt content, and improve nutritional profiles, all while addressing sustainability and health concerns. Key technological strides include the development of enzymatic hydrolysis methods, like papain extraction from Yanbian cattle bones, which preserved native collagen structures and achieved superior yield, gel strength, viscosity, and thermal stability compared to traditional acid or alkaline approaches. Additionally, the integration of transglutaminase (TGase) and polysaccharides, such as cassia bean gum, fortified gel networks in surimi and meat products, boosting their water-holding capacity, reducing their porosity, and mitigating salt reduction effects through enhanced protein crosslinking. Non-thermal processing technologies, including high-pressure processing, ultrasonication, and microfluidization, optimized emulsion stability in milk-based systems by reducing fat globule size and improving sensory properties without nutrient degradation. Low-salt formulations utilizing amino acids (e.g., *L*-arginine) and hydrocolloids maintained gel integrity and functionality, addressing health risks linked to sodium intake, while the structural remodeling of milk emulsions (single, mixed, and double types) through homogenization and membrane emulsification enabled functional foods with improved nutrient delivery, oxidative stability, and tailored textures for specific consumer needs.

Looking ahead, future research should prioritize scaling up innovative processing techniques, ensuring cost-effectiveness and reproducibility in large-scale food production. This entails developing standardized protocols for evaluating the safety and regulatory compliance of new additives to facilitate faster market entry. Advancing molecular-level understanding of protein–polysaccharide–lipid interactions during gelation will allow for tailored textural customization, particularly for demographic-specific products like those for elderly populations or health-focused diets. Exploring sustainable raw material sources, such as underutilized fish species and agricultural byproducts, can reduce environmental impact and promote circular economy practices. Comprehensive in vivo and clinical studies are needed to validate health benefits like antioxidant or antihypertensive effects, strengthening evidence-based nutritional claims. Leveraging digital technologies, including artificial intelligence and machine learning, will optimize formulation design and predict gel behavior under varying conditions, enabling personalized nutrition solutions.

The interdisciplinary collaboration among food scientists, engineers, nutritionists, and regulatory specialists remains crucial for translating laboratory innovations into commercially viable products that meet evolving consumer demands for healthy, sustainable, and high-quality food options, ultimately contributing to global food security and public health.

Funding: This research received no external funding.

Acknowledgments: The editors sincerely acknowledge the authors' valuable contributions to this field and hope that their work will garner growing recognition and impact in the future. We also thank the Editorial Office of *Gels* for their support throughout the review process.

Conflicts of Interest: The authors declare no conflicts of interest.

List of Contributions:

1. Gray, E.T.; Huang, W.; Zhou, Z.; Cheng, H.; Liang, L. A Comparative Study of Soy Protein Isolate-κ-Carrageenan Emulsion Gels and Bigels for the Encapsulation, Protection, and Delivery of Curcumin. *Gels* **2025**, *11*, 782. https://doi.org/10.3390/gels11100782.
2. Wu, Z.; Qu, Y.; Li, O.; Benjakul, S.; Zhou, A. Improvement of Gel Properties of Nemipterus virgatus Myofibrillar Protein Emulsion Gels by Curdlan: Development and Application to Emulsified Surimi. *Gels* **2025**, *11*, 753. https://doi.org/10.3390/gels11090753.
3. Cid-Córdoba, A.Y.; Calderón-Domínguez, G.; Perea-Flores, M.d.J.; Peña-Barrientos, A.; Serrano-Villa, F.S.; Barrios-Francisco, R.; González-Vázquez, M.; Rentería-Ortega, M. Encapsulation of *Lactobacillus reuteri* in Chia–Alginate Hydrogels for Whey-Based Functional Powders. *Gels* **2025**, *11*, 613. https://doi.org/10.3390/gels11080613.
4. Naves, B.V.B.; da Silva, T.L.T.; Nunes, C.A.; Haddad, F.F.; Bastos, S.C. Development, Characterization, and Stability of Margarine Containing Oleogels Based on Olive Oil, Coconut Oil, Starch, and Beeswax. *Gels* **2025**, *11*, 513. https://doi.org/10.3390/gels11070513.
5. Zhi, J.; Xu, F.; Yu, S.; Hao, J.; Wang, J.; Fan, Z. Development and Application of Anthocyanin-Based Complex Polysaccharide Gels Based on Blueberry Pomace for Monitoring Beef Freshness. *Gels* **2025**, *11*, 385. https://doi.org/10.3390/gels11060385.
6. Nakamizo, R.; Hayashi, T.; Kominami, Y.; Ushio, H. A Mathematical Model of Myosin Heavy Chain Dynamics in the Disintegration of Golden Threadfin Bream *Nemipterus virgatus* Surimi Gel. *Gels* **2025**, *11*, 348. https://doi.org/10.3390/gels11050348.
7. Wang, H.; Li, Q.; Yang, M.; Wang, H.; Wang, M.; Lin, L.; Lu, J. High-Quality Application of Crayfish Muscle in Surimi Gels: Fortification of Blended Gels by Transglutaminase. *Gels* **2025**, *11*, 204. https://doi.org/10.3390/gels11030204.
8. Zhang, S.; Zhao, D.; Yin, L.; Wang, R.; Jin, Z.; Xu, H.; Xia, G. Physicochemical and Functional Properties of Yanbian Cattle Bone Gelatin Extracted Using Acid, Alkaline, and Enzymatic Hydrolysis Methods. *Gels* **2025**, *11*, 186. https://doi.org/10.3390/gels11030186.
9. Zou, Q.; Zheng, Y.; Liu, Y.; Luo, L.; Chen, Y.; Ran, G.; Liu, D. Preparation of Cassia Bean Gum/Soy Protein Isolate Composite Matrix Emulsion Gel and Its Effect on the Stability of Meat Sausage. *Gels* **2024**, *10*, 643. https://doi.org/10.3390/gels10100643.
10. Walayat, N.; Blanch, M.; Moreno, H.M. Surimi and Low-Salt Surimi Gelation: Key Components to Enhance the Physicochemical Properties of Gels. *Gels* **2025**, *11*, 142. https://doi.org/10.3390/gels11020142.
11. Yao, D.; Sun, L.-C.; Zhang, L.-J.; Chen, Y.-L.; Miao, S.; Cao, M.-J.; Lin, D. Emulsion Structural Remodeling in Milk and Its Gelling Products: A Review. *Gels* **2024**, *10*, 671. https://doi.org/10.3390/gels10100671.

References

1. Mao, L.; Lu, Y.; Cui, M.; Miao, S.; Gao, Y. Design of gel structures in water and oil phases for improved delivery of bioactive food ingredients. *Crit. Rev. Food Sci. Nutr.* **2020**, *60*, 1651–1666. [CrossRef] [PubMed]

2. Cao, Y.; Mezzenga, R. Design principles of food gels. *Nat. Food* **2020**, *1*, 106–118. [CrossRef] [PubMed]
3. Liu, F.; Liang, X.; Yan, J.; Zhao, S.; Li, S.; Liu, X.; Ngai, T.; McClements, D.J. Tailoring the properties of double-crosslinked emulsion gels using structural design principles: Physical characteristics, stability, and delivery of lycopene. *Biomaterials* **2022**, *280*, 121265. [CrossRef] [PubMed]

gels

MDPI

Article

Preparation of Cassia Bean Gum/Soy Protein Isolate Composite Matrix Emulsion Gel and Its Effect on the Stability of Meat Sausage

Qiang Zou [1,2,†], Yuhan Zheng [1,2,†], Yudie Liu [1,2], Linghui Luo [1,2], Yuyou Chen [1,2], Guilian Ran [1,2] and Dayu Liu [1,2,*]

1 School of Food and Biological Engineering, Chengdu University, Chengdu 610106, China; zouqiang@cdu.edu.cn (Q.Z.); zhengyuhan@stu.cdu.edu.cn (Y.Z.); liuyudie@stu.cdu.edu.cn (Y.L.); luolinghui@stu.cdu.edu.cn (L.L.); chenyuyou@stu.cdu.edu.cn (Y.C.); ranguilian@stu.cdu.edu.cn (G.R.)
2 Meat Processing Key Laboratory of Sichuan Province, School of Food and Biological Engineering, Chengdu University, Chengdu 610106, China
* Correspondence: liudayu0912@163.com
† These authors contributed equally to this work.

Abstract: The use of plant-derived emulsified gel systems as fat substitutes for meat products has always been an important direction in the development of healthy foods. In this study, a composite matrix emulsion gel was prepared with soy protein isolate (SPI) and different concentrations of cassia bean gum (CG), and then the selected emulsion gel was applied to meat sausage as a fat substitute to explore its stability. Our results showed that the hardness, chewiness, viscosity, shear stress, and G′ and G″ moduli of the emulsion gel increased considerably with the cassia bean gum concentration, the thickness of the emulsion gel increased, and the pore size decreased. The gel strength of the 1.75% CG/SPI emulsion gel was the highest, which was 586.91 g. The elasticity was 0.94 mm, the masticability was 452.94 mJ, and the water-holding capacity (WHC) was 98.45%. Then, the 1.75% CG/SPI emulsion gel obtained via screening was applied as a fat substitute in meat sausage. With an increase in the substitution amount, the cooking loss, emulsification stability, pH, color difference, texture, and antioxidant activity of the meat sausage before and after freezing and thawing increased first and then decreased. The indexes of meat sausage with 50% fat replacement were not considerably different from those of full-fat meat sausage. This study can provide a theoretical basis for the application of plant-derived emulsified gel systems as fat substitutes in meat sausage.

Keywords: composite matrix emulsion gel; cassia bean gum; meat and sausage; fat substitute; freeze–thaw stability

1. Introduction

Meat sausage is a popular and frequently eaten processed meat product because of its high nutritional value, convenience, and unique flavor. Pig back fat is one of the main raw materials of meat sausage (accounting for approximately 30%). It not only improves the flavor of meat sausage but also affects its texture and sensory quality. However, excessive intake will increase the risk of diseases, such as coronary heart disease, cardiovascular disease, diabetes, and obesity [1], adversely affecting human health. In addition, traditional meat sausage retains oil and has poor freeze–thaw and oxidation stability because of excessive animal fat [2]. Thus, the meat industry has studied different lipid reduction technology strategies to produce healthier and more nutritious products while improving product stability.

Emulsion gels as animal fat substitutes have gradually attracted interest. They are gel-like solid materials with spatial network structures that are formed through an emulsion that is initiated using induction methods, which can effectively emulsify and disperse oil droplets [3]. The preparation of emulsion gels with proteins or polysaccharides often has limited practical applications because of the poor emulsification and thermal stability of the products [4]. Hence, many studies have focused on preparing composite matrix emulsion gels with excellent properties, not only enhancing the emulsifying effect of the emulsions but also improving the stability of the emulsion gels [5].

Soy protein isolate is one of the high-quality plant proteins that can replace animal protein, mainly consisting of soy globulin and β-conglycinin, and can be applied to foods to increase their protein content and provide desired functions such as gelation, emulsification, and foaming properties, where the ability to gel is a key functional property of SPI. The gelation process begins with protein denaturation, resulting in structural changes, the exposure of functional groups including sulfhydryl or hydrophobic groups, and then various interactions that trigger aggregation [6]. Moreover, the incorporation of polysaccharides can alter the intermolecular and/or intramolecular interactions involved in the complex, thus affecting the gelling function of the protein [7]. Zhang et al. [8] prepared emulsion gels containing soy protein isolate and pectin using the thermal sensitivity method, which effectively improved the texture, rheology, and microscopic and other functional properties of composite gels. For novel food development, protein–polysaccharide interactions are widely used to achieve certain food functions. Wang et al. [9] added an emulsion gel, which was prepared via the composite preparation of soy protein isolate and κ-carrageenan, to sausage in different proportions instead of pig backfat to reduce the content of fat and cholesterol, and the difference in texture and product yield was not significant compared with the control group. In this regard, there are few reports on emulsion gels being prepared using the combination of soybean protein isolate and cassia bean gum. Cassia bean gum is the polysaccharide of cassia seeds. It is a water-soluble polysaccharide that is extracted and isolated from the endosperm of mature seeds of Cassia obtusifolia or small cassia in the legumes. Its main structure is composed of galactomannan. It has been reported that cassia bean gum can be used as a fat substitute in food and can achieve thickening, freeze–thaw stability, synergistic gelatinization, film formation, etc. [10]. Cao et al. [11] made an edible oil packaging film from cassia bean gum, which has a higher barrier property and heat-sealing property. There are few reports on the application of cassia bean gum as a fat substitute. Generally, meat sausage is a frozen storage food, and its freeze–thaw stability and storage stability have an important impact on the product quality. However, the application of soy protein isolate/cassia bean gum composite matrix emulsion gel as a fat substitute in meat sausage has not been studied.

Therefore, in this study, soybean protein isolate and cassia bean gum were prepared into a protein–polysaccharide composite matrix emulsion gel using heat treatment, and the effects of cassia bean gum (0%, 0.85%, 1.15%, 1.45%, 1.75%, and 2%) at different concentrations on the rheological behavior, texture properties, microstructure, WHC, and thermal phase transformation temperature of the composite matrix emulsion gel were investigated. Then, the selected emulsion gel was used as a fat substitute and added into meat sausage at different replacement amounts (0%, 25%, 50%, 75%, and 100%). The texture, emulsification stability, color difference, and antioxidant activity during the storage of meat sausage before and after freezing and thawing were used as indicators to screen out the meat sausage with the best stable fat replacement amount. This study serves as a reference for the research and development of low-fat meat sausage products.

2. Results and Discussion

2.1. Texture Characteristics of the Emulsion Gel

As shown in Table 1, the hardness and chewiness of the emulsion gel increased with the CG concentration, increasing from 43.67 N to 586.91 N and 43.53 mJ to 452.94 mJ, respectively. When the CG concentration was greater than 1.75%, the properties decreased to 558.94 N and 339.89 mJ, respectively. Cassia bean gum may have promoted the interaction between the proteins and polysaccharides, forming a dense gel network structure [12]. Cassia bean gum at 2% reduced firmness and chewiness because the overfilling effect inhibited the aggregation of SPIs. This result is consistent with the previous results of Jiang et al.'s study [13] on the effect of κ-carrageenan on the texture of oyster protein treated with high pressure homogenization.

Table 1. Texture characteristics of emulsion gels.

Groups	Hardness (N)	Elasticity	Cohesiveness	Chewiness (mJ)
SPI/CG 0%	43.67 ± 0.16 f	0.99 ± 0.02 a	0.98 ± 0.01 a	43.53 ± 0.16 f
SPI/CG 0.85%	131.68 ± 1.22 e	0.98 ± 0.01 ab	0.91 ± 0.01 b	115.35 ± 4.16 e
SPI/CG 1.15%	221.50 ± 3.10 d	0.95 ± 0.01 c	0.82 ± 0.01 c	167.26 ± 8.16 d
SPI/CG 1.45%	384.67 ± 6.34 c	0.94 ± 0.01 c	0.81 ± 0.01 c	294.79 ± 4.41 c
SPI/CG 1.75%	586.91 ± 11.84 b	0.94 ± 0.01 bc	0.81 ± 0.01 c	452.94 ± 9.77 a
SPI/CG 2%	558.94 ± 9.14 a	0.93 ± 0.01 bc	0.80 ± 0.01 d	399.89 ± 7.23 b

Data are expressed as the mean ± standard deviation. a–f: different lowercase letters indicate significant differences ($p < 0.05$). SPI/CG0%: soy protein isolate, no cassia bean gum added; SPI/CG 0.85%: SPI + 0.85% cassia bean gum; SPI/CG 1.15%: SPI + 1.15% cassia bean gum; SPI/CG 1.45%: SPI + 1.45% cassia bean gum; SPI/CG 1.75%: SPI + 1.75% cassia bean gum; SPI/CG 2%: SPI + 2% cassia bean gum.

The initial elasticity and cohesion of the emulsion gel were 0.99 and 0.98, respectively (Table 1). As the CG concentration increased, the values gradually declined. The lowest values were 0.93 and 0.80, respectively, and the gradual decrease in cohesiveness indicated that damage to the irreversible structure of the sample increased after compression. The elasticity decreased, but the change was not considerable. The high chewiness of the sample indicated that the emulsion gel was mainly elastic. This may be because the electrostatic attraction between soy protein isolate increases the interface adsorption, protein aggregation enhances the viscoelasticity of the gel, and the addition of an appropriate polysaccharide increases the viscosity of the continuous phase; the structure of the emulsion gel is more stable, and the polysaccharide does not affect the protein-led gel behavior in the emulsion gel [14].

2.2. *Rheological Behavior of the Emulsion Gel*

The shear rate ranged from 0.01 s^{-1} to 10 s^{-1}, the viscosity of the emulsion gel decreased with an increasing shear rate, and the emulsion gel showed typical pseudoplastic fluid characteristics [15]. Different concentrations of CG showed similar effects on the emulsion gel's viscosity, and shear thinning behavior was observed at all concentrations possibly because the gel network structure in the emulsion was destroyed by shear force. Moreover, droplet migration resistance was reduced, and this effect prevented the dispersion and aggregation of oil droplets and decreased the apparent viscosity [16]. At the same shear rate, compared with the 0% CG/SPI emulsion gel, the emulsion gel with a higher CG concentration had higher viscosity, and the 2% CG/SPI emulsion gel had the highest viscosity possibly due to the cross-linking of CG and SPI after heating. This process increased the viscosity of the gel network structure and, thus, reduced the degree of shear thinning [17].

The energy storage modulus (G') and consumption modulus (G'') of emulsion gels with different concentrations of CG varied within a frequency range of 10–100 rad/s (Figure 1B). Different concentrations of cassia bean gum increased the G' values of the emulsion gels to different degrees, and the G' value increased with the cassia bean gum concentration. These changes indicated that the G' value of an emulsion gel was less dependent on the application frequency, and the network structure of the gel was stable and not easily destroyed [18]. In addition, the G' values of the emulsion gels with different concentrations of CG were higher than the G'' values, indicating that the emulsion gels were mainly elastic gel network structures. The 2% CG/SPI emulsion gels had large G' and G'' values and prominent viscoelastic properties possibly due to the noncovalent interactions between cassia bean gum and proteins as anion hydrophilic colloids. The flow of free water in the gel network was restricted, and the gel network was enhanced [19].

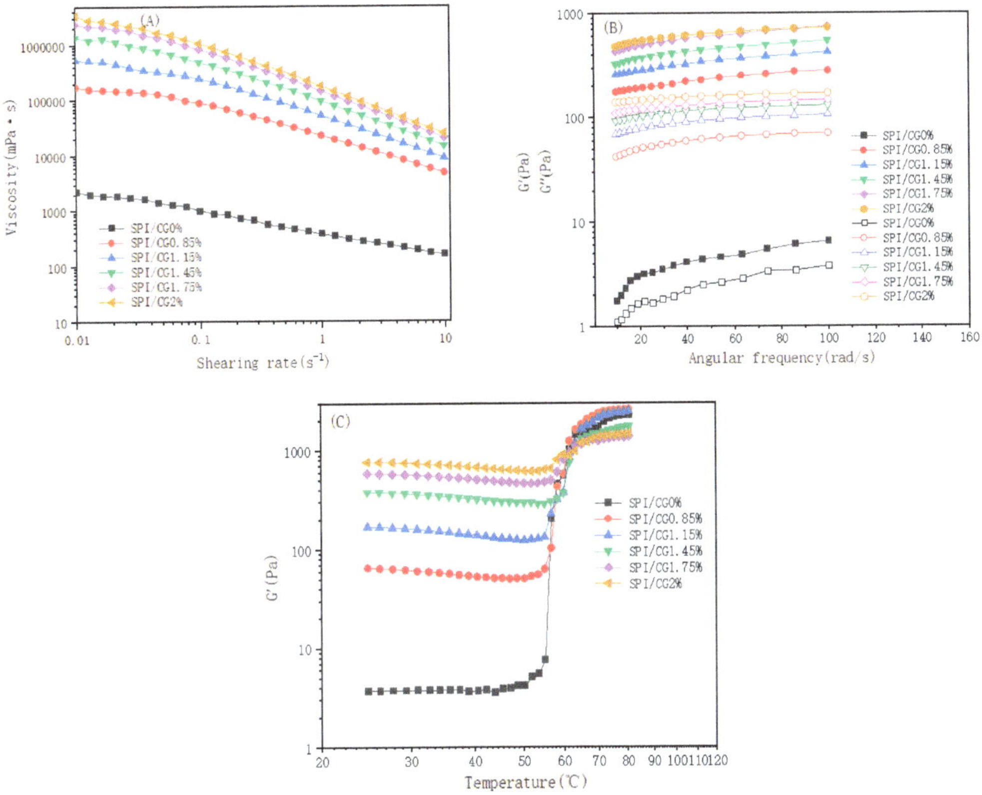

Figure 1. Rheological characteristics of emulsion gels with different concentrations of cassia bean gum. Effects of cassia bean gum on the viscosity (**A**), frequency scan (**B**), and temperature scan (**C**). Data are expressed as the mean ± standard deviation. SPI/CG 0%: soy protein isolate, no cassia bean gum added; SPI/CG 0.85%: SPI + 0.85% cassia bean gum; SPI/CG 1.15%: SPI + 1.15% cassia bean gum; SPI/CG 1.45%: SPI + 1.45% cassia bean gum; SPI/CG 1.75%: SPI + 1.75% cassia bean gum; SPI/CG 2%: SPI + 2% cassia bean gum. Different letters indicate significant differences in the data ($p < 0.05$).

Figure 1C shows the change in the G′ value of the SPI/CG emulsion gel at varying temperatures. The change in the G′ values of the composite gels presented two stages during heating. At 25–53 °C, the G′ value tended to be stable at an increasing temperature. The G′ value of the emulsion gels at this stage was much higher than that of the 0% CG/SPI emulsion gels, and the G′ value increased with the CG concentration. The interaction of hydrogen or ionic bonds between CG and SPI in the gel network enhanced the elasticity and hardness of the emulsion gel [20]. When the temperature was higher than 56 °C, the G′ value of the emulsion gel with different concentrations of CG increased considerably, and different degrees of "kinking" occurred on the G′ curve, that is, the melting and gelation behaviors of the gel, indicating that CG exerted a synergistic effect that enhanced the gel strength of the emulsion and helped to improve the final gel strength [21].

2.3. Water-Holding Capacity and Strength of the Emulsion Gel

The gel strength can directly reflect the gelling degree of a complex gel, is a key index for evaluating the gel structures of proteins and protein-based products, and is closely related to the WHC. As shown in Figure 2, the gel strength and WHC increased first and then decreased with an increasing CG concentration. The gel strength increased from 43.67 g to 586.91 g, and the highest gel strength of the emulsion gel (586.91 g) was obtained at a 1.75% CG concentration. The presence of polysaccharides could affect the microstructure and texture properties of the gel, which might be related to an exclusion effect of incompatible biopolymers in a mixed solution [22]. Moreover, an increase in the CG concentrations led to the increase in hydrogen-bonding groups, thus enhancing the internal network structure of the emulsion gels.

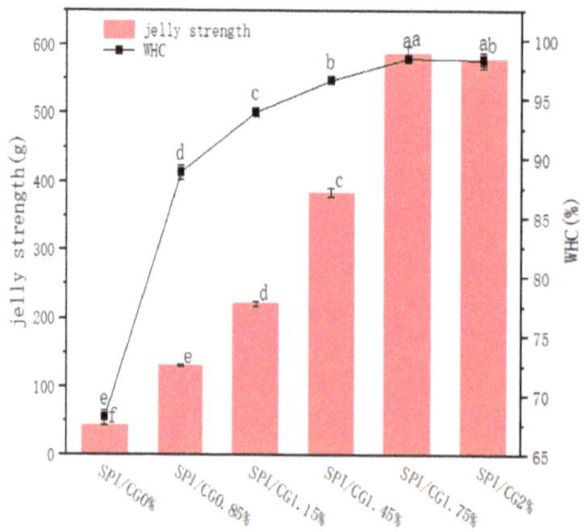

Figure 2. Effects of different concentrations of cassia bean gum on the water holding capacity and strength of emulsion gel. SPI/CG 0%: soy protein isolate, no cassia bean gum added; SPI/CG 0.85%: SPI + 0.85% cassia bean gum; SPI/CG 1.15%: SPI + 1.15% cassia bean gum; SPI/CG 1.45%: SPI + 1.45% cassia bean gum; SPI/CG 1.75%: SPI + 1.75% cassia bean gum; SPI/CG 2%: SPI + 2% cassia bean gum. Different letters indicate significant differences in the data ($p < 0.05$).

The WHC is one of the most important indexes for evaluating the structural quality of protein gel networks. The WHC increased from 68.17% to 98.45%, and a 30% increase in water retention rate was observed. These results are similar to those in Table 1. On the one hand, more and more CG filled the void space of the SPI, thus reducing the pore size of the gel network; on the other hand, the increase in CG concentrations led to the tighter adsorption of SPI with CG onto the oil–water interface, thus improving the compactness of the gel structure [23].

2.4. Observation of the Emulsion Gel Structure

As shown in Figure 3, the emulsion gels with different concentrations of CG were white, and the addition of 1.45% cassia bean gum resulted in the formation of a line dividing the different morphology of the composite gel. The addition of 1.45% or more CG can form a complete semisolid composite gel in positive and negative positions, whereas the composite emulsion gels with less than 1.45% CG were semifluid and semisolid. These results demonstrated that the cassia bean gum concentration plays an important role in emulsion gel formation, and smooth and delicate gels with hard surfaces can be obtained by increasing the cassia bean gum concentration.

The microstructure of the gels was observed using scanning electron microscopy. The 0% CG/SPI emulsion gel presented a relatively smooth and irregular network composed of cavities of different sizes. After cassia bean gum was added, the emulsion gel presented a network structure composed of pores of different sizes and flaky edges. The pores were a result of the synergistic effect of SPI and cassia bean gum, which formed a dense gel with a lamellar pore structure [24]. As the CG concentration increased, the size of the structural pores in the emulsion gel gradually decreased, whereas the thickness of the flake increased. Thus, the stability of the emulsion increased. This result was consistent with the observed appearance of the gel. The highest thickness of the emulsion gel and the smallest pores were obtained by adding 2% CG. This may be because during the gelation process of soy protein isolate, protein structures aggregate to form a three-dimensional gel network, and some denatalized, exposed functional groups interact with cassia gum to form a dense, saturated spatial network structure [25].

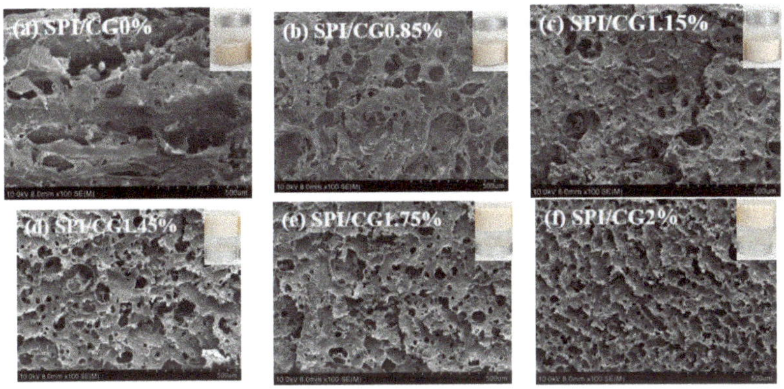

Figure 3. The appearance and scanning electron microscopy (SEM) images of emulsion gels with different concentrations of cassia bean gum. (**a**) SPI/CG 0%: soy protein isolate, no cassia bean gum added, (**b**) SPI/CG 0.85%: SPI + 0.85% cassia bean gum, (**c**) SPI/CG 1.15%: SPI + 1.15% cassia bean gum, (**d**) SPI/CG 1.45%: SPI + 1.45% cassia bean gum, (**e**) SPI/CG 1.75%: SPI + 1.75% cassia bean gum, and (**f**) SPI/CG 2%: SPI + 2% cassia bean gum.

2.5. FTIR Spectroscopy of Emulsion Gel

The chemical interaction information of emulsion gels with different concentrations of cassia bean gum were investigated using FTIR spectroscopy (Figure 4). The wide absorption peak between 3100 and 3500 cm^{-1} was a response to hydrogen bonds (O-H and N-H; Figure 4). As the CG concentration increased, the peak value of the hydrogen bond gradually increased, and the strength of the hydrogen bonds was enhanced during gel formation [26]. These observations were consistent with previous reports on cassia bean gum and κC/KGM composite gels [27].

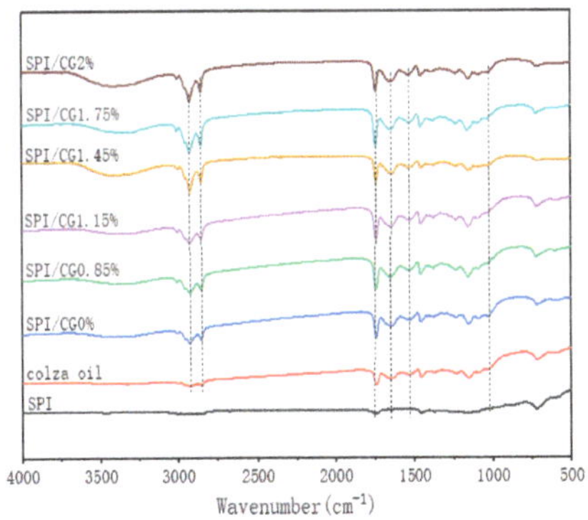

Figure 4. FTIR spectra of emulsion gels with different concentrations of cassia bean gum. SPI: pure soy protein isolate solution; SPI/CG 0%: soy protein isolate, no added cassia bean gum; SPI/CG 0.85%: SPI + 0.85% cassia bean gum; SPI/CG 1.15%: SPI + 1.15% cassia bean gum; SPI/CG 1.45%: SPI + 1.45% cassia bean gum; SPI/CG 1.75%: SPI + 1.75% cassia bean gum; SPI/CG 2%: SPI + 2% cassia bean gum.

The strong absorption bands at 2928, 2854, 1748, and 1026 cm^{-1} were the characteristic signals of rapeseed oil, indicating that rapeseed oil only served as a filler in the emulsion gel, but the particle size of the oil droplets changed. The amide I band (1635.69 cm^{-1}) was mainly caused by the C=O stretching vibration in the peptide bond, and the amide II band (1532.03 cm^{-1}) represented the bending vibration of the N-H group and the stretching vibration of the C-N group. All SPI/CG emulsion gels showed an increased absorption peak strength at amides I and II possibly due to the large number of hydrogen bonds and cross-linked isopeptide bonds that were generated during gel formation [28].

2.6. Analysis of the Thermal Properties of the Emulsion Gel

The influence of DSC on the thermal properties of the cassia bean gum emulsions of different concentrations was studied. The absorption peak on the thermal characteristics curve of the 0% CG/SPI emulsion gel was not obvious (Figure 5), and the denaturation temperature was 69.25 °C. As the CG concentration increased, the absorption peaks on the thermal characteristic curves of the cassia bean gum emulsion gels were evident at 69–81 °C, which was the denaturation temperature range of the emulsion gel. The absorption peak gradually moved to the right, and the denaturation temperature increased. When the CG concentration was 2%, the denaturation temperature of the emulsion gels reached 80.66 °C possibly because the interaction between cassia bean gum at a high concentration and SPI was strengthened, the oil–water interface area increased, and a stable emulsion gel structure formed. The complete denaturation of the emulsion gel required an increase in the heat absorption rate, thus improving the thermal stability of the gels [29].

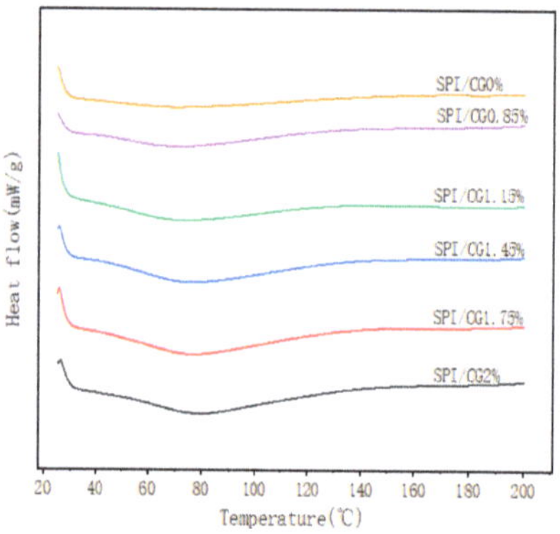

Figure 5. Effect of different concentrations of cassia bean gum on the thermal properties (DSC) of emulsion gels. SPI/CG0%: soy protein isolate, no cassia bean gum added; SPI/CG 0.85%: SPI + 0.85% cassia bean gum; SPI/CG 1.15%: SPI + 1.15% cassia bean gum; SPI/CG 1.45%: SPI + 1.45% cassia bean gum; SPI/CG 1.75%: SPI + 1.75% cassia bean gum; SPI/CG 2%: SPI + 2% cassia bean gum.

2.7. Cooking Loss, Emulsification Stability, and pH of Meat Sausage before and after FT Treatment

The cooking loss, pH, and emulsion stability of the meat sausage before and after freezing and thawing are shown in Table 2. The pH of the meat sausage was not significantly affected by the addition of different fat replacement ratios before and after freezing and thawing ($p \leq 0.05$), and the average pH of the emulsion gels ranged from 6.53 to 6.68. Panagiotopoulou et al. [30] demonstrated that substituting part of or all the animal fat with a Pickering emulsion did not affect the average pH value of the pork sausage; this result was consistent with the results of the present study. As the amount of emulsion gel increased, the cooking loss of meat sausage before and after freezing and thawing increased. When the amount of fat replacement was less than 50%, the cooking loss of S1 (0.04%) and S2 (0.06%) was significantly lower than that of the control group (0.07%) possibly due to the tight network structure and texture of the emulsion gel during cooking, which locked oil and water [31]. However, the rates of cooking loss in S3 and S4 were greater than the rate in the control group because water and fat were bound during heating. When the amount of fat replacement was 50%, the cooking loss was similar to that in the control group. The same trend was observed for the water seepage rate and oil permeability in all the experimental groups possibly because an appropriate amount of emulsion gel and meat protein formed the dense spatial network structure, which effectively locked water and oil. Paglarini et al. [32] found that the emulsification stability and cooking loss of emulsified sausage with a certain proportion of composite gel were close to those of the control group.

Table 2. Cooking loss, emulsification stability, and pH of meat sausage before and after FT treatment.

Groups	Cooking Loss	Before FT Treatment			After FT Treatment		
		PH	W	F	PH	W	F
C	0.07 ± 0.002 c	6.56 ± 0.02 c	0.046 ± 0.004 c	0.01 ± 0.008 c	6.53 ± 0.04 d	0.058 ± 0.001 c	0.028 ± 0.004 c
S1	0.04 ± 0.002 e	6.61 ± 0.02 b	0.041 ± 0.007 c	0.011 ± 0.003 bc	6.55 ± 0.01 c	0.046 ± 0.003 d	0.026 ± 0.003 d
S2	0.06 ± 0.002 d	6.62 ± 0.04 d	0.047 ± 0.005 c	0.017 ± 0.006 ab	6.57 ± 0.02 c	0.056 ± 0.004 c	0.031 ± 0.005 c
S3	0.09 ± 0.003 b	6.63 ± 0.01 b	0.073 ± 0.006 b	0.02 ± 0.005 ab	6.59 ± 0.01 b	0.074 ± 0.008 b	0.04 ± 0.005 b
S4	0.1 ± 0.001 a	6.68 ± 0.02 a	0.085 ± 0.001 a	0.022 ± 0.002 a	6.63 ± 0.02 a	0.096 ± 0.002 a	0.053 ± 0.002 a

Data are expressed as the mean ± standard deviation. FT, freeze–thaw; cook loss; W, water permeability; F, oil permeability. a–e: different lowercase letters indicate significant differences ($p < 0.05$). C stands for (emulsion gel 1.75% CG/SPI) minced sausage with 0% fat replacement; S1 represents (emulsion gel 1.75% CG/SPI) meat sausage with 25% fat replacement; S2 stands for (emulsion gel 1.75% CG/SPI) minced sausage with 50% fat replacement; S3 stands for (emulsion gel 1.75% CG/SPI) minced meat with 75% fat replacement; S4 stands for (emulsion gel 1.75% CG/SPI) meat sausage with 100% fat replacement.

2.8. Color Difference Characteristics of Meat Sausage before and after Freeze–Thaw Treatment

Color is one of the most important factors for determining consumers' preferences regarding meat products. As shown in Table 3, the color parameters brightness (L*), red (a*), and yellow (b*) of the meat sausage before and after freezing and thawing were higher than those of the control group ($p < 0.05$). The highest L* values were 63.74 and 63.81 in the S4 group. This may be because the oil droplet diameter of a meat sausage containing the emulsified gel is smaller than that produced by animal fat, which creates a greater light reflection, resulting in a higher L* [33]. The a* value of the meat sausage was not considerably different from that of the control group before and after the freeze–thaw treatment, indicating that the addition of the emulsion gel had no adverse effect on the redness value of the meat sausage. The b* value of the meat sausage before and after the freeze–thaw treatment was lower than that of the control group, showing a downward trend. The lowest values were 1.78 and 2.71 (S4), which may have been influenced by changes in the vegetable oil pigment and the emulsification of SPI after the emulsion gel was heated [34]. Li et al. [35] added laminaria polysaccharides at different concentrations into chicken sausage as a fat substitute. The L* value increased with the proportion of the substitute, and the a* value gradually decreased. Moreover, after the freeze–thaw treatment, the L* and a* values increased, and the b* values showed no considerable difference before and after the freeze–thaw treatment.

Table 3. Color difference in minced meat before and after FT treatment.

Groups	Before FT Treatment			After FT Treatment		
	L*	a*	b*	L*	a*	b*
C	60.55 ± 0.17 e	15.78 ± 0.13 a	3.11 ± 0.03 a	61.59 ± 0.16 e	15.86 ± 0.11 a	3.77 ± 0.12 a
S1	61.45 ± 0.16 d	15.56 ± 0.08 b	2.84 ± 0.13 b	62.04 ± 0.04 d	15.64 ± 0.05 c	3.63 ± 0.04 ab
S2	61.78 ± 0.05 c	15.49 ± 0.02 bc	2.34 ± 0.09 c	62.81 ± 0.14 c	15.61 ± 0.26 ab	3.49 ± 0.11 b
S3	62.63 ± 0.13 b	15.37 ± 0.08 c	1.90 ± 0.09 d	63.51 ± 0.10 b	15.42 ± 0.17 b	2.92 ± 0.04 c
S4	63.74 ± 0.12 a	15.31 ± 0.14 c	1.78 ± 0.06 d	63.81 ± 0.06 a	15.40 ± 0.19 b	2.71 ± 0.15 d

Data are expressed as the mean ± standard deviation. FT, freeze–thaw; L*, brightness; a*, red; b*, yellow. a–e: different lowercase letters indicate significant differences ($p < 0.05$). C stands for (emulsion gel 1.75% CG/SPI) minced sausage with 0% fat replacement; S1 represents (emulsion gel 1.75% CG/SPI) meat sausage with 25% fat replacement; S2 stands for (emulsion gel 1.75% CG/SPI) minced sausage with 50% fat replacement; S3 stands for (emulsion gel 1.75% CG/SPI) minced meat with 75% fat replacement; S4 stands for (emulsion gel 1.75% CG/SPI) meat sausage with 100% fat replacement.

2.9. Texture Characteristics of Meat Sausage before and after Freeze–Thaw Treatment

TPA is an important index for evaluating the quality and overall acceptability of meat products. As shown in Table 4, as the amount of emulsion gel replacement increased, the hardness of the meat sausage before and after the freeze–thaw treatment increased and then decreased, and the hardnesses of S1 and S2 before the treatment were 9097.83 and 9745.53 N, respectively. After freezing and thawing, the hardnesses of S1 and S2 were 7643.38 and 7911.79 N. When the amount of emulsion gel was greater than 50%, the hardness decreased considerably, and the lowest hardness of S4 was obtained (5959.92 and 4850.13 N). The possible reason is that animal fat is a solid fat with a hard texture. By contrast, the emulsion gel has a soft texture. The lowest hardness of S4 was observed when the emulsion gel partially or completely replaced animal fat in meat sausage. The quality of meat products can be negatively affected [36]. Pintado et al. [37] found that when different proportions of a soybean emulsion gel were used to replace animal fat in sausage, its hardness and chewiness were be substantially reduced. The hardness of S2 before and after the freeze–thaw treatment was close to that of the control group, showing excellent freeze–thaw stability. In all the experimental groups, the chewiness, viscosity, and hardness showed the same change trend; that is, the chewiness and viscosity of S3 and S4 were significantly lower than those of C, S1, and S2 ($p < 0.05$), and the chewiness and viscosity of S2 and C were similar. This also proves that meat sausage has a lower hardness, and S2 showed excellent freeze–thaw stability. There was no significant difference in the elasticity and cohesiveness between the two groups (0.91–0.96) ($p < 0.05$).

Table 4. Texture characteristics of meat sausage before and after FT treatment.

Groups	Before FT Treatment					After FT Treatment				
	Hardness (N)	Elasticity	Cohesiveness	Adhesiveness (N-mm)	Chewiness (mJ)	Hardness (N)	Elasticity	Cohesiveness	Adhesiveness (N-mm)	Chewiness (mJ)
C	9897.42 ± 107.7 a	0.95 ± 0.01	0.94 ± 0.02	9413.59 ± 76.09 a	8831.74 ± 107.64 a	7643.38 ± 57.28 b	0.94 ± 0.02	0.94 ± 0.02	7248.75 ± 131.14 a	6905.06 ± 73.94 a
S1	9097.83 ± 65.34 b	0.96 ± 0.01	0.95 ± 0.01	8612.65 ± 44.14 b	8268.27 ± 121.65 c	7911.79 ± 111.77 a	0.95 ± 0.01	0.91 ± 0.06	6532.66 ± 92.26 b	6108.69 ± 72.63 b
S2	9745.53 ± 55.31 c	0.96 ± 0.01	0.95 ± 0.01	9131.36 ± 83.81 c	8989.2 ± 83.95 b	7860.28 ± 99.01 a	0.94 ± 0.01	0.91 ± 0.05	7169.74 ± 292.5 a	6878.2 ± 84.31 a
S3	7236.98 ± 22.04 d	0.95 ± 0.02	0.95 ± 0.01	6907.28 ± 64.29 d	6566.53 ± 53.83 d	5236.98 ± 22.04 c	0.93 ± 0.01	0.92 ± 0.01	5207.28 ± 64.29 c	5033.2 ± 33.25 d
S4	5969.92 ± 43.49 e	0.95 ± 0.01	0.92 ± 0.01	5471.42 ± 113.59 e	5195.68 ± 101.17 e	4850.13 ± 60.66 d	0.93 ± 0.01	0.92 ± 0.01	4552.31 ± 55.05 d	4394.4 ± 65.19 c

Data are expressed as the mean ± standard deviation. FT, freeze–thaw. a–e: different lowercase letters indicate significant differences ($p < 0.05$). C stands for (emulsion gel 1.75% CG/SPI) minced sausage with 0% fat replacement; S1 represents (emulsion gel 1.75% CG/SPI) meat sausage with 25% fat replacement; S2 stands for (emulsion gel 1.75% CG/SPI) minced sausage with 50% fat replacement; S3 stands for (emulsion gel 1.75% CG/SPI) minced meat with 75% fat replacement; S4 stands for (emulsion gel 1.75% CG/SPI) meat sausage with 100% fat replacement.

2.10. Antioxidant Activity of Meat Sausage

As shown in Figure 6, as the amount of emulsion gel added increased, the DPPH and ABTS free radical clearance rates of the meat sausage increased and then decreased. When the amount of fat replacement was 0–50%, the DPPH free radical scavenging rate increased from 83% to 92%, and the ABTS free radical scavenging rate increased from 86% to 95% possibly because the dense networks in the emulsion gel hindered the movement of the liquid oil phase and the transfer of oxidation products while inhibiting the penetration and diffusion of oxygen within the meat sausage meat sausage, thus inhibiting oil oxidation [38]. Millao et al. [39] reported that due to the strengthening of the gel structure, the antioxidant capacity of oleogel gradually increased with an increasing EC concentration. In addition, the antioxidant capacities of raw materials enhanced the antioxidant capacity of oil; that is, rapeseed oil was rich in antioxidant materials, such as vitamin E, which can remove free radicals. Therefore, the free radical scavenging rates of DPPH and ABTS were high in all the groups and increased with the amount of the emulsion gel. However, when the amount of fat replacement was greater than 50%, the free radical scavenging rates for DPPH and ABTS significantly decreased ($p < 0.05$). The scavenging rates for DPPH and ABTS decreased to 63% and 65%, respectively, because of the large amount of water lost from the meat sausage during processing after the addition of the emulsion gel. Moreover, the structure of the gel network was damaged, and the rapeseed oil was degraded by heat.

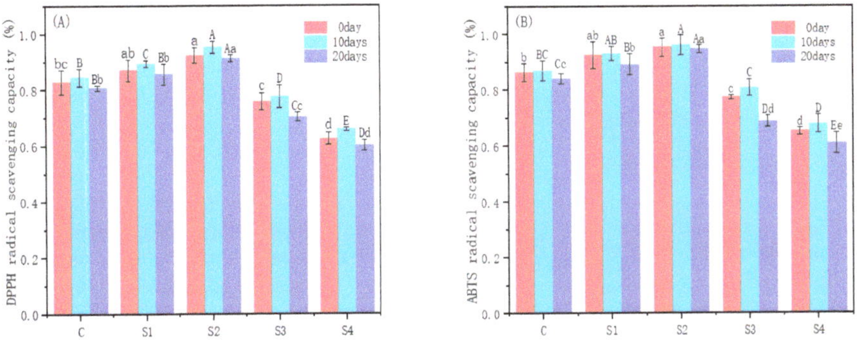

Figure 6. DPPH (**A**) and ABTS (**B**) free radical clearance of meat sausage during storage with different fat replacement amounts. C stands for (emulsion gel 1.75% CG/SPI) minced sausage with 0% fat replacement; S1 represents (emulsion gel 1.75% CG/SPI) meat sausage with 25% fat replacement; S2 stands for (emulsion gel 1.75% CG/SPI) minced sausage with 50% fat replacement; S3 stands for (emulsion gel 1.75% CG/SPI) minced meat with 75% fat replacement; S4 stands for (emulsion gel 1.75% CG/SPI) meat sausage with 100% fat replacement. Different letters indicate significant differences in the data ($p < 0.05$).

As the storage time increased, the free radical scavenging rates for DPPH and ABTS in the meat sausage increased and then decreased because of the hydrolysis of some antioxidants at the later stage of storage. The DPPH and ABTS free radical scavenging rates in S2 were the most stable during storage.

3. Conclusions

The effects of cassia bean gum on the texture, rheology, stability, and microstructure of the emulsion gels of a protein–polysaccharide composite matrix were investigated. As the cassia bean gum concentration increased, the texture of the emulsion gels became harder, and the surface was smooth and delicate. The size of the internal structural pores gradually decreased and was evenly distributed, and the thickness of the gel increased. Thus, the stability of the emulsion increased. During gel formation, the interaction between CG and SPI produced a large number of hydrogen and isopeptide bonds, increasing the absorption peak intensity at amides I and II in the infrared region. The presence of CG considerably improved the hardness, elasticity, and WHC of the emulsion gel. The viscosity of the emulsion gels with different concentrations of CG decreased with an increasing shear rate, showing typical pseudoplastic fluid characteristics. The G' values of emulsion gels in frequency scanning were higher than the G'' values, and the G' and G'' values were less dependent on the frequency. Thus, the gel network structure was stable and resilient. DSC analysis showed that the denaturation temperature of emulsion gel reached to 80.66 °C after the addition of different concentrations of CG. The 1.75% CG/SPI emulsion gel was applied to replace animal fat in meat sausage. As the amount of the replacement increased, the texture, emulsification stability, color difference, and DPPH and ABTS free radical clearance rates of the meat sausage increased first and then decreased. The freeze–thaw stability and oxidation stability of S2 were stronger than those of the control group. Therefore, the emulsion gel can replace 50% of fat for low-fat meat sausage production. This study provides a novel method for reducing the content of animal fat in meat sausage and maintaining its quality.

4. Materials and Methods

4.1. Materials

The materials used were as follows: soybean protein isolate (purity $\geq$ 90%): Shandong Zeenda Food Raw Materials Co., LTD. (Shandong, China); rapeseed Oil: Hunan Baling Oil

Co., LTD. (Hunan, China); cassia bean gum: Shaanxi Chenming Biotechnology Co., LTD. (Shaanxi, China); potassium bromide (analytical grade): Guangdong Qianjin Chemical Reagent Co., LTD. (Guangdong, China); pork lean meat, pig backfat, and other ingredients were purchased from local shopping malls (Chengdu, China); all other reagents used in this study are analytical-grade reagents.

4.2. Preparation of Emulsion Gel

The emulsion gel was prepared by referring to the method of Gao et al. [40], with modifications. The soybean protein isolate powder was dispersed in deionized water (5 g/100 mL) and magnetically stirred at 6 °C at room temperature for 1.5 h to achieve a fully hydrated solution. The soybean protein isolate solution was heated to 75 °C for 15 min and then rapidly cooled in an ice bath. The concentrations of cassia bean gum used were 0%, 0.85%, 1.15%, 1.45%, 1.75%, and 2%; these were mixed with soybean protein isolate solution, and then the composite solution was mixed with rapeseed oil (19%, *w*/*w*). A high-pressure homogenizer (XHF-DY, Xinzhi Biotechnology Co., LTD., Ningbo, China) was used for homogenization at 20,000 rpm for 8 min, and the sample was then heated at 50 °C for 30 min and cooled to a gel. The final emulsion gel was left to rest overnight at 4 °C for analysis.

4.3. Observation of Emulsion Gel Structure

Our appearance observation occurred as follows: 30 g of the prepared fresh sample was transferred into a bottle and placed in the refrigerator at 5 °C overnight, and then it was left to stand at room temperature for 1 h before shooting, and the inverted non-flowing form was used as the standard for forming the gel.

We performed scanning electron microscopy (SEM) as follows: According to the method of Lei et al. [41], with some changes, the sample was first frozen in a freezer (−80 °C) for 8 h to fix its structure, was then put in a freeze-drying machine for 48 h for removal, and then the sample was soaked in petroleum ether for 12 h; this process was repeated 6 times, and then the degreased sample was placed in a vacuum drying oven at 70 °C for 4 h to evaporate the petroleum ether. The microstructure of the samples was observed using SEM (JSM-5800 LV, JEOL Ltd., Tokyo, Japan). The acceleration voltage was 10.0 kV, and the micromorphology of the sample was observed at 500×.

4.4. Determination of Physical and Chemical Properties of Emulsion Gel

4.4.1. Texture Characteristics

Here, we referred to the method of Meng et al. [42] and modified it. The physical property tester (Vector 33, Bruker Optics, Ettlingen, Germany) and P36 R probe were used to determine the emulsion gel's texture properties. Before the experiment, the emulsion gel was left to stand at room temperature at 5 °C for 2 h. The emulsion gels (3.2 cm in diameter) were tested in parallel 4 times, the strain level was 50%, and the following parameters were used for determination: the pre-test velocity, the lateral center velocity, and the post-test velocity; these were 5.0, 5.0, and 6.0 mm/s, respectively, and the contact force was 15 g. The hardness (N), elasticity, cohesiveness, and chewiness (mJ) of the emulsion gel were recorded.

4.4.2. Rheological Behavior

Referring to the method of You et al. [43], with modifications, the rheological properties of the sample were determined using a rheological analyzer (MCR-101, Anton Paar Co. Ltd., Graz, Austria) with a measuring gap of 1 mm and a parallel plate diameter of 50 mm.

Temperature scanning was conducted as follows: The gel sample was placed in the rheological plate, the excess sample around the pressed plate was scraped off, and the plate was sealed with silicone oil and covered to prevent water loss at high temperature. The fixed scanning frequency was 1 Hz, the strain was 0.1%, and the change of the energy storage modulus (G') when the gel sample was heated from 20 °C to 90 °C was recorded.

The shear rate was determined as follows: in the shear rate range of 0.01–10 s^{-1}, the temperature is 25 °C, and the change of the apparent viscosity of the sample with the shear rate is recorded to determine the fluid type.

Frequency scanning was conducted as follows: The angular frequency range is 10~100 rad/s and the strain is 0.1%. Frequency scanning is performed on the emulsion gel at 25 °C, and the changes in the energy storage modulus (G′) and loss modulus (G″) with the angular frequency are recorded.

4.4.3. Water Holding Capacity

According to Zhao et al.'s method [44], with slight modifications, 20 g samples were placed in a 50 mL centrifuge tube, the gels were centrifuged at 4 °C at 12,000× *g* for 15 min to remove excess water, and the surface water of the emulsion gel was absorbed using filter paper before being weighed. The water holding capacity (WHC) of the emulsion gel is calculated according to the formula:

$$WHC(\%) = \frac{W2}{W1} \times 100$$

where $W1$ is the mass of emulsion gel before centrifugation, and $W2$ is the mass of emulsion gel after centrifugation.

4.4.4. FTIR Spectroscopy

Using a modified method with reference to Meng et al. [45], freeze-dried and crushed samples were mixed with potassium bromide at 1:120 and pressed into tablets for further FTIR determination (Vector 33, Bruker Optics, Ettlingen, Germany). The scanning range of infrared spectrum is 4000–400 cm^{-1}, the resolution is 4 cm^{-1}, and the scanning times is 32.

4.4.5. Differential Scanning Calorimetry (DSC) Analysis

Referring to Zhu et al.'s [46] method, with modifications, a differential scanning calorimeter (Q200M, TA instrument, New Castle, DE, USA) was used to analyze the thermal properties of the samples. Each sample of about 5–6 mg was sealed in an aluminum crucible with a hole cover, and an empty aluminum crucible without samples was used as a reference group. The heating temperature range was 25–200 °C, and the heat distribution was measured under inert nitrogen gas at a rate of 5 °C/min and a flow rate of 10 mL/min.

4.5. Preparation of Meat Sausage

Here, we referred to the method of Qi et al. [47] and modified it. Each experiment consisted of five groups of meat sausages, as shown in Table 5. All laboratory meat sausages utilized the same lean meat from the front leg of the pig, along with consistent backfat and toppings, to minimize variations between batches. The pork front leg meat underwent a pre-treatment process in which the visible fasciae were removed and the meat was minced. Subsequently, the minced meat was combined with an appropriate amount of salt, sodium ascorbate, and sodium tripolyphosphate to facilitate the dissolution of the meat proteins. After marinating for eight hours, the mixture was blended at a high speed for approximately 20 min until a strong texture was achieved, during which 30 g of cold water was added in batches. Following this, the cured minced meat, pork fat, or emulsion gel (with fat replacement ratios of 0%, 25%, 50%, 75%, and 100%) was stirred in a food processor for five hours. The resulting meat sausage mixture was then steamed in a pot at 60 °C for 30 min, rapidly cooled in an ice bath, and stored at 4 °C for further analysis.

4.6. Quality Determination of Meat Sausage before and after FT Treatment

The samples were frozen in a constant temperature freezer at −18 °C for 24 h. After freezing, the samples were thawed in a constant temperature refrigerator at 4 °C for testing.

Table 5. List of ingredients for meat sausage preparation.

Component	Groups (Unit: g/100 g)				
	C	S1	S2	S3	S4
Forehock	70	70	70	70	70
Pork backfat	30	22.5	15	7.5	0
Emulsion gel	0	7.5	15	22.5	30
Ice water	25	25	25	25	25
Salt	3	3	3	3	3
Sugar	1	1	1	1	1
Sodium tripolyphosphate	0.2	0.2	0.2	0.2	0.2
natrascorb	0.1	0.1	0.1	0.1	0.1
Other ingredients	0.7	0.7	0.7	0.7	0.7

C stands for (emulsion gel 2% CG/SPI) minced sausage with 0% fat replacement; S1 represents (emulsion gel 2% CG/SPI) meat sausage with 25% fat replacement; S2 represents (emulsion gel 2% CG/SPI) minced meat with 50% fat replacement; S3 stands for (emulsion gel 2% CG/SPI) minced meat with 75% fat replacement; S4 stands for (emulsion gel 2% CG/SPI) meat sausage with 100% fat replacement.

4.6.1. Determination of Cooking Loss

About 15 g of minced meat samples were weighed and heated at 90 °C for 30 min. After cooling in a cold bath for 30 min, the surface moisture of the sample was absorbed with filter paper to record the quality of the minced meat before and after cooking. The cooking loss of minced meat was calculated according to the formula:

$$Cooking\ loss\ (\%) = \frac{M1 - M2}{M1} \times 100$$

where $M1$ is the weight of the minced meat before cooking, and $M2$ is the weight of the minced meat after cooking.

4.6.2. Determination of Emulsification Stability

In accordance with the modifications suggested by Pan et al. [48], 20 g of raw minced meat was placed into a 50 mL centrifuge tube, which was then heated in a water bath at 80 °C for 20 min. After heating, the tube was removed and allowed to cool before being centrifuged at a centrifugal force of 2500× g for 3 min. The centrifuge tube was subsequently inverted and placed in a beaker for 1 h. Following this, the solid portion of the sample was extracted, weighed, and dried in an oven at 100 °C for 10 h to achieve a constant weight. The water permeability (W) and fat permeability (F) of the minced meat were calculated using the appropriate formulas:

$$W(\%) = \frac{M2 - M3}{M1} \times 100$$

$$F(\%) = \frac{M3 - M4}{M1} \times 100$$

where $M1$ is the quality of the raw meat, $M2$ is the quality before drying, $M3$ is the quality after drying, and $M4$ is the quality of the empty bottle.

4.6.3. Measurement of pH Value

We referred to the method of Yu et al. [49] to determine the pH value of meat sausage. We took 5 g of the cut sausage sample, added 50 mL distilled water to homogenize it for 5 min, then let it stand for 30 min, and removed the supernatant to determine the pH value.

4.6.4. Determination of Color Difference

According to the method of Cheng et al. [50], with modifications, the color of the meat sausage was determined. The meat sausage was cut into a flat and smooth cylinder with a section height of 2 cm, corrected with a standard blackboard and whiteboard, and the observer angle was measured to be 0°. Using the three color values (brightness, L* value; redness, a* value; yellowness, b* value), each sample was measured 6 times in parallel.

4.6.5. Determination of Texture

The method of Wang et al. [51], with slight modifications, was used to determine the texture of the meat sausage. Meat sausage samples were cut into cylinders with a diameter of 20 mm and a height of 20 mm, and the texture analysis of meat sausage samples was performed with a physical property tester (Vector 33, Bruker Optics, Ettlingen, Germany). A p/50 probe was equipped to perform a two-cycle axial compression test with a stress of 30%. The test speeds before, during, and after the test were 5.0 mm/s, 5.0 mm/s, 1.0 mm/s, and the trigger force was 5.0 g, respectively. The parameters measured were hardness (g), elasticity, cohesion, chewiness (mJ), and adhesion.

4.7. Determination of Antioxidant Activity of Meat Sausage

According to the method of Feng et al. [52], although slightly modified, 10 g of cut meat sausage was placed into 50 mL of distilled water and homogenized in a water bath at 30 °C for 1 h, and 0.2 mL of filtrate was mixed with 1 mL of DPPH solution (0.1 mM, diluted with anhydrous ethanol). After 30 min of light-avoidance reaction at room temperature, the absorbance at 517 nm was measured as the experimental group. Also, 0.2 mL of distilled water and 1 mL of DPPH solution were used as the control group, and 0.2 mL of filtrate and 1 mL of anhydrous ethanol were used as the blank group. The DPPH radical scavenging capacity of the meat sausage was calculated according to formula:

$$DPPH\ radical\ scavenging\ capacity\ (\%) = \frac{B1 - (B2 - B3)}{B1} \times 100$$

where $B1$ is the absorbance at 517 nm of the control group, $B2$ is the absorbance at 517 nm of the experimental group, and $B3$ is the absorbance at 517 nm of the blank group.

According to the method of Kong et al. [53], although slightly modified, the mother liquor of ABTS was obtained by mixing 200 mg of ABTS with 34 mg of potassium persulfate and 50 mL of distilled water, shaking well, and leaving overnight at room temperature away from light. The ABTS solution, with an absorbance 0.7 ± 0.02 at 734 nm, was prepared by diluting the ABTS mother solution to a certain multiple with 95% anhydrous ethanol. The absorbance at 734 nm was measured by mixing 25 μL of filtrate and 1 mL of ABTS solution at room temperature for 10 min after the reaction. Here, 25 μL of distilled water and 1 mL of ABTS solution were used as the blank group. The ABTS radical scavenging capacity of the meat sausage was calculated according to formula:

$$ABTS\ radical\ scavenging\ capacity\ (\%) = \frac{Bblank - Bsample}{Bsample} \times 100$$

where *Bsample* is the absorbance of the experimental group at 734 nm, and *Bblank* is the absorbance of the blank group at 734 nm.

4.8. Data Analysis

All tests were carried out in triplicate, and the data were presented as the mean ± SD (standard deviation). SPSS software (26.0 for Mac, IBM SPSS Statistical software Inc., Chicago, IL, USA) was used for statistical analysis. ANOVA and Duncan's multiple range tests were used for statistical analysis. The significance level for all tests was established at $p \leq 0.05$.

Author Contributions: Q.Z.: Conceptualization, Writing—review and editing, Project administration, Validation, Formal analysis, Funding acquisition, Methodology, and Writing—original draft; Y.Z.: Writing—review and editing, Writing—original draft, Methodology, Formal analysis, Data curation, and Validation; Y.L.: Investigation, Formal analysis, and Methodology; L.L.: Investigation; Y.C.: Formal analysis; G.R.: Data curation; D.L.: Methodology, Writing—review and editing, Resources, Supervision, and Project administration. All authors have read and agreed to the published version of the manuscript.

Funding: This research was funded by the Sichuan Provincial Science and Technology Achievement Transfer (2022ZHCG0072) and Sichuan Provincial Key Research and Development Project (2022YFN0050).

Institutional Review Board Statement: The study was conducted in accordance with the Declaration of Helsinki and approved by the Ethics Committee of Chengdu University (date of approval: 30 June 2024).

Informed Consent Statement: Informed consent was obtained from all subjects involved in the study.

Data Availability Statement: The data presented in this study are available on request from the corresponding author.

Acknowledgments: The authors would like to thank Chengdu University for providing experimental instrumentation support for this study and also thank the team members for their time.

Conflicts of Interest: The authors declare no conflicts of interest.

References

1. Leeuwis, C.; Boogaard, B.K.; Atta-Krah, K. How Food Systems Change (or Not): Governance Implications for System Transformation Processes. *Food Secur.* **2021**, *13*, 761–780. [CrossRef] [PubMed]
2. Tremonte, P.; Pannella, G.; Lombardi, S.J.; Iorizzo, M.; Vergalito, F.; Cozzolino, A.; Maiuro, L.; Succi, M.; Sorrentino, E.; Coppola, R. Low-Fat and High-Quality Fermented Sausages. *Microorganisms* **2020**, *8*, 1025. [CrossRef] [PubMed]
3. Huang, X.; Liu, B.; Li, Y.; Huang, D.; Zhu, S. Influence Mechanism of Components and Characteristics on Structural and Oxidative Stability of Emulsion Gel. *Food Hydrocoll.* **2024**, *151*, 109852. [CrossRef]
4. Wang, S.; Wu, Z.; Jia, L.; Wang, X.; He, T.; Wang, L.; Yao, G.; Xie, F. Soybean Protein Isolate-Sodium Alginate Double Network Emulsion Gels: Mechanism of Formation and Improved Freeze-Thaw Stability. *Int. J. Biol. Macromol.* **2024**, *274*, 133296. [CrossRef]
5. Li, M.; Feng, L.; Dai, Z.; Li, D.; Zhang, Z.; Zhou, C.; Yu, D. Improvement of 3D Printing Performance of Whey Protein Isolate Emulsion Gels by Regulating Rheological Properties: Effect of Polysaccharides Incorporation. *Food Bioprocess Technol.* **2024**, 1–15. [CrossRef]
6. Matos, M.E.; Sanz, T.; Rosell, C.M. Establishing the Function of Proteins on the Rheological and Quality Properties of Rice Based Gluten Free Muffins. *Food Hydrocoll.* **2014**, *35*, 150–158. [CrossRef]
7. Yang, Q.; Wang, Y.-R.; Li-Sha, Y.-J.; Chen, H.-Q. The Effects of Basil Seed Gum on the Physicochemical and Structural Properties of Arachin Gel. *Food Hydrocoll.* **2021**, *110*, 106189. [CrossRef]
8. Zhang, X.; Chen, X.; Gong, Y.; Li, Z.; Guo, Y.; Yu, D.; Pan, M. Emulsion Gels Stabilized by Soybean Protein Isolate and Pectin: Effects of High Intensity Ultrasound on the Gel Properties, Stability and β-Carotene Digestive Characteristics. *Ultrason. Sonochemistry* **2021**, *79*, 105756. [CrossRef]
9. Wang, X.; Luo, K.; Liu, S.; Adhikari, B.; Chen, J. Improvement of Gelation Properties of Soy Protein Isolate Emulsion Induced by Calcium Cooperated with Magnesium. *J. Food Eng.* **2019**, *244*, 32–39. [CrossRef]
10. Sharma, D.; Sharma, P. Synergistic Studies of *Cassia Tora* Gum with Xanthan and Guar Gum: Carboxymethyl Synthesis of Cassia Gum-Xanthan Synergistic Blend and Characterization. *Carbohydr. Res.* **2023**, *523*, 108723. [CrossRef]
11. Cao, L.; Ge, T.; Meng, F.; Xu, S.; Li, J.; Wang, L. An Edible Oil Packaging Film with Improved Barrier Properties and Heat Sealability from Cassia Gum Incorporating Carboxylated Cellulose Nano Crystal Whisker. *Food Hydrocoll.* **2020**, *98*, 105251. [CrossRef]
12. Li, X.; Chen, X.; Cheng, H. Impact of κ-Carrageenan on the Cold-Set Pea Protein Isolate Emulsion-Filled Gels: Mechanical Property, Microstructure, and In Vitro Digestive Behavior. *Foods* **2024**, *13*, 483. [CrossRef]
13. Jiang, S.; Ma, Y.; Wang, Y.; Wang, R.; Zeng, M. Effect of κ-Carrageenan on the Gelation Properties of Oyster Protein. *Food Chem.* **2022**, *382*, 132329. [CrossRef]
14. Hou, W.; Long, J.; Hua, Y.; Chen, Y.; Kong, X.; Zhang, C.; Li, X. Formation and Characterization of Solid Fat Mimetic Based on Pea Protein Isolate/Polysaccharide Emulsion Gels. *Front. Nutr.* **2022**, *9*, 1053469. [CrossRef]
15. Lu, Y.; Ma, Y.; Zhang, Y.; Gao, Y.; Mao, L. Facile Synthesis of Zein-Based Emulsion Gels with Adjustable Texture, Rheology and Stability by Adding β-Carotene in Different Phases. *Food Hydrocoll.* **2022**, *124*, 107178. [CrossRef]
16. Wang, X.-Y.; Wang, J.; Rousseau, D.; Tang, C.-H. Chitosan-Stabilized Emulsion Gels *via* pH-Induced Droplet Flocculation. *Food Hydrocoll.* **2020**, *105*, 105811. [CrossRef]
17. Li, A.; Gong, T.; Hou, Y.; Yang, X.; Guo, Y. Alginate-Stabilized Thixotropic Emulsion Gels and Their Applications in Fabrication of Low-Fat Mayonnaise Alternatives. *Int. J. Biol. Macromol.* **2020**, *146*, 821–831. [CrossRef]
18. Bi, C.; Chi, S.; Wang, X.; Alkhatib, A.; Huang, Z.; Liu, Y. Effect of Flax Gum on the Functional Properties of Soy Protein Isolate Emulsion Gel. *LWT* **2021**, *149*, 111846. [CrossRef]
19. Çakır, E.; Foegeding, E.A. Combining Protein Micro-Phase Separation and Protein–Polysaccharide Segregative Phase Separation to Produce Gel Structures. *Food Hydrocoll.* **2011**, *25*, 1538–1546. [CrossRef]
20. Zhao, H.; Chen, J.; Hemar, Y.; Cui, B. Improvement of the Rheological and Textural Properties of Calcium Sulfate-Induced Soy Protein Isolate Gels by the Incorporation of Different Polysaccharides. *Food Chem.* **2020**, *310*, 125983. [CrossRef]

21. Neiser, S.; Draget, K.I.; Smidsrød, O. Gel Formation in Heat-Treated Bovine Serum Albumin–κ-Carrageenan Systems. *Food Hydrocoll.* **2000**, *14*, 95–110. [CrossRef]
22. Hu, Y.; Liang, H.; Xu, W.; Wang, Y.; An, Y.; Yan, X.; Ye, S.; Huang, Q.; Liu, J.; Li, B. Synergistic Effects of Small Amounts of Konjac Glucomannan on Functional Properties of Egg White Protein. *Food Hydrocoll.* **2016**, *52*, 213–220. [CrossRef]
23. Huang, H.; Belwal, T.; Aalim, H.; Li, L.; Lin, X.; Liu, S.; Ma, C.; Li, Q.; Zou, Y.; Luo, Z. Protein-Polysaccharide Complex Coated W/O/W Emulsion as Secondary Microcapsule for Hydrophilic Arbutin and Hydrophobic Coumaric Acid. *Food Chem.* **2019**, *300*, 125171. [CrossRef] [PubMed]
24. Wei, Z.; Dong, Y.; Si, J. Ovotransferrin Fibril—Gum Arabic Complexes as Stabilizers for Oleogel-in-Water Pickering Emulsions: Formation Mechanism, Physicochemical Properties, and Curcumin Delivery. *Foods* **2024**, *13*, 1323. [CrossRef]
25. Jiang, Y.; Zhang, C.; Yuan, J.; Wu, Y.; Li, F.; Waterhouse, G.I.N.; Li, D.; Huang, Q. Exploiting the Robust Network Structure of Zein/Low-Acyl Gellan Gum Nanocomplexes to Create Pickering Emulsion Gels with Favorable Properties. *Food Chem.* **2021**, *349*, 129112. [CrossRef] [PubMed]
26. Li, P.; Guo, C.; Li, X.; Yuan, K.; Yang, X.; Guo, Y.; Yang, X. Preparation and Structural Characteristics of Composite Alginate/Casein Emulsion Gels: A Microscopy and Rheology Study. *Food Hydrocoll.* **2021**, *118*, 106792. [CrossRef]
27. Cheng, Z.; Zhang, B.; Qiao, D.; Yan, X.; Zhao, S.; Jia, C.; Niu, M.; Xu, Y. Addition of κ-Carrageenan Increases the Strength and Chewiness of Gelatin-Based Composite Gel. *Food Hydrocoll.* **2022**, *128*, 107565. [CrossRef]
28. Zhang, B.; Meng, R.; Li, X.-L.; Liu, W.-J.; Cheng, J.-S.; Wang, W. Preparation of Pickering Emulsion Gels Based on κ-Carrageenan and Covalent Crosslinking with EDC: Gelation Mechanism and Bioaccessibility of Curcumin. *Food Chem.* **2021**, *357*, 129726. [CrossRef]
29. Zhang, X.; Wang, W.; Wang, Y.; Wang, Y.; Wang, X.; Gao, G.; Chen, G.; Liu, A. Effects of Nanofiber Cellulose on Functional Properties of Heat-Induced Chicken Salt-Soluble Meat Protein Gel Enhanced with Microbial Transglutaminase. *Food Hydrocoll.* **2018**, *84*, 1–8. [CrossRef]
30. Panagiotopoulou, E.; Moschakis, T.; Katsanidis, E. Sunflower Oil Organogels and Organogel-in-Water Emulsions (Part II): Implementation in Frankfurter Sausages. *LWT* **2016**, *73*, 351–356. [CrossRef]
31. Abbasi, E.; Amini Sarteshnizi, R.; Ahmadi Gavlighi, H.; Nikoo, M.; Azizi, M.H.; Sadeghinejad, N. Effect of Partial Replacement of Fat with Added Water and Tragacanth Gum (*Astragalus Gossypinus* and *Astragalus Compactus*) on the Physicochemical, Texture, Oxidative Stability, and Sensory Property of Reduced Fat Emulsion Type Sausage. *Meat Sci.* **2019**, *147*, 135–143. [CrossRef]
32. Paglarini, C.d.S.; Martini, S.; Pollonio, M.A.R. Using Emulsion Gels Made with Sonicated Soy Protein Isolate Dispersions to Replace Fat in Frankfurters. *LWT* **2019**, *99*, 453–459. [CrossRef]
33. Wu, Y.; Sun, S.; Li, X.; Li, X.; Huang, Y.; An, F.; Huang, Q.; Song, H. Effect of Gel Composition Interaction on Rheological, Physicochemical and Textural Properties of Methyl Cellulose Oleogels and Lard Replacement in Ham Sausage. *Int. J. Biol. Macromol.* **2024**, *280*, 135902. [CrossRef]
34. Cui, B.; Mao, Y.; Liang, H.; Li, Y.; Li, J.; Ye, S.; Chen, W.; Li, B. Properties of Soybean Protein Isolate/Curdlan Based Emulsion Gel for Fat Analogue: Comparison with Pork Backfat. *Int. J. Biol. Macromol.* **2022**, *206*, 481–488. [CrossRef]
35. Li, Y.; Wang, S.; Liu, X.; Zhao, G.; Yang, L.; Zhu, L.; Liu, H. Improvement in Texture and Color of Soy Protein Isolate Gel Containing Capsorubin and Carotenoid Emulsions Following Microwave Heating. *Food Chem.* **2023**, *428*, 136743. [CrossRef]
36. Kılıç, B.; Özer, C.O. Potential Use of Interesterified Palm Kernel Oil to Replace Animal Fat in Frankfurters. *Meat Sci.* **2019**, *148*, 206–212. [CrossRef]
37. Pintado, T.; Muñoz-González, I.; Salvador, M.; Ruiz-Capillas, C.; Herrero, A.M. Phenolic Compounds in Emulsion Gel-Based Delivery Systems Applied as Animal Fat Replacers in Frankfurters: Physico-Chemical, Structural and Microbiological Approach. *Food Chem.* **2021**, *340*, 128095. [CrossRef]
38. Liu, N.; Lu, Y.; Zhang, Y.; Gao, Y.; Mao, L. Surfactant Addition to Modify the Structures of Ethylcellulose Oleogels for Higher Solubility and Stability of Curcumin. *Int. J. Biol. Macromol.* **2020**, *165*, 2286–2294. [CrossRef]
39. Millao, S.; Iturra, N.; Contardo, I.; Morales, E.; Quilaqueo, M.; Rubilar, M. Structuring of Oils with High PUFA Content: Evaluation of the Formulation Conditions on the Oxidative Stability and Structural Properties of Ethylcellulose Oleogels. *Food Chem.* **2023**, *405*, 134772. [CrossRef]
40. Gao, T.; Wu, X.; Gao, Y.; Teng, F.; Li, Y. Construction of Emulsion Gel Based on the Interaction of Anionic Polysaccharide and Soy Protein Isolate: Focusing on Structural, Emulsification and Functional Properties. *Food Chem. X* **2024**, *22*, 101377. [CrossRef]
41. Lei, D.; Qin, L.; Wang, M.; Li, H.; Lei, Z.; Dong, N.; Liu, J. Insights into the Acid-Induced Gelation of Original Pectin from Potato Cell Walls by Gluconic Acid-δ-Lactone. *Foods* **2023**, *12*, 3427. [CrossRef] [PubMed]
42. Meng, Z.; Qi, K.; Guo, Y.; Wang, Y.; Liu, Y. Macro-Micro Structure Characterization and Molecular Properties of Emulsion-Templated Polysaccharide Oleogels. *Food Hydrocoll.* **2018**, *77*, 17–29. [CrossRef]
43. You, K.-M.; Murray, B.S.; Sarkar, A. Tribology and Rheology of Water-in-Water Emulsions Stabilized by Whey Protein Microgels. *Food Hydrocoll.* **2023**, *134*, 108009. [CrossRef]
44. Zhao, Q.; Zheng, B.; Li, J.; Cheong, K.L.; Li, R.; Chen, J.; Liu, X.; Jia, X.; Song, B.; Wang, Z.; et al. Emulsion-Filled Surimi Gel: A Promising Approach for Enhancing Gel Properties, Water Holding Capacity, and Flavor. *Trends Food Sci. Technol.* **2024**, *152*, 104663. [CrossRef]
45. Meng, Z.; Qi, K.; Guo, Y.; Wang, Y.; Liu, Y. Effects of Thickening Agents on the Formation and Properties of Edible Oleogels Based on Hydroxypropyl Methyl Cellulose. *Food Chem.* **2018**, *246*, 137–149. [CrossRef]

46. Zhu, Q.; Gao, J.; Han, L.; Han, K.; Wei, W.; Wu, T.; Li, J.; Zhang, M. Development and Characterization of Novel Bigels Based on Monoglyceride-Beeswax Oleogel and High Acyl Gellan Gum Hydrogel for Lycopene Delivery. *Food Chem.* **2021**, *365*, 130419. [CrossRef]
47. Qi, W.; Wu, J.; Shu, Y.; Wang, H.; Rao, W.; Xu, H.-N.; Zhang, Z. Microstructure and Physiochemical Properties of Meat Sausages Based on Nanocellulose-Stabilized Emulsions. *Int. J. Biol. Macromol.* **2020**, *152*, 567–575. [CrossRef] [PubMed]
48. Pan, Y.; Liu, S.; Han, Z.; Shen, L.; Lan, W.; Shao, J.-H.; Cheng, K.; Liu, Y.; Xia, Q.; Wang, Z.; et al. The Underlying Mechanism between Emulsification Stability and in Vitro Digestion in Golden Pompano (*Trachinotus Ovatus*) Myofibrillar Protein-Fish Oil Oleogel Emulsion under Ultrasonic Treatments. *Food Hydrocoll.* **2024**, *154*, 110015. [CrossRef]
49. Yu, D.; Feng, M.; Sun, J. Influence of Mixed Starters on the Degradation of Proteins and the Formation of Peptides with Antioxidant Activities in Dry Fermented Sausages. *Food Control* **2021**, *123*, 107743. [CrossRef]
50. Cheng, J.-R.; Liu, X.-M.; Zhang, Y.-S.; Zhang, Y.-H.; Chen, Z.-Y.; Tang, D.-B.; Wang, J.-Y. Protective Effects of *Momordica Grosvenori* Extract against Lipid and Protein Oxidation-Induced Damage in Dried Minced Pork Slices. *Meat Sci.* **2017**, *133*, 26–35. [CrossRef]
51. Wang, J.; Lu, J.; Zhang, X.; Kong, B.; Li, Y.; Chen, Q.; Wen, R. Effect of Inoculation with Autochthonous Lactic Acid Bacteria on Flavor, Texture, and Color Formation of Dry Sausages with NaCl Partly Substituted by KCl. *Foods* **2024**, *13*, 1747. [CrossRef]
52. Feng, Y.; Zhang, B.; Fu, X.; Huang, Q. Starch-Lauric Acid Complex-Stabilised Pickering Emulsion Gels Enhance the Thermo-Oxidative Resistance of Flaxseed Oil. *Carbohydr. Polym.* **2022**, *292*, 119715. [CrossRef]
53. Kong, Y.; Feng, M.; Sun, J. Effects of *Lactobacillus Plantarum* CD101 and *Staphylococcus Simulans* NJ201 on Proteolytic Changes and Bioactivities (Antioxidant and Antihypertensive Activities) in Fermented Pork Sausage. *LWT* **2020**, *133*, 109985. [CrossRef]

gels

Review

Emulsion Structural Remodeling in Milk and Its Gelling Products: A Review

Dexing Yao [1], Le-Chang Sun [1], Ling-Jing Zhang [1], Yu-Lei Chen [1], Song Miao [2], Ming-Jie Cao [1] and Duanquan Lin [1,*]

[1] College of Ocean Food and Biological Engineering, Jimei University, Xiamen 361021, China
[2] Teagasc Food Research Centre, Moorepark, P61 C996 Fermoy, Co. Cork, Ireland; song.miao@teagasc.ie
* Correspondence: dq.lin@jmu.edu.cn

Abstract: The fat covered by fat globule membrane is scattered in a water phase rich in lactose and milky protein, forming the original emulsion structure of milk. In order to develop low-fat milk products with good performance or dairy products with nutritional reinforcement, the original emulsion structure of milk can be restructured. According to the type of lipid and emulsion structure in milk, the remolded emulsion structure can be divided into three types: restructured single emulsion structure, mixed emulsion structure, and double emulsion structure. The restructured single emulsion structure refers to the introduction of another kind of lipid to skim milk, and the mixed emulsion structure refers to adding another type of oil or oil-in-water (O/W) emulsion to milk containing certain levels of milk fat, whose final emulsion structure is still O/W emulsion. In contrast, the double emulsion structure of milk is a more complicated structural remodeling method, which is usually performed by introducing W/O emulsion into skim milk (W_2) to obtain milk containing (water-in-oil-in-water) $W_1/O/W_2$ emulsion structure in order to encapsulate more diverse nutrients. Causal statistical analysis was used in this review, based on previous studies on remodeling the emulsion structures in milk and its gelling products. In addition, some common processing technologies (including heat treatment, high-pressure treatment, homogenization, ultrasonic treatment, micro-fluidization, freezing and membrane emulsification) may also have a certain impact on the microstructure and properties of milk and its gelling products with four different emulsion structures. These processing technologies can change the size of the dispersed phase of milk, the composition and structure of the interfacial layer, and the composition and morphology of the aqueous phase substance, so as to regulate the shelf-life, stability, and sensory properties of the final milk products. This research on the restructuring of the emulsion structure of milk is not only a cutting-edge topic in the field of food science, but also a powerful driving force in promoting the transformation and upgrading of the dairy industry to achieve high-quality and multi-functional dairy products, in order to meet the diversified needs of consumers for health and taste.

Keywords: milk; emulsion structure; milk fat globule membrane; protein; nutrient

1. Introduction

Milk is rich in macro- and micronutrients, including lipids, proteins, carbohydrates, vitamins and minerals [1]. These nutrients contribute to the growth and development of mammalian newborns. Among them, the lipids in milk and other dairy products are an important source of energy and nutrition for infants and even adults. For example, lipids in breast milk account for about 40–50% of an infant's energy needs. Milk lipids are usually in the form of fat globules, which are natural colloidal particles that release energy-rich lipids (e.g., triglycerides) and various bioactive molecules (e.g., essential fatty acids, conjugated linoleic acid, phospholipids, sphingolipids, cholesterol, carotenoids, and fat-soluble vitamins A, D, E, and K). After infancy, humans continue to consume milk from non-human sources to supplement various nutrients, such as cow's milk and ewe's milk [2]. However, due to the potential microbiological risks of milk (e.g., the presence of pathogens, such as *Campylobacter*, *Salmonella*, and *E. coli*), as well as the instability of milk because of the large size of natural milk fat globules, commercially available milk often undergoes multiple processing operations to improve its safety, stability, and shelf-life, the most common of which is thermal treatment and homogenization [2].

Processing operations, such as homogenization or heating, can improve the stability and shelf-life of milk, and one of the most important reasons is their ability to alter the fat globule structure in raw milk and remodel the emulsion structure. In fact, milk, including cow's milk, is itself an unstable oil-in-water (O/W) emulsion, in which milk fat globules (about 3.25 wt% in milk, ranging from 1 to 10 μm in diameter) encapsulated by a milk fat globule membrane (MFGM, containing highly polar lipids and proteins) are dispersed in ~90% water. Over time, this emulsion system may experience physical instability, such as changes in the arrangement or size distribution of milk fat globules, which can eventually lead to problems, such as flocculation, coalescence, and creaming [3]. However, after the processing of milk, its stability is mainly improved by reducing the size of emulsion droplets and reducing the number of microorganisms. The reduction in the size of the emulsion droplets leads to an increase in the interfacial area, but the newly formed interface is not completely covered by the MFGM, and casein can then be incorporated into the newly produced fat surface and thus provide stability against coalescence [1,2]. In addition, a lower microbial count reduces microbial spoilage and thus improves the safety of milk [4].

The structural remodeling of milk involves not only the modification of the particle size of the original milk fat globules, but also the change of the structure and configuration of the milk emulsion. Firstly, another oil/bubble can be used to replace the original milk fat globules to obtain milk with the structure of remolded single emulsion, which is generally based on skim milk for structural remodeling with the purpose of improving its taste and nutritional value. Second, a small amount of another O/W emulsion or a small amount of another oil can be mixed into the original O/W structure of the milk, such as fish oil [5,6], flaxseed oil [7], rapeseed oil [8] and sunflower oil [9], obtaining milk with a mixed emulsion structure, which is eventually still a kind of O/W emulsion. Finally, the original O/W structure of milk can also be transformed into water-in-oil-water ($W_1/O/W_2$) or oil-in-water-in-oil ($O_1/W/O_2$). For example, after removing the milk fat globules, another W/O emulsion system is introduced, and the $W_1/O/W_2$ composite/double emulsion structure is fabricated after mixing and homogenization. The purpose of this kind of milk emulsion structure remodeling is mainly to improve the physical properties of milk and construct milk with a specific functional property or suitable for a certain type of special population (e.g., consumers looking for functional foods and lactose intolerant patients), so as to produce new milk-based products and enhance their market competitiveness.

In recent years, there have been many in-depth studies on the structure of milk, but there is a lack of a systematic overview of the construction and application of milk with different emulsion structures and milk emulsion structure remodeling during processing. Therefore, the content of this review mainly focuses on the following aspects: (i) the original emulsion structure of milk and its remodeled single emulsion structure, (ii) the mixed emulsion structure of milk, (iii) the double emulsion structure of milk, and (iv) the changes of original/remodeled emulsion structures in milk during food processing, in order to provide researchers with a comprehensive understanding of the emulsion structure remodeling of milk.

2. Original Emulsion Structure and Restructured Single Emulsion Structure of Milk

Milk is an emulsion composed of aqueous and oil phases, which is a chemically and physically unstable system. The aqueous and oil phases in an emulsion have different compositions and functions. The aqueous phase of milk is the continuous phase that is mainly composed of water, emulsifiers, and other water-soluble components, such as lactose and whey protein [10]. Water, as the main component of milk, mainly acts as a solvent in the milk emulsion, dissolving some water-soluble components, such as proteins, sugars, and vitamins, so that they are evenly dispersed in the milk emulsion system (Figure 1). At the same time, it also ensures the fluidity and taste of the milk. The oil phase of milk is the dispersed phase, which is usually composed of oil, emulsifiers, and other fat-soluble components, such as carotenoids, fat-soluble vitamins (A, D, E, K), and a variety of volatile flavor compounds [2] (Figure 1). The oil phase provides nutrients and taste, as the oils provide nutrients, such as energy and fat-soluble vitamins, while also imparting a rich mouthfeel and smoothness to the milk.

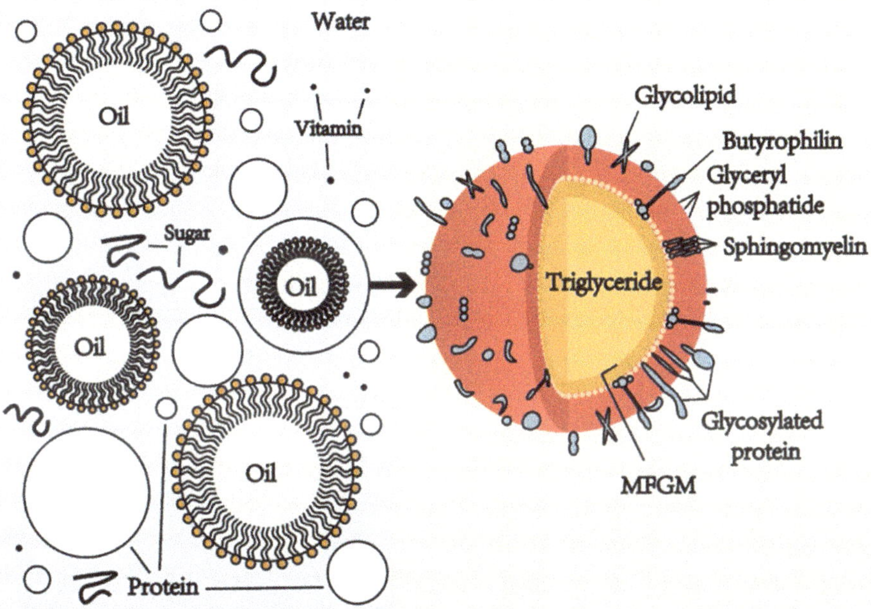

Figure 1. The original emulsion structure of milk. The left part of the figure shows the O/W emulsion structure of milk, in which the aqueous phase contains proteins, sugars, vitamins, etc., and the oil phase is composed of fat-soluble components, such as carotenoids, fat-soluble vitamins (A, D, E, K) and a variety of volatile flavor compounds. The right part is the structure of the milk fat globules, which are formed by a triglyceride core and coated by a milk fat globule membrane (MFGM) with a three-layer structure, where MFGM is composed of phospholipids, sphingolipids, cholesterol, glycoproteins and enzymes.

In milk, lipids account for about 3.5–5.2% of its total composition, mainly composed of triglycerides, accounting for more than 98% of the total lipids of milk, and the remaining, approximately 2%, of milk fats can be subdivided into different subcategories, mainly including diglycerides, monoglycerides, free fatty acids, phospholipids, and cholesterol [11]. The triglycerides in milk are in the form of MFGM-encapsulated spherical droplets, also known as milk fat globules. Structurally, milk fat globules are formed by a triglyceride-based core and surrounded by MFGMs with a three-layer structure, where MFGMs are composed of phospholipids, sphingolipids, cholesterol, glycoproteins, and enzymes [2] (Figure 1). MFGM plays a key role in the physical stability of triglycerides in milk because, as a natural emulsifier, it is able to form an interfacial adsorption layer on the surface of the lipid droplets, reducing the physical attraction between fat globules and reducing their tendency to aggregate, thereby reducing the flocculation and coalescence of fat globules [11–13]. In addition, MFGM can also protect fat globules from the action of lipase and slow down the decomposition of milk fat, because the MFGM membrane can act as a physical barrier to block the contact between lipase and fat globules, and MFGM also contains lipase inhibitors, which can slow down the action of lipase, thereby delaying the hydrolysis process of fat [2,11,12,14,15].

Although the presence of MFGMs in milk can reduce the surface tension between the aqueous and oil phases to a certain extent, the emulsified colloidal structure is still susceptible to interference and damage by external factors, which leads to the deterioration of the stability and quality of milk. There are two main triggers for this instability. On the one hand is the creaming caused by gravity, due to the different densities of oil droplets and water droplets, during which the oil may float to the top layer of the emulsion, leading to instability of the milk emulsion structure; on the other hand, the changes in temperature and pH can also affect the stability of milk. High-temperature treatment may melt emulsifiers with high melting points, losing the ability to stabilize the emulsion, and extreme acid-based conditions may alter the structure and properties of milk proteins, thereby disrupting the emulsion structure. In order to improve the stability of milk, some measures can be taken, such as adding appropriate emulsifiers, reasonably adjusting the pH and temperature of the emulsion system and adopting appropriate processing methods to prevent the occurrence of unstable phenomena, such as emulsion creaming [16–18].

In addition, most of the fat in fresh milk can be removed by physical methods, such as high-speed centrifugation, and skim milk can be obtained. Skim milk is easy to store because it contains almost no fat, is less prone to oxidation, and is also beneficial for lowering cholesterol, blood pressure and triglycerides. However, skim milk does not taste as good as whole milk, and fat-soluble substances (vitamins A, D, E, K) are lost during the skimming process. Therefore, there are also a small number of studies on the structural remodeling of skim milk, mainly by introducing a small amount of another oil/bubble into skim milk to obtain milk with a restructured single emulsion structure, in order to improve the taste and nutritional value of skim milk (Figure 2). For example, the replacing of fat globules with bubbles in acidified milk matrices has been studied to develop better-tasting and low-fat products [19].

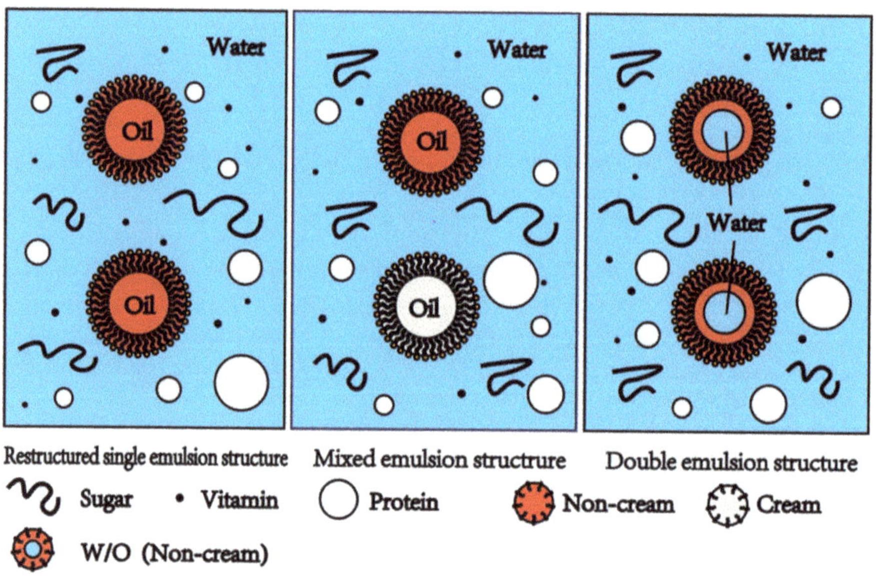

Figure 2. Milk with three different remodeled emulsion structures: restructured single emulsion structure, mixed emulsion structure, and double emulsion structure. Milk with the restructured single emulsion structure contains non-milk lipids. Milk with the mixed emulsion structure contains both milk fats and non-milk lipids. Milk with the double emulsion structure contains W/O emulsion droplets.

3. Mixed Emulsion Structure in Milk

Milk with mixed emulsion structure refers to the original structure of milk (i.e., the O/W emulsion structure) mixed with a small amount of another O/W or a small amount of another lipid substance (such as fish oil and flaxseed oil), and the final milk emulsion contains multiple lipids but it is still an O/W emulsion (Figure 2). The first step in fabricating the mixed emulsion structure is to add various admixtures to the original emulsion structure to obtain a mixed dispersion after mixing. These admixtures can be prepared with O/W emulsions and single or mixed oil phases (Table 1). Then, various processing methods are used, and the above mixed dispersion is emulsified to obtain a mixed emulsion. The aim of this kind of structural remodeling is to nurture the milk without affecting its stability, integrity and organoleptic properties, with the expectation of enhancing its beneficial effects on human health.

Table 1. Examples of milk containing mixed emulsion structures.

The Type of Admixture	Original Milk	Added Oil Phase	References
O/W emulsion	Whole milk	8–12 wt% flaxseed oil emulsion (addition amount of 5 wt%)	[20]
	Whole milk or milk with 2% fat	25 wt% algal oil emulsion (final concentration with 5 wt% algal oil added)	[21]
Oil	Whole milk	Flaxseed oil (addition amount of 10 wt%)	[7]
	Whole milk	Cod liver oil (addition amount of 5 wt%)	[6]
	Whole milk	Ghee and canola oil blends (addition amount of 20–50 wt%)	[8]
	Milk with 1.5% or 3.5% fat	Cod liver oil (addition amount of 0.5 wt%)	[22]
	Milk with 3.5% fat	Buttermilk powder (addition amount of 0.3–1.5%, *w/v*)	[15]
	Milk with 4.5% fat	Chia oil and α-lipoic acid nanoliposomes (addition amount of 20 wt%)	[23]
	Milk with 0.5 wt% and 1.5 wt % fat (1:1)	Cod liver oil (addition amount of 0.5 wt%)	[5]
	Milk with 0.5 wt% and 1.5 wt % fat (1:1)	Cod liver oil (addition amount of 0.5 wt%)	[24]
	Remix concentrated milk	Phospholipids (addition amount of 0.0–0.2 wt%)	[25]
	Compound evaporated milk	Cream residuum powder and sweet buttermilk powder (addition amount of 0–6 wt%)	[26]
	Full-fat donkey milk	Sunflower oil (addition amount of 1.6%, *v/v*)	[9]

w/v mass solubility, *v/v* volume ratio, and wt% mass percentage.

The addition of different admixtures may have different fortifying effects on milk. In terms of adding O/W emulsions, the nutritional value of milk has been enhanced by the addition of emulsions containing flaxseed oil or algae oil (Table 1), as they are both rich in omega-3 fatty acids. However, this method of structural remodeling may cause some adverse effects on milk. On the one hand, studies have shown that, with the increase of flaxseed oil addition (i.e., 8 wt%, 10 wt%, 12 wt%), the sensory score of milk decreases; therefore, an appropriate added concentration of oil should be selected and controlled to guarantee the nutritional fortification effect on milk without significant effects on its sensory scores [20]. On the other hand, the addition of oils rich in polyunsaturated fatty acids may reduce the oxidative stability of milk; therefore, it is necessary to add some antioxidants (such as ascorbic acid, ethylenediaminetetraacetic acid (EDTA) and sodium caseinate) to improve the oxidative stability of the mixed emulsion system [21].

In terms of adding lipids directly, many studies have been reported on remodeling the structure of milk by adding oils, such as fish oil, flaxseed oil, and sunflower oil. On the one hand, these enhance the nutritional value of milk. For example, fish oil and flaxseed oil are a class of oils rich in omega-3 polyunsaturated fatty acids (PUFAs) [6,7]; the addition of long-chain unsaturated lipids (e.g., canola oil) can balance lipid chain length and saturation levels to replicate the structure and nutrition of human milk [8]; the addition of sunflower oil to donkey milk lowers energy intake and improves its texture and health properties [9]; the introduction of lipids with specific biological activities, such as α-linolenic acid (ALA) and α-lipoic acid (LA), can also confer specific health effects on milk [23]. On the other hand, added lipids may also be beneficial in enhancing the physical stability of milk. For example, buttermilk powder is added to homogenized milk as an emulsifier for dairy products to produce milk emulsions containing a stable colloidal phase [15]; the addition of substances such as sweet buttermilk powder and cream residue powder, which are rich in natural emulsifier phospholipids, may also enhance the physical stability of milk [25,26].

Although, in most cases, the direct addition of lipids will reduce the oxidative stability of milk, this problem can be effectively solved by encapsulating the oil in advance (i.e., using spray drying to encapsulate the oil in food-grade wall materials, such as sodium caseinate, maltodextrin, and soy protein) or encapsulating it in liposomes to form nanoliposomes [6,23]. There are also studies of cow's milk and soy milk supplemented with fish oil containing gallates, which have been shown to maintain the tocopherol content in the emulsion and inhibit lipid oxidation [22]. In addition, the direct addition of lipids with special flavors (such as fish oil) can adversely affect the flavor of milk. It has been found that the use of different homogenization temperatures and pressures to blend fish oil into commercial homogenized milk affected the fishy smell residue of fish oil [5]. The mixed milk emulsion after high-temperature and high-pressure homogenization (72 °C, 22.5 Mpa) is less fishy than that after low-temperature and low-pressure homogenization (50 °C, 5 Mpa); high-temperature and low-pressure (72 °C, 5 Mpa) emulsification also lead to a stronger fishy smell than high-temperature and high-pressure (72 °C, 22.5 Mpa) homogeneous emulsification, because the high temperature and high pressure can reduce the formation of volatiles [5].

4. Double Emulsion Structure in Milk

Liquid emulsion with a double emulsion structure refers to the transformation of the original O/W structure of milk into an $O_1/W/O_2$ or $W_1/O/W_2$ structure, which may involve multiple steps, including the removal of milk fat globules from milk and the introduction of another W/O emulsion system [16,27,28] (Figure 2). Table 2 shows the composition of the inner aqueous phase, oil phase, outer aqueous phase and emulsifier of milk with a double emulsion structure in several previous studies. Studies on introducing prepared $W_1/O/W_2$ emulsions into whole milk have also been reported [29]. However, such emulsion structures are complex, and precious research has mainly focused on skim milk with a double emulsion structure, which is therefore also the main focus of this review. unless otherwise specified.

Table 2. Examples of milk containing a double emulsion structure.

Inner Aqueous Phase W_1	Aliphatic Phase O	Outer Aqueous Phase W_2	Emulsifiers	References
Distilled water (20 g)	(LT, BF, HS, or SO, 80 g)	Reconstituted milk (80 g, containing 10.0 wt% skimmed milk powder)	PGPR	[30]
Distilled water (20 wt%)	Olive oil (71.9 wt%)	Aqueous solution of biopolymers (80 wt%)	WE, OE, and GG	[31]
Ferrous sulfate (1%, *w*/*v*) in distilled water (40 wt% to W_1/O emulsion)	MCT (60 wt%)	30% WPI solution (75 wt% to $W_1/O/W_2$)	PSML, and PGPR	[32]
0.001 M phosphate buffered saline containing 0.2%, *w*/*v* vitamin B_{12} (10–30% oil-based)	Sunflower Oil (70–90%)	Skim milk (75–95%)	PGPR	[17]

Table 2. *Cont.*

Inner Aqueous Phase W_1	Aliphatic Phase O	Outer Aqueous Phase W_2	Emulsifiers	References
0.001 M phosphate buffered saline containing 0.2%, *w*/*v* vitamin B_{12} (10–30% oil-based)	Sunflower Oil (70–90%)	Skim milk (75–95%)	PGPR	[18]
Skimmed milk (30 wt%)	Sunflower Oil (70 wt%)	Skim milk (95 wt%)	PGPR and lecithin mixture	[28]
Skimmed milk (10 wt% oil-based)	Sunflower Oil (90 wt%)	Skim milk (80–95 wt%)	Span 80	[16]
Skim milk (40 wt%)	Canola oil or anhydrous milk fat (60 wt%)	Skim milk (80 wt%)	Sunflower lecithin, and PGPR	[27]

[Abbreviation: low trans vegetable fat (LTVF), refined bovine fat (BF), partially hydrogenated soybean oil (HS), refined sunflower oil (SO), hydrophilic emulsifier (WE), hydrophobic emulsifier (OE), gellan gum (GG), medium-chain triglyceride (MCT), whey protein isolate (WPI), poly-oxyethylene sorbitan monolaurate (PSML), polyglycerol poly-ricinoleate (PGPR), Span 80: lipophilic surfactant], [*w*/*v* mass solubility, *v*/*v* volume ratio, wt% mass percentage].

Preparation of milk with a double emulsion structure enables fat replacement and the development of low-fat products. For example, $W_1/O/W_2$ milk has been prepared with skim milk, PGPR, and different alternative non-dairy fats, including LTVF, BF, HS, and SO, which may serve as a potential lipid-lowering alternative to full-fat dairy products [30]. However, this fat substitution may affect the subsequent processing performance of milk. For example, a study on how olive oil-based ($W_1/O/W_2$) composite milk emulsion affects curdling behavior indicated that the restructured milk had the lowest cheese yield compared to whole and low-fat milk [31]. Although olive oil-based restructured milk is not ideal in terms of cheese yield, this study provides a reference for the replacement of cheese milk fat with olive oil emulsions, and also provides a basis for the application of milk with a double emulsion structure in the development of other milk fat substitute products [31].

In addition, the double emulsion structure can also be fabricated for milk fortification. Some studies have used water-in-oil (W/O) emulsification to encapsulate iron, glycyrrhizic acid (GA) or vitamin B_{12} to prepare nutrient-fortified milk [17,18,32]. It is important to note that iron in milk may catalyze lipid oxidation, which subsequently leads to rancidity, producing unpleasant odors and tastes. Therefore, in order to prevent lipid oxidation in iron-micro-capsulated milk, the amount of added iron needs to be controlled within a small range (i.e., 0.1–0.3%, *w*/*v*) [32].

5. Remodeled Emulsion Structures in Gelled Dairy Products

Some studies have mainly focused on the effect of the original structures on the properties of gelled dairy products with the purpose of improving their taste and nutritive quality [33–35]. For example, it has been indicated that buffalo set-yoghurts made with unhomogenized milk exhibited higher syneresis and poor stability upon shear-induced breakdown, mainly due to the porous gel structure containing a large number of bigger fat globules [36]. Generally, the purpose of fabricating the emulsion structures of gelled dairy products is to use health-beneficial lipids, like those derived from plants, to replace saturated fatty acids [37]. Therefore, in most cases, remodeled emulsion structures in gelled dairy products include restructured single, mixed and double emulsion structures [31]. Milk can be processed along with other gelled/semi-gelled dairy products, including fermented (e.g., cheese and yogurt) and condensed milk products. Therefore, the first strategy in fabricating the emulsion structures in gelled dairy products is implying milk with pre-fabricated emulsion structures to produce dairy products with different emulsion structures [31]. For example, recombined milks were formulated using two different concentrations (4 g/100 g and 6 g/100 g) of spray-dried emulsions containing PUFA-rich oils encapsulated with buttermilk, followed by fermentation to produce yogurt [38]. However, milk with different fabricated emulsion structures can affect the processing and quality of its gelling products. It has been found that skim milk containing olive oil-based $W_1/O/W_2$ emulsions showed a shorter reticulation phase and a lower cheese-making yield than full-fat milk [31].

Another strategy is processing skim milk or whole-fat milk followed by directly fabricating the emulsion structures in gelled dairy products [39]. For instance, the replacement of milk fat by rapeseed oil stabilized emulsion with an oil content of 25%, 50% or 75% (*w*/*w* basis) in full-fat or free fat commercial yogurt has been investigated in a previous study [37]. The findings indicated that incorporating rapeseed oil stabilized emulsion with 75% oil would seemingly reduce oil droplet size without much compromise to bacterial viability, sensory, or texture, and that such a fat replacement strategy shows promise in dairy products [37].

6. Remodeling Emulsion Structures of Milk During Food Processing

Milk is highly nutritious, which also means that it is prone to spoilage and is an excellent substrate for the growth of disease-causing microorganisms. This poses a huge challenge for the dairy industry in providing safe, shelf-stable, and affordable milk [40]. As a result, various processing techniques are applied to extend the shelf-life, improve the taste and texture, enhance the stability, and improve the nutritional value and functionality of milk. The common processing methods include heat treatment [41,42], high-pressure processing [43,44], homogenization [34,45], ultrasound [10,46], micro-fluidization [35,47], freezing [48–50] and membrane emulsification [51–53]. These processing technologies are widely used in the manufacture of milk, dairy products, and dairy ingredients.

6.1. Thermal Treatment

Thermal treatment is one of the major traditional food processing methods. Heat treatment of milk reduces microbial spoilage by destroying the proteins and nucleic acids in the cells, cell wall and cell membrane structure of microorganisms, and thus provides safer products with an extended shelf-life [40]. Although intense heat treatment of milk components can improve the safety and shelf-life of milk, the flavor and nutritional value are negatively affected, accompanied by many other drawbacks, such as protein denaturation and modification, reduced nutritional value, and the formation of Maillard products. Heat treatment also affects mineral and vitamin balance, leading to undesirable changes in the flavor and color of milk [4,26,54].

Heat treatment may adversely affect the stability of the original emulsion structure and restructured single emulsion structure in milk for the following reasons. Firstly, coagulation in milk during thermal processing is one of the biggest challenges in dairy sterilization. Several reactions have been found to occur during heating, including denaturation of whey protein and the formation of complexes between denatured whey proteins, casein micelles, and milk fat globules, leading to the occurrence of thermocoagulation [12,26,55]. Therefore, in milk emulsion systems with high concentrations of proteins, such as (recombinant) evaporated milk, the thermal stability of proteins during heating, especially whey proteins, is very important for milk products [12,13]. The mechanism of thermal instability of dairy proteins is that heat treatment leads to the unfolding of the natural dense structure of globulin, which exposes buried hydrophobic residues and leads to the aggregation of denatured proteins [41]. Studies have shown that thermally induced protein instability and calcium phosphate thermo-precipitation are the main causes of scale on some parts during the thermal processing of milk. Under certain conditions, whey protein, casein, calcium phosphate and fat can be contained in the scale. When milk is heated, κ-casein may undergo heat-induced dissociation at temperatures greater than 60 °C, reducing the stability of casein micelles. In addition, pH (from 6.6 to 5.5) is a major factor in heat-induced κ-casein dissociation, which collapses casein micelles during acidification due to charge neutralization [40].

Second, heat treatment of milk can also affect lipid digestion. Studies have shown that heat treatment may alter the structure of the interfacial coating surrounding the fat globules, such as protein denaturation and cross-linking, making lipase molecules more difficult to adsorb. In addition, heat treatment may promote a greater degree of fat globule flocculation, which again limits the ability of lipase to adsorb on the surface of fat globules, thereby reducing the initial rate of lipid digestion [2]. Temperature also has a significant impact on the stability and integrity of MFGMs. Studies have shown that, in milk, heat treatment between 60 °C and 95 °C affects the surface structure of milk fat globules, mainly by whey protein binding to MFGM protein [11,56].

Third, during thermal processing, the degradation of lipids, proteins, and other emulsion components leads to change in the flavor profile of milk, which is associated with the formation of volatile compounds [12]. It has also been shown that heat treatment of milk also produces methyl ketones, such as 2-pentanone, 2-heptanone, and 2-nonanone, and 2-heptanone, which has been identified as one of the compounds with the strongest volatile flavor in hot milk [12].

Ultra-high-temperature heat-treated (UHT) milk has a long shelf-life at room temperature, making it a nutritionally, technically and economically important food. Studies have shown that certain chemical and physical changes caused by UHT treatment can lead to storage instability, including the formation of protein precipitates at the bottom of the storage container or the formation of precipitates or gels throughout the milk (mainly involving the aggregation of milk proteins), which limits the shelf-life and market potential of UHT milk [42]. However, the heat load after UHT (including evaporation and spray drying) generally has little effect on the physicochemical and functional properties of the production of infant formula (IMF) powders [57].

Indeed, consumer acceptance of milk is largely determined by its organoleptic characteristics and nutritional value, so there is a huge demand for new technologies to replace conventional thermal processing with technologies that cause minimal damage to nutrients and provide a longer shelf-life. To achieve this, a number of non-thermal techniques have been explored, such as high-pressure processing and ultrasonics [4,54].

There are currently no studies on the effect of heat treatment on the structure of milk with mixed or double emulsion structures. Heat treatment may affect the stability, taste and nutritional value of this kind of milk, which however needs further investigation.

6.2. High-Pressure (HP) Processing

HP processing is a non-thermal processing method that can replace traditional thermal processing and is widely used in the food processing industry. HP treatment is able to kill microorganisms by disrupting the fluidity and integrity of their cell membranes and inactivate specific enzymes by changing their spatial structures under high pressures at room temperature to extend the shelf-life of raw materials and food products in order to ensure food safety while reducing quality loss in the processed products. HP treatment does not lead to thermal degradation of food ingredients, does not impair the organoleptic, nutritional and physicochemical properties of food, does not affect the composition of fatty acids, and minimizes the use of chemical additives [43,44].

HP treatment is also widely used in the pre-treatment of milk. HP treatment requires the use of extremely high pressures (typically between 100 and 1000 Mpa) for a specific period of time [58]. This allows HP treatment to achieve a reduction in the population of microorganisms in milk, such as *E. coli*, without the aid of heating [44].

In addition, the pressure during HP treatment also affects the properties of the milk itself, such as temperature, pH, fat globule size, casein micelle particle size and protein distribution and hydrophobicity [46]. Studies have shown that the diameter of the particle size of untreated goat's milk increases after 14 days of refrigeration while, in HP-treated goat's milk, the increase in the above parameters is inhibited or eliminated after higher pressure (>200 MPa) treatment, and even 400 Mpa HP-treated goat's milk has no significant difference in particle size during storage [43].

In terms of effects on lipids, HP treatment has been shown to alter the size of the fat globules and the composition of the MFGM, with pressure slightly affecting the size of the fat globules and temperature parameters having a greater effect on the size [44]. Processing milk at temperatures above 25 °C produces smaller fat globules, while the effect is opposite at lower temperatures. However, the HP treatment itself has been shown to have little to no effect on the lipid composition of milk. In addition, under mild pressure (250 MPa) conditions without heating, enzymatic lipolysis is facilitated and, due to the release of free fatty acids, as well as the production of mono- and ditriglycerides and additional volatile compounds, leads to undesirable flavors and aromas. Therefore, the effects of high pressure on milk fat and fat globules need to be studied in more depth, especially since these changes are related to the stability of the emulsion system and quality changes in the final products [44].

HP treatment also alters casein micelle size. The effect of HP treatment on the electrostatic and hydrophobic interactions leads to the breakdown of submicelles, followed by the dissociation of casein moieties from micellar casein, after which they re-aggregate and rebind, causing changes in casein micelle size [43]. HP treatment can also cause protein denaturation, mainly by causing the unfolding and destruction of different bonds and interactions, which changes the native structure of proteins. These changes in milk proteins also lead to alterations in their functions and properties. During HP treatment, the size, composition and hydration of casein micelles undergo structural changes. However, the specific variation and the extent of these changes depend on the processing conditions. During HP treatment, the water is compressed, which causes the hydrophobic bonds of the casein micellar components to be broken and then changes the light transmission of the milk, and the minerals are also dissolved (mainly micellar phosphate). A study indicated that HP treatment exacerbated the change in casein micelle size [44]. HP treatment after the application of 500 MP severely affected and changed the size and structure of casein micelles (the spherical structure became irregular, the structural integrity was lost, and the size decreased), and these changes were thought to be responsible for additional effects on the properties of milk, such as changes in color and stability, rheological properties, and pH [44].

HP treatment also has an effect on whey protein. The whey protein in HP-treated milk undergoes reversible unfolding due to pressure, which allows the water molecules in the medium to penetrate the hydrophobic region of molecules, leading to conformational change of the protein and eventually the formation of aggregates. This structural change promotes the improvement of some specific functions of the milk, such as hydrophobicity, solubility, gelatinity, firmness and emulsifying properties, which can improve the quality of other dairy products, such as cheese and yogurt. However, the specific variation and the extent of these changes depend on the processing conditions. For instance, due to the interaction between denatured β-lactoglobulin and κ-casein, a lighter pressure of about 250 MPa may increase the size of the micelles; however, these changes are reversible [44]. Temperature and pH also affect the change in micelle size during processing, and it was found that, as the pH increased, the temperature increased the micelle size [44]. At the same time, HP treatment has a dual effect on enzymes, which can be activated or inhibited depending on the intensity of the pressure, the type of enzyme, and the temperature. Pressures below 350 MPa can increase the activity of the enzyme because the partial unfolding of the enzyme and protein-based substrate is conformationally flexible, thus facilitating the interaction between them. Inactivation is known to begin at pressures above 400 MPa, and inactivation may increase as pressure increases. However, the degree of inactivation is affected not only by pressure level, but also by processing time, enzyme type, milk composition, and pH level [44].

During the processing of milk, the degradation of lipids, proteins, and other milk components leads to changes in the flavor profile of final products, which is associated with the formation of volatile compounds, such as aldehydes and sulfur compounds [12]. Overall, however, HP treatment offers an alternative to heat preservation, pasteurization, and sterilization in dairy technology, and allows for the development of new products with the desired texture, flavor, and functional properties [43].

6.3. Homogenization

In the food industry, homogenization usually refers to the more even dispersion of fat, protein and other components in food through physical methods (such as high-pressure homogenizers), so as to improve the taste, texture and stability of food. Homogenization is widely used in the dairy industry to prevent creaming of milk during storage [34]. For example, in milk processing, homogenization can cause the fat globules to break into smaller particles, preventing the fat from floating and making the milk more delicate. Homogenization is also used to produce reformulated milk with skimmed milk powder and milk fat, as well as to produce filled milk with non-dairy lipids, such as vegetable oils [1].

6.3.1. Effect of Homogenization on Milk with Original or Restructured Single Emulsion Structures

Homogenization can inhibit emulsion creaming during storage of liquid dairy products by reducing the size of milk fat globules [45]. The main mechanism is that the milk is forced through a narrow gap under high pressure in the high-pressure homogenizer, creating a strong shear force, impact force, and cavitation effect, which work together to break the fat globule. Homogenization also prevents milk fat globules from clumping together by converting them into smaller particles. Thus, homogenization is able to promote fat dispersion [33]. For example, homogenization can reduce the size of milk fat globules, thereby improving milk stability [59]. Studies have also shown that homogenization can reduce the size of goat milk fat globules, thereby increasing the interaction between fat globules and casein, which is beneficial to the stability of the emulsion structure of goat milk [60]. Turbulence, shear, and cavitation during homogenization are the main forces that lead to fat globule rupture [11,12]. However, the extent of reduced size of fat droplets by homogenization (5–25 MPa) highly depends on the pressure. Sometimes, fat globule aggregation may occur during homogenization, but this can be eliminated by two-stage homogenization [56].

Studies have shown that homogenization reduced the size of milk fat globules in milk, increased their number and surface area, and altered MFGM properties [12,13,15,56]. As a result, the milk fat globules are not completely covered by the MFGMs. Therefore, the surface-active components in the aqueous phase of milk emulsion, such as casein and whey protein, spontaneously adsorb to the surface of the fat globules, form new interfaces, and provide stability against coalescence [1,2,56]. However, at the time of homogenization to reduce the size of milk fat globules, proteins can be partially replaced by phospholipids, which is determined by the polarity of phospholipids and their concentration on the outer bilayer of MFGMs [34].

The above-mentioned changes in emulsion structure caused by homogenization may affect the digestibility of milk. The reduced size of the milk fat globules may result in a faster digestion rate. However, homogenization of natural milk fat globules does not appear to improve the lipolysis rate, as the effect of increased surface area may be offset by the presence of proteins at the interface of lipid droplets, and homogenized droplets tend to aggregate more quickly than natural milk fat globules under gastric conditions [11,13,56]. Studies have shown that homogenization does not appear to have a significant effect on the final degree of milk fat digestion, but following thermal treatment appears to reduce the overall degree of lipid digestion of homogenized samples [2]. The probable reason was that heating above the thermal denaturation temperature of the adsorbed whey protein can promote the unfolding and interfacial cross-linking of the whey protein. As a result, the interfacial structure and fat globule aggregation are significantly changed, which reduces the speed and degree of fat globule digestion [2]. However, higher temperatures during homogenization can improve the efficiency of homogenization [12], but previous studies showed that milk fat globules (MFGs) after homogenization and thermal treatment had higher flocculation rates than natural MFGs [36].

In addition, different homogenization methods (e.g., shear homogenization, atmospheric homogenization, high-pressure homogenization (HPH), and ultra-high-pressure homogenization (UHPH)) may have different effects on the structural changes of the milk described above.

The application of HPH can increase the temperature of milk [15,34]. As a result, milk can be homogenized at a lower initial temperature compared to atmospheric pressure homogenization, which eliminates the heating phase and simplifies the process [34]. HPH is more effective in generating small emulsion droplets [28]. This technique can replace atmospheric homogenization or pasteurization with a lower pasteurization temperature. At the same time, it effectively eliminates pathogenic microorganisms and improves the quality of milk emulsions [15,34].

HPH treatment is also able to affect the adsorption of proteins on the surface of fat globules, lead to the denaturation of whey proteins, and alter the structural properties of casein micelles, thereby altering the functional properties of proteins and their condensation [45]. It was found that the adsorption of milk proteins on the surface of milk fat globules was promoted by the application of HPH in the pressure range of 100–200 MPa and the temperature range of 20–40 °C, and the adsorption capacity increased with the increase of temperature [34]. The properties of these new interfaces (i.e., stability and viscosity) can also be significantly altered during thermal treatment, due to changes in the conformation and interactions (i.e., protein–protein and protein–fat interactions) of the different types of proteins present (e.g., casein and whey protein) [12]. In the temperature range of 20 °C to 40 °C, the molecular movement of milk protein intensified with the increase in temperature [12], which was conducive to its adsorption and rearrangement on the surface of milk fat globules. In addition, elevated temperatures may also promote protein–protein and protein–fat interactions, further enhancing the stability of the emulsion. However, excessively high temperatures can also cause protein denaturation or aggregation, which is detrimental to the stability of the emulsion.

UHPH is a homogenization method based on the same principles as atmospheric pressure homogenization, but working at higher pressures. UHPH can operate at pressures as high as 400 Mpa, compared to 20–50 MPa homogenized at atmospheric pressure [38,54]. Cavitation, friction, turbulence, and shear stresses are the forces that arise during the UHPH process. UHPH is also able to reduce milk microbial counts, denature whey protein, inactivate intrinsic lactases, such as plasmin and alkaline phosphatase, and drastically reduce fat globule size. Studies have shown that UHPH could reduce the size of fat globules (down to a fraction or even less) and enhance the activity of natural lactoprotein lipase [54]. However, the study also showed that high-temperature short-time (HTST) thermal treatment (72 °C, 15 s) was able to completely inactivate natural lipases that cause liquid lipolysis, such as natural lactoprotein lipase [54].

Shear homogenization refers to crushing and refining materials through extrusion, impact and shear, so that the dispersed phase is distributed in the immiscible continuous phase to achieve the purpose of uniform mixing. Studies have shown that shear homogenization of buffalo milk improved its texture and made its gel firmer with a higher storage modulus, a smaller hysteresis area, and better deformation recovery ability. In addition, the newly formed MFGMs after homogenization are different from the original MFGMs, and the former are mainly composed of membrane fragments and casein, in which a newly formed dense protein layer is covered by the MFGMs. These newly formed MFGMs are cross-linked with the surrounding milk protein matrix, resulting in higher gel strength. However, homogenization also has side effects, with studies showing that shear homogenization increased the total content of free fatty acids in buffalo milk compared to untreated buffalo milk, which may lead to fat-soluble rancidity [36]. Compared with raw milk, shear homogenization could increase free saturated fatty acids by 2.75~3 g per 100 g of fat [36].

6.3.2. Effect of Homogenization on Milk with Mixed Emulsion Structure

In the study of milk with mixed emulsion structure, after homogenization under atmospheric pressure, milk is rich in substances that can accumulate on the surface of newly formed fat globules, accompanied by a reduction in the size of fat globules. HPH can also reduce the diameter of milk fat globules, increase the surface area of milk fat globules, and promote the emulsification of excess fat in dairy products. Both conventional homogenization and HPH can improve the thermal stability (HS) of milk emulsions, while HPH can increase the viscosity of emulsions [15]. In a study on the effect of homogenization conditions on the oxidative stability of fish oil-fortified milk, it was found that homogenization conditions (e.g., temperature and pressure) had no effect on the oxidative stability of milk emulsions containing stable fish oil, and that the actual composition of the oil–water interface (e.g., protein composition) was more important than the total surface area itself [5].

6.3.3. Effect of Homogenization on Milk with Double Emulsion Structure

In the study of the formation of a double $W_1/O/W_2$ emulsion in skim milk using minimal food-grade emulsifiers (i.e., skim milk in the inner and outer aqueous phases and sunflower oil in the oil phase), ultrasonic waves were used in the first emulsification step and different high-shear techniques (i.e., ultrasonic and HPH) were used in the second emulsification step [28]. Results showed that an increase in HPH pressure or an increase in the duration of ultrasound resulted in a smaller droplet diameter, but HPH was more efficient at producing small emulsion droplets because the mechanism of HPH is not to produce droplets larger than the valve gap. However, for a given energy load, either ultrasound or HPH could form a double emulsion with similar size and similar degree of encapsulation [28]. Therefore, it can be concluded that homogenization can reduce the size of the fat globules in milk with a double emulsion structure and, at the same time, increase the surface area of the milk fat globules.

6.4. Ultrasonic Treatment

Ultrasound treatment is a promising technology for disrupting particles and can be applied to food processing [4,61]. Ultrasound refers to sound waves with frequencies above 20 kHz that are imperceptible to the human ear and are divided into two categories: power ultrasound and diagnostic ultrasound. Diagnostic ultrasound ranges from 2 MHz to about 15 MHz and is widely used in the medical field. In contrary, power ultrasound, which ranges from 20 kHz to about 1 MHz, has been applied in food processing [4]. For example, powder ultrasound has been used to form stable functional skim milk with the bioactive substance (i.e., black seed oil) [10].

6.4.1. Effect of Ultrasonic Treatment on Milk with Original or Restructured Single Emulsion Structures

Ultrasonic treatment is considered to be an effective processing method, because it can lead to acoustic cavitation in liquids. When energy passes through a liquid medium, it causes the formation, growth, and rupture of bubbles, resulting in strong local shear forces and turbulence with an increase in temperature. These stresses and repeated disintegration of bubbles can disrupt the MFGMs, creating spatial and transient stresses on the surface of the fat globules and particles, which can lead to particle breakage. Many studies have proved that ultrasonic waves mechanically create air bubbles in fluids through cavitation, resulting in a reduction in particle size [4,28,33,36,62].

For milk, the turbulence associated with acoustic cavitation enhances the partial mobility of milk fat globule particles, thereby improving the aggregation between co-aggregated protein and the newly formed MFG-protein complex. As a result, milk gels produced from ultrasonicated milk showed improved gel properties, such as gel strength, elasticity and hardness, viscosity, and water holding capacity, while reducing gelling time [36]. However, prolonged ultrasonic treatment has been reported to have adverse effects on milk gel because whey proteins dissociate from micellar aggregates and MFG flocculates to form homogeneous clusters, resulting in a weak gel network with high synergistic effects [36].

During ultrasonic treatment, the shear forces generated by acoustic cavitation can also cause changes in the protein particles in the milk emulsion. The strong cavitation force usually leads to partial cleavage of the hydrophobic interaction between whey protein molecules, resulting in reversible or irreversible denaturation, depending on the ultrasound intensity [36]. It has been found that whey proteins and whey–whey aggregates in milk were denatured and soluble whey–whey/whey–casein aggregates were formed, which further interacted with casein micelles to form micellar aggregates during the first 30 min ultrasound treatment; however, prolonged treatment with ultrasound results in the destruction of some of the whey protein from these aggregates [4]. Partially denatured whey protein can provide stability to emulsified droplets, allowing for the production of emulsion droplets with smaller sizes [63]. In addition, the physical forces of acoustic cavitation have no effect on the integrity of the casein micelles but result in a slight reduction in their size. However, studies have shown that small changes to the protein by ultrasound do not alter the viscosity of milk [4].

Ultrasonic treatment can also be used to assist in producing milk with a restructured single emulsion structure. Studies have shown that ultrasound-assisted emulsification can be used to prepare stable turmeric oil-loaded milk emulsions [64]. Ultrasound-assisted emulsification has considerable advantages over conventional methods, such as obtaining submicron-sized milk fat globule particles, narrow particle size distributions and more stable emulsions, and low energy requirements, and can be used as a simple and low-cost processing technique [16,64]. Ultrasound-assisted emulsification is a two-step process in which Rayleigh–Taylor instability drives the formation of droplets of the dispersed phase in a continuous phase, followed by resulting droplets in the second step, which are broken into tiny droplets by a shock wave generated by instantaneous cavitation. The formation and final size of the droplets are influenced by a number of operational factors, including the presence of surfactants. The main role of surfactants is to reduce the interfacial tension so that less energy is required to produce the droplets, and also to adsorb on the surface of the newly prepared droplets to prevent droplet recoalescence, resulting in a stable emulsion [64].

Ultrasonic treatment can also be used in combination with homogenization to process milk. The main purpose of ultrasonic homogenization is to reduce the size of fat globules, but also to increase the stability and consistency of the resulting milk emulsion, and to reduce the separation of fat and aqueous phases during consumption and storage. By controlling the temperature and shortening the homogenization duration, the negative effects of ultrasonic treatment, such as lipid oxidation, can be avoided [46].

6.4.2. Effect of Ultrasonic Treatment on Milk with Double Emulsion Structure

The physical and chemical effects produced by cavitation bubbles have many practical applications in milk with a double emulsion structure. In the study of the preparation of $W_1/O/W_2$ emulsions containing skim milk and sunflower seed oil with low-frequency ultrasonic treatment, sonication was used in the preparation of both W_1/O and $W_1/O/W_2$ emulsions, and the results showed that the resulted double emulsion droplets were relatively stable within 7 days [16]. The results of above study suggested that, for a given energy load, either ultrasound or HP treatment can form a similar size and similar degree of encapsulation of milk with a double emulsion structure [28]. As expected, an increase in the duration of ultrasound or HP treatment pressure could result in a smaller droplet diameter. However, increasing the ultrasonic power may lead to greater shear forces, which results in more emulsion droplets being destroyed and then reduces the retention of internalized aqueous droplets [16]. Compared with HP homogenization, ultrasonic treatment has an inherent randomness due to its mechanism of acoustic cavitation, and the reduction in droplet size is random, resulting in a larger size distribution. Therefore, in this case, the presence of larger droplets may lead to a higher encapsulation efficiency at the time of encapsulation, but this may also reduce the stability of the double milk emulsion [28].

Another application example is the encapsulation of glycyrrhizic acid (GA) with an ultrasound-assisted milk-based double emulsion, in which ultrasonic treatment was used in the preparation of both W_1/O and $W_1/O/W_2$ emulsions [17]. Studies have shown that sonication was able to reduce droplet size; However, prolonged ultrasonic treatment could lead to droplet rupture in the original emulsion, a decrease in pH and an increase in temperature, resulting in uncontrolled release of the encapsulated compound and lower encapsulation efficiency [17]. Therefore, an appropriate sonication time should be selected to control the encapsulation efficiency [16–18].

6.5. Membrane Emulsification (ME)

ME involves the use of low compression to allow the dispersed phase to permeate through the membrane with a specific pore size under the action of different pressures into the continuous phase. In contrast to homogenization, the resulting droplet size is mainly controlled by the choice of membrane rather than by the generation of turbulent droplet breakage. The technique is very attractive due to its simplicity, potentially lower energy requirements, the need for less surfactant and the resulting narrow droplet size distribution [1,51]. The advantages of ME in the food processing industry may come from its low shear properties and mild processing conditions, especially for the preparation of double emulsions, structural phases with fine droplets, and microcapsules. Therefore, in contrast to traditional emulsification methods (e.g., homogenization), small and monodispersed droplets can be produced without the use of high shear stresses that cause internal droplets to escape, which is more effective in maintaining the microstructures and nutrients in emulsion food products [51]. Another advantage of ME is the amplification capability of the membrane device. By adding more fibers to the unit, increasing the fiber length, and using the unit in parallel, the flow rate can be easily increased to achieve the desired conversion rate and productivity. The limitations of the ME process may be related to the low flux and fouling of monodisperse emulsions. These shortcomings can be addressed by other operations, such as premixed processing, as well as rotating or vibrating membrane devices [51].

ME technology is divided into two types: direct ME and premixed ME: (i) in direct ME, a continuous phase flows tangentially to the membrane surface, while the dispersed phase is squeezed through the pores of the membrane, and the droplets of the dispersed phase grow at the openings of the pores in the membrane and separate when they reach a certain size; (ii) premixed ME is based on the formation of a coarse emulsion by traditional mechanical techniques and then passing it through a membrane to obtain a narrow droplet size distribution [1,65].

In ME, different types of membranes can be used, such as Shirasu-porous glass (SPG) membranes, ceramic membranes, and silicon and silicon nitride membranes [1,51]. In the food industry, different membranes have their own specific application scenarios. Inorganic membranes (e.g., ceramic and SPG membranes) are suitable for the preparation of food emulsions that require strict hygiene, high-temperature sterilization or chemical resistance. Silicone-based membranes may be more suitable for some food emulsions that require precise control of droplet size. ME is a suitable technology for emulsified oils and milk. It has been reported that milk proteins could stabilize emulsions without the use of any external reagents as surfactants, achieve encapsulation at an oil concentration of 30 wt%, and obtain droplet sizes that are approximately 2 to 6 times the size of the membrane pore [1].

In the study of the preparation of O/W emulsion containing fish oil by premixed ME method, the study showed that the emulsion produced by premixed ME has a smaller droplet size distribution and is more monodisperse than the emulsion produced by the rotor-stator [53]. Another study indicated that ME could be successfully used to produce astaxanthin-containing O/W emulsions with significantly narrow droplet size distribution, and that the average size of the droplets could be adjusted by varying experimental parameters, such as pressure and dispersed phase fraction, and small mean diameter values were obtained at high pressure and low disperse phase fractions [65].

At present, there are no studies on the effect of membrane emulsification on the structure of mixed milk emulsions. However, the predicted effect of membrane emulsification on the structure of the mixed milk emulsion is mainly in terms of droplet size, which can produce relatively small droplets and may have a higher encapsulation efficiency. In future, understanding the effect of different material-based membranes with different pore sizes on mixed milk emulsions can help to provide a basis for selecting the right membrane. The experimental parameters in the membrane emulsification process, such as pressure, temperature, flow rate, emulsifier type and concentration, need to be investigated for regulating the structure and properties of mixed milk emulsions. In terms of milk with a double emulsion structure (i.e., $W_1/O/W_2$), primary emulsions can be prepared by conventional methods or by ME. The mild conditions of ME are particularly useful for the second emulsion step, which prevents the rupture of the double emulsion droplets [51].

6.6. Microfluidization

Microfluidization is an emerging technique and the most effective method for generating uniformly distributed droplets [35,66]. Due to its high cost and production limitations, its applicability on a larger scale is limited, but microfluidization has an important role in various studies due to its advantages such as short processing time, low thermal effect, and almost no nutrient loss. Therefore, microfluidization technology can produce emulsified products with uniform particle size, good nutrient retention and high stability, which meets the production requirements of high value-added products. In such cases, the advantages of microfluidics may indeed outweigh its high costs. Microfluidization is not only used in the preparation and production of emulsions, nanoparticles, and beverages, but is also used to process some food materials to improve their specific properties and thus is suitable for new applications [67]. In principle, the production of emulsions using microfluidizers is the most energy-efficient and suitable for high-throughput processing, though the cost of maintenance is high [16].

The formation of milk emulsion using a microfluidizer usually involves two steps. First, a high-shear agitator is used to mix the oil, emulsifier, and water phase together to form a coarse emulsion. Secondly, pneumatic pressure is used to force the coarse emulsion through the microfluidizer. After the coarse emulsion enters the microfluidizer, it is split into two separate streams, causing them to collide with each other at high speed. This process generates strong destructive forces, such as cavitation, turbulence, and shear, which effectively disrupt the large droplets in the coarse emulsion, resulting in the formation of very tiny droplets [66–71]. In addition, the frequent collisions occurring between the particles (e.g., fat globules and protein aggregates) and between the particles and the microchannel walls are also able to break up larger particles and reduce their size, thus improving the uniformity and stability of milk. Crude emulsions prepared by high-shear mixers can reach a particle size of 1.5 μm, while emulsions prepared by microfluidization processes have a smaller particle size (≥200 nm) [68]. The results of a study in the production of emulsion containing fish oil showed that microfluidization could form stable nano=emulsions, and that, when the average droplet size of nano=emulsions was around 200 nm, it could be used to improve their absorption in the digestive tract [68]. Another study showed that microfluidization techniques allowed emulsions to have droplet sizes in the nanometer range (<500 nm) with a single-modal particle size distribution and good stability [70].

In general, the size of the emulsion decreases with increasing pressure and duration within the optimal microfluidization pressure range. However, after exceeding the optimal treatment conditions, the emulsion size may increase or barely change, because the emulsion droplets break and aggregate [69,72]. A study indicated that that, during the process of emulsification homogenization, the shear, turbulence and collision caused by microfluidization could simultaneously change the size of oil droplets in an O/W emulsion stabilized by natural pea globulin and the structure of the pea globulin-based interfacial membrane [73]. In addition, when the microfluidization pressure increased, the emulsion stability of the liquid emulsion had two opposite effects, either an improvement in the emulsion stability or a decrease in the emulsion stability of the liquid emulsion, which might be a consequence of flocculation [73]. Further research should be carried out to determine the specific effects of different pressure gradients, temperatures, and processing flow on the droplet size in milk. However, it has been indicated that, when the microfluidization pressure was held constant, the particle sizes increased as the milk fat concentration was increased, indicating that higher microfluidization pressure might be needed for milk containing higher levels of fat to obtain stable final products [74].

In terms of the effect of microfluidization on the digestive behavior of milk, it has been indicated that microfluidization treatment could promote in vitro and in vivo protein digestibility of milk [75]. This may be due to the breakage of fat, as well as protein molecules, into a very small size with increased surface area as a result of microfluidization, thereby interacting more fully with digestive enzymes and accelerating the release of nutrients, such as fatty acids and amino acids. However, more research is needed to further clarify this relationship and provide a basis for the development of more efficient nutrient delivery systems.

6.7. Freezing

Freezing is a common technique in the food industry to preserve food and extend its shelf-life. Food is exposed to extremely low temperatures, usually below freezing, which slows down bacterial growth and chemical changes, thus preserving the freshness and quality of food. Freezing can also be applied to the storage of milk, and freezing has a certain effect on the original emulsion structure in milk. When milk is frozen, several reactions occur, which are related to the freezing temperature or freezing time. There are two unstable phenomena in milk when it is frozen. One is fat separation, which occurs during the cooling process of milk in the early stage of freezing. The other is protein flocculation, which depends on the final storage temperature and the time of frozen storage [48].

When milk is stored at low temperatures, the fat globules may undergo partial coalescence. The possible mechanism is that freezing can lead to the crystallization of triglyceride nuclei, and the fat crystals can protrude from the surface of the fat globules and pierce the neighboring MFGMs, partially coalescing to produce large, irregularly shaped particles or a continuous network of aggregated fats, which tend to rapidly form fat creaming. The partially coalesced fat globules themselves can further coalesce as the temperature rises and the fat crystals melt. During the partial coalescence, some components in MFGMs can be released into the water phase [11,56,60].

In addition to the changes in fat, freezing also has an effect on the non-fat compositions in milk, mainly the balance of salt. This can be explained by the partial freezing of water and the formation of a supersaturated salt solution, increasing the ionic strength and thus the osmotic pressure of the system [60].

7. Future Outlook

The future development of milk with different emulsion structures may focus on the following areas:

(i) The design of functional milk: Future research may focus on the design of multifunctional milk, which can give milk more functionality by changing the surface structures of milk fat globules, remodeling its emulsion structures, and introducing other functional components into milk. Moreover, in the relevant practical applications, it is necessary to select the appropriate additives (e.g., oil from different sources with various properties) and processing processes according to the specific production needs and product characteristics to achieve the best effects, which also needs more in-depth and systematic research.

(ii) The innovation in emulsifying technology: The innovation of emulsification technology helps to improve the stability and texture of milk with remodeled emulsion structures. On the one hand, although some emulsifying technologies (e.g., homogenization and ultrasonic treatment) have been used for preparing milk with original, restructured single, mixed, or double emulsion structures, some novel emulsifying technologies (e.g., ME and microfluidization) are still limited to remodeling the original emulsion structures in milk. These novel emulsifying technologies may show good performance to improve the properties of milk, which however needs further investigation for confirmation. On the other hand, the combination of different emulsifying technologies (e.g., the combination of homogenization and ultrasonic treatment) may show synergetic effects. Therefore, more diverse combinations of different emulsifying technologies with different working mechanisms to overcome their respective shortcomings in the production of milk with different emulsion structures may also be an interesting research topic in future.

(iii) The development of gelled dairy products with different emulsion structures and functions: Previous studies mainly focus on the fabrication of emulsion structures in milk for various purposes, but research on the processing properties of milk with remodeled emulsion structures and the functional properties of the relevant processed products are still insufficient. Therefore, more efforts are needed to understand how remodeling emulsion structures in milk affects its subsequent processing performance (e.g., fermentation, concentration, and curdling) and the sensory and functional properties of the resultant final dairy products (e.g., yogurt and cheese).

8. Conclusions

Milk with its original emulsion structure has problems, such as short shelf-life and high number of microorganisms, which limit its economic value. The treatment of milk using different processing technologies, including thermal treatment, high-pressure treatment, homogenization, sonication, micro-fluidization, freezing and membrane emulsification, can improve the stability of milk and reduce microbial counts, thereby extending its shelf-life. Among them, heat treatment and UHP technologies have been most widely used in reducing the number of microorganisms in milk, due to their advantages in low cost and good maturity. In addition, these processing methods can have an impact on the microstructure of milk, such as altering the structural and functional properties of milk proteins, the size and surface area of milk fat globules, and the structural and digestive properties of milk lipids. In addition, in order to improve the taste and texture of milk (especially skim milk) and improve its nutritional value and functionality, the original emulsion structure of milk can be remolded by using cod liver oil, flaxseed oil and sunflower oil to prepare final products containing restructured single, mixed or double emulsion structures. Similarly, different processing methods can also affect the microstructure of liquid emulsions containing the three remodeled emulsion structures above, but current research mainly focuses on homogenization and sonication. In addition, changes of the structural properties of milk during a single processing method have been widely investigated in previous studies, while the effect of multiple processing methods on milk is rarely reported. In summary, the structural remodeling of milk can improve the texture, taste, stability and nutritional value of dairy products. Further research on remodeling the emulsion structure of milk will promote the application and development of milk emulsion, and provide more innovative and improved directions for the food industry.

Author Contributions: D.Y.: Writing—original draft (in Chinese). L.-C.S.: Methodology, writing—review and editing. L.-J.Z.: Funding acquisition, writing—review & editing. Y.-L.C.: Methodology, writing—review and editing. S.M.: Conceptualization, writing—review and editing. M.-J.C.: Conceptualization, writing—review and editing. D.L.: Conceptualization, methodology, writing—original draft, visualization, project administration, funding acquisition, supervision. All authors have read and agreed to the published version of the manuscript.

Funding: This work was supported by the Xiamen Natural Science Foundation, China [3502Z20227046] and the Foundation of Fujian Educational Committee, China [JAT220177]. The support from the opening project of National & Local Joint Engineering Research Center of Deep Processing Technology for Aquatic Products is highly appreciated.

Conflicts of Interest: The authors declare that they have no known competing financial interests or personal relationships that could have appeared to influence the work reported in this paper.

References

1. Gutiérrez, G.; Rayner, M.; Dejmek, P. Production of vegetable oil in milk emulsions using membrane emulsification. *Desalination* **2009**, *245*, 631–638. [CrossRef]
2. Liang, L.; Qi, C.; Wang, X.; Jin, Q.; McClements, D.J. Influence of homogenization and thermal processing on the gastrointestinal fate of bovine milk fat: In vitro digestion study. *J. Agric. Food Chem.* **2017**, *65*, 11109–11117. [CrossRef] [PubMed]
3. Dos Santos Morais, R.; Louvet, N.; Borges, F.; Dumas, D.; Cvetkovska-Ben Mohamed, L.; Barrau, S.; Scher, J.; Gaiani, C.; Burgain, J. Impact of lacticaseibacillus rhamnosus GG on the emulsion stability of raw milk. *Foods* **2021**, *10*, 991. [CrossRef] [PubMed]
4. Shanmugam, A.; Chandrapala, J.; Ashokkumar, M. The effect of ultrasound on the physical and functional properties of skim milk. *Innov. Food Sci. Emerg. Technol.* **2012**, *16*, 251–258. [CrossRef]
5. Let, M.B.; Jacobsen, C.; Sorensen, A.D.; Meyer, A.S. Homogenization conditions affect the oxidative stability of fish oil enriched milk emulsions: Lipid oxidation. *J. Agric. Food Chem.* **2007**, *55*, 1773–1780. [CrossRef]
6. Santhanam, A.K.; Lekshmi, M.; Chouksey, M.K.; Tripathi, G.; Gudipati, V. Delivery of omega-3 fatty acids into cake through emulsification of fish oil-in-milk and encapsulation by spray drying with added polymers. *Dry. Technol.* **2014**, *33*, 83–91. [CrossRef]
7. Lamothe, S.; Guérette, C.; Britten, M. Nutrient release and oxidative stability during in vitro digestion of linseed oil emulsions produced from cow milk, soy drink, and green tea extract. *LWT-Food Sci. Technol.* **2020**, *134*, 110137. [CrossRef]
8. Clulow, A.J.; Binte Abu Bakar, S.Y.; Salim, M.; Nowell, C.J.; Hawley, A.; Boyd, B.J. Emulsions containing optimum cow milk fat and canola oil mixtures replicate the lipid self-assembly of human breast milk during digestion. *J. Colloid Interface Sci.* **2021**, *588*, 680–691. [CrossRef]
9. Tidona, F.; Charfi, I.; Povolo, M.; Pelizzola, V.; Carminati, D.; Contarini, G.; Giraffa, G. Fermented beverage emulsion based on donkey milk with sunflower oil. *Int. J. Food Sci. Technol.* **2015**, *50*, 2644–2652. [CrossRef]
10. Anandan, S.; Keerthiga, M.; Vijaya, S.; Asiri, A.M.; Bogush, V.; Krasulyaa, O. Physicochemical characterization of black seed oil-milk emulsions through ultrasonication. *Ultrason. Sonochem.* **2017**, *38*, 766–771. [CrossRef]
11. Singh, H.; Gallier, S. Nature's complex emulsion: The fat globules of milk. *Food Hydrocoll.* **2017**, *68*, 81–89. [CrossRef]
12. Reis, M.G.; Harris, P.; Berry, C.; Nguyen, H.; Maclean, P.; Weeks, M. Tracking changes in volatile components and lipids after homogenisation and thermal processing of milk. *Int. Dairy J.* **2020**, *103*, 104624. [CrossRef]
13. Wade, T.; Beattie, J.K. Electroacoustic determination of size and zeta potential of fat globules in milk and cream emulsions. *Colloids Surf. B Biointerfaces* **1997**, *10*, 73–85. [CrossRef]
14. Augustin, M.A.; Sanguansri, L.; Oliver, C.M. Functional properties of milk constituents: Application for microencapsulation of oils in spray-dried emulsions—A minireview. *Dairy Sci. Technol.* **2009**, *90*, 137–146. [CrossRef]
15. Garczewska-Murzyn, A.; Kiełczewska, K.; Smoczyński, M. The influence of buttermilk powder on the stability of emulsion and colloidal phases of homogenized milk. *Eur. Food Res. Technol.* **2022**, *248*, 2629–2636. [CrossRef]
16. Leong, T.S.H.; Zhou, M.; Kukan, N.; Ashokkumar, M.; Martin, G.J.O. Preparation of water-in-oil-in-water emulsions by low frequency ultrasound using skim milk and sunflower oil. *Food Hydrocoll.* **2017**, *63*, 685–695. [CrossRef]
17. Maghamian, N.; Goli, M.; Najarian, A. Ultrasound-assisted preparation of double nano-emulsions loaded with glycyrrhizic acid in the internal aqueous phase and skim milk as the external aqueous phase. *LWT-Food Sci. Technol.* **2021**, *141*, 110850. [CrossRef]
18. Zaghian, N.; Goli, M. Optimization of the production conditions of primary (W_1/O) and double ($W_1/O/W_2$) nano-emulsions containing vitamin B_{12} in skim milk using ultrasound wave by response surface methodology. *J. Food Meas. Charact.* **2020**, *14*, 3216–3226. [CrossRef]
19. Tritschler, D.; Kinting, N.; Braig, A.; Sala, G.; Hinrichs, J. Critical consideration of gas bubbles to replace fat globules in acidified milk matrices. *Int. Dairy J.* **2024**, *151*, 105858. [CrossRef]
20. Nagarajappa, V.; Battula, S.N. Effect of fortification of milk with omega-3 fatty acids, phytosterols and soluble fibre on the sensory, physicochemical and microbiological properties of milk. *J. Sci. Food Agric.* **2017**, *97*, 4160–4168. [CrossRef]
21. Gallaher, J.J.; Hollender, R.; Peterson, D.G.; Roberts, R.F.; Coupland, J.N. Effect of composition and antioxidants on the oxidative stability of fluid milk supplemented with an algae oil emulsion. *Int. Dairy J.* **2005**, *15*, 333–341. [CrossRef]

22. Delfanian, M.; Sahari, M.A.; Barba, F.J. Effect of lipophilized gallic acid on the oxidative stability of omega-3 fatty acids rich soy and cow milk. *LWT-Food Sci. Technol.* **2023**, *190*, 115475. [CrossRef]
23. Choudhary, P.; Dutta, S.; Moses, J.A.; Anandharamakrishnan, C. Liposomal encapsulation of omega-3 fatty acid and α-lipoic acid conjugate for cow milk fortification. *J. Food Process. Preserv.* **2021**, *46*, 16082. [CrossRef]
24. Venkateshwarlu, G.; Let, M.B.; Meyer, A.S.; Jacobsen, C. Chemical and olfactometric characterization of volatile flavor compounds in a fish oil enriched milk emulsion. *J. Agric. Food Chem.* **2004**, *52*, 311–317. [CrossRef] [PubMed]
25. Kasinos, M.; Mukarukundo, F.; De Beuf, K.; Van der Meeren, P. Anionic and zwitterionic phospholipids differently affect the heat coagulation of recombined concentrated milk emulsions. *Int. Dairy J.* **2014**, *39*, 131–138. [CrossRef]
26. Kasinos, M.; Tran Le, T.; Van der Meeren, P. Improved heat stability of recombined evaporated milk emulsions upon addition of phospholipid enriched dairy by-products. *Food Hydrocoll.* **2014**, *34*, 112–118. [CrossRef]
27. Klojdová, I.; Troshchynska, Y.; Štětina, J. Influence of carrageenan on the preparation and stability of w/o/w double milk emulsions. *Int. Dairy J.* **2018**, *87*, 54–59. [CrossRef]
28. Leong, T.S.H.; Zhou, M.; Zhou, D.; Ashokkumar, M.; Martin, G.J.O. The formation of double emulsions in skim milk using minimal food-grade emulsifiers—A comparison between ultrasonic and high pressure homogenisation efficiencies. *J. Food Eng.* **2018**, *219*, 81–92. [CrossRef]
29. Molet-Rodríguez, A.; Ramezani, M.L.; Martín-Belloso, O. Impact of the lipid phase composition and state on the in vitro digestibility and chlorophyllin bioaccessibility of $W_1/O/W_2$ emulsions into whole milk. *Food Res. Int.* **2023**, *173*, 113455. [CrossRef]
30. Pérez, M.P.; Wagner, J.R.; Márquez, A.L. Partial coalescence in double ($W_1/O/W_2$) emulsions prepared with skimmed milk, polyglycerol polyricinoleate, and different fats. *Eur. J. Lipid Sci. Technol.* **2017**, *119*, 1600447. [CrossRef]
31. Felfoul, I.; Bornaz, S.; Belhadj Hmida, W.; Sahli, A.; Attia, H.; Blecker, C. Effect of milk fat substitution of rennet milk induced coagulation on physico-Chemical properties. *J. Chem.* **2013**, *2013*, 732024. [CrossRef]
32. Chang, Y.H.; Lee, S.Y.; Kwak, H.S. Physicochemical and sensory properties of milk fortified with iron microcapsules prepared with water-in-oil-in-water emulsion during storage. *Int. J. Dairy Technol.* **2016**, *69*, 452–459. [CrossRef]
33. Kashaninejad, M.; Razavi, S.M.A. Influence of thermosonication treatment on the average size of fat globules, emulsion stability, rheological properties and color of camel milk cream. *LWT-Food Sci. Technol.* **2020**, *132*, 109852. [CrossRef]
34. Kiełczewska, K.; Ambroziak, K.; Krzykowska, D.; Aljewicz, M. The effect of high-pressure homogenisation on the size of milk fat globules and MFGM composition in sweet buttermilk and milk. *Int. Dairy J.* **2021**, *113*, 104898. [CrossRef]
35. Verma, K.; Tarafdar, A.; Mishra, V.; Dilbaghi, N.; Kondepudi, K.K.; Badgujar, P.C. Nanoencapsulated curcumin emulsion utilizing milk cream as a potential vehicle by microfluidization: Bioaccessibility, cytotoxicity and physico-functional properties. *Food Res. Int.* **2021**, *l148*, 110611. [CrossRef] [PubMed]
36. Abesinghe, A.M.N.L.; Vidanarachchi, J.K.; Islam, N.; Prakash, S.; Silva, K.F.S.T.; Bhandari, B.; Karim, M.A. Effects of ultrasonication on the physicochemical properties of milk fat globules of Bubalus bubalis (water buffalo) under processing conditions: A comparison with shear-homogenization. *Innov. Food Sci. Emerg. Technol.* **2020**, *59*, 102237. [CrossRef]
37. Kasprzak, M.M.; Sady, M.; Kruk, J.; Bartkova, S.; Sanka, I.; Scheler, O.; Jamróz, E.; Berski, W.; Onacik-Gür, S.; Szram, R.; et al. Replacement of milk fat by rapeseed oil stabilised emulsion in commercial yogurt. *PeerJ* **2023**, *11*, 16441. [CrossRef]
38. Aghababaei, F.; Trujillo, A.J.; Juan, B.; Capellas, M.; Ferragut, V. Encapsulating capacity of ultra-high-pressure homogenization (UHPH): Replacement of milk fat by vegetable oils using buttermilk as a functional ingredient in yogurt processing. *LWT-Food Sci. Technol.* **2023**, *187*, 115304. [CrossRef]
39. Trujillo-Ramirez, D.; Olivares-Martinez, I.; Lobato-Calleros, C.; Rodriguez-Huezo, E.; Jaime Vernon-Carter, E.; Alvarez-Ramirez, J. Impact of the droplet size of canola oil-in-water emulsions on the rheology and sensory acceptability of reduced-milk fat stirred yogurt. *J. Food Sci. Technol.* **2022**, *59*, 4853–4862. [CrossRef]
40. Huppertz, T.; Nieuwenhuijse, H. Constituent fouling during heat treatment of milk: A review. *Int. Dairy J.* **2022**, *126*, 105236. [CrossRef]
41. Ben-Harb, S.; Panouillé, M.; Huc-Mathis, D.; Moulin, G.; Saint-Eve, A.; Irlinger, F.; Bonnarme, P.; Michon, C.; Souchon, I. The rheological and microstructural properties of pea, milk, mixed pea/milk gels and gelled emulsions designed by thermal, acid, and enzyme treatments. *Food Hydrocoll.* **2018**, *77*, 75–84. [CrossRef]
42. Grewal, M.K.; Chandrapala, J.; Donkor, O.; Apostolopoulos, V.; Stojanovska, L.; Vasiljevic, T. Fourier transform infrared spectroscopy analysis of physicochemical changes in UHT milk during accelerated storage. *Int. Dairy J.* **2017**, *66*, 99–107. [CrossRef]
43. Kiełczewska, K.; Jankowska, A.; Dąbrowska, A.; Wachowska, M.; Ziajka, J. The effect of high pressure treatment on the dispersion of fat globules and the fatty acid profile of caprine milk. *Int. Dairy J.* **2020**, *102*, 104607. [CrossRef]
44. Serna-Hernandez, S.O.; Escobedo-Avellaneda, Z.; García-García, R.; Rostro-Alanis, M.d.J.; Welti-Chanes, J. High hydrostatic pressure induced changes in the physicochemical and functional properties of milk and dairy products: A review. *Foods* **2021**, *10*, 1867. [CrossRef]
45. Kiełczewska, K.; Brożek, O.; Rudkowska, P.; Garczewska-Murzyn, A.; Smoczyński, M. The effects of full-stream and partial high-pressure homogenisation on the properties of milk emulsion. *Int. J. Food Sci. Technol.* **2022**, *57*, 7167–7174. [CrossRef]
46. Karlović, S. Reducing Fat Globules Particle-Size in Goat Milk: Ultrasound and high hydrostatic pressures approach. *Chem. Biochem. Eng. Q. J.* **2015**, *28*, 499–507. [CrossRef]

47. McCrae, C.H. Homogenization of milk emulsions:use of microfluidizer. *Int. J. Dairy Technol.* **1994**, *47*, 28–31. [CrossRef]
48. Bottiroli, R.; Zhang, C.; Aprea, E.; Fogliano, V.; Hettinga, K.; Gasperi, F. Short communication: Short-time freezing does not alter the sensory properties or the physical stability of ultra-high-temperature hydrolyzed-lactose milk. *J. Dairy Sci.* **2020**, *103*, 8822–8828. [CrossRef]
49. Nurliyani; Suranindyah, Y.; Pretiwi, P. Quality and emulsion stability of milk from ettawah crossed bred goat during frozen storage. *Procedia Food Sci.* **2015**, *3*, 142–149. [CrossRef]
50. Zhang, L.; Qu, J.; Huppertz, T.; Liu, J.; Sun, Z.; Zhou, P. Effects of different freeze-thaw processes on the bioactivity and digestibility of human milk. *LWT-Food Sci. Technol.* **2022**, *156*, 113025. [CrossRef]
51. Charcosset, C. Preparation of emulsions and particles by membrane emulsification for the food processing industry. *J. Food Eng.* **2009**, *92*, 241–249. [CrossRef]
52. Luo, X.; Ramchandran, L.; Vasiljevic, T. Lower ultrafiltration temperature improves membrane performance and emulsifying properties of milk protein concentrates. *Dairy Sci. Technol.* **2015**, *95*, 15–31. [CrossRef]
53. Ramakrishnan, S.; Ferrando, M.; Aceña-Muñoz, L.; De Lamo-Castellví, S.; Güell, C. Fish oil microcapsules from O/W emulsions produced by premix membrane emulsification. *Food Bioprocess Technol.* **2012**, *6*, 3088–3101. [CrossRef]
54. Pereda, J.; Ferragut, V.; Quevedo, J.M.; Guamis, B.; Trujillo, A.J. Effects of ultra-high-pressure homogenization treatment on the lipolysis and lipid oxidation of milk during refrigerated storage. *J. Agric. Food Chem.* **2008**, *56*, 7125–7130. [CrossRef] [PubMed]
55. Kasinos, M.; Goñi, M.L.; Nguyen, M.T.; Sabatino, P.; Martins, J.C.; Dewettinck, K.; Van der Meeren, P. Effect of hydrolysed sunflower lecithin on the heat-induced coagulation of recombined concentrated milk emulsions. *Int. Dairy J.* **2014**, *38*, 187–194. [CrossRef]
56. Gallier, S.; Acton, D.; Garg, M.; Singh, H. Natural and processed milk and oil body emulsions: Bioavailability, bioaccessibility and functionality. *Food Struct.* **2017**, *13*, 13–23. [CrossRef]
57. Masum, A.K.M.; Chandrapala, J.; Huppertz, T.; Adhikari, B.; Zisu, B. Production and characterization of infant milk formula powders: A review. *Dry. Technol.* **2020**, *39*, 1492–1512. [CrossRef]
58. San Martín-González, M.F.; Welti-Chanes, J.; Barbosa-Cánovas, G.V. Cheese manufacture assisted by high pressure. *Food Rev. Int.* **2006**, *22*, 275–289. [CrossRef]
59. Shao, Y.; Yuan, Y.; Xi, Y.; Zhao, T.; Ai, N. Effects of homogenization on organoleptic quality and stability of pasteurized milk samples. *Agriculture* **2023**, *13*, 205. [CrossRef]
60. Tribst, A.A.L.; Falcade, L.T.P.; Carvalho, N.S.; Cristianini, M.; Leite Júnior, B.R.d.C.; Oliveira, M.M.d. Using physical processes to improve physicochemical and structural characteristics of fresh and frozen/thawed sheep milk. *Innov. Food Sci. Emerg. Technol.* **2020**, *59*, 102247. [CrossRef]
61. Juliano, P.; Kutter, A.; Cheng, L.J.; Swiergon, P.; Mawson, R.; Augustin, M.A. Enhanced creaming of milk fat globules in milk emulsions by the application of ultrasound and detection by means of optical methods. *Ultrason. Sonochem.* **2011**, *18*, 963–973. [CrossRef] [PubMed]
62. Astráin-Redín, L.; Skipnes, D.; Cebrián, G.; Álvarez-Lanzarote, I.; Rode, T.M. Effect of the application of ultrasound to homogenize milk and the subsequent pasteurization by pulsed electric field, High hydrostatic pressure, and microwaves. *Foods* **2023**, *12*, 1457. [CrossRef] [PubMed]
63. Alemán, M.; Bou, R.; Guardiola, F.; Durand, E.; Villeneuve, P.; Jacobsen, C.; Sørensen, A.-D.M. Antioxidative effect of lipophilized caffeic acid in fish oil enriched mayonnaise and milk. *Food Chem.* **2015**, *167*, 236–244. [CrossRef] [PubMed]
64. Patil, L.; Gogate, P.R. Ultrasound assisted synthesis of stable oil in milk emulsion: Study of operating parameters and scale-up aspects. *Ultrason. Sonochem.* **2018**, *40*, 135–146. [CrossRef] [PubMed]
65. Ribeiro, H.S.; Rico, L.G.; Badolato, G.G.; Schubert, H. Production of O/W emulsions containing astaxanthin by repeated premix membrane emulsification. *J. Food Sci.* **2005**, *70*, E117–E123. [CrossRef]
66. Bai, L.; Lv, S.; Xiang, W.; Huan, S.; McClements, D.J.; Rojas, O.J. Oil-in-water Pickering emulsions via microfluidization with cellulose nanocrystals: 1. Formation and stability. *Food Hydrocoll.* **2019**, *96*, 699–708. [CrossRef]
67. Ozturk, O.K.; Turasan, H. Applications of microfluidization in emulsion-based systems, nanoparticle formation, and beverages. *Trends Food Sci. Technol.* **2021**, *116*, 609–625. [CrossRef]
68. García-Márquez, E.; Higuera-Ciapara, I.; Espinosa-Andrews, H. Design of fish oil-in-water nanoemulsion by microfluidization. *Innov. Food Sci. Emerg. Technol.* **2017**, *40*, 87–91. [CrossRef]
69. Jafari, S.M.; He, Y.; Bhandari, B. Production of sub-micron emulsions by ultrasound and microfluidization techniques. *J. Food Eng.* **2007**, *82*, 478–488. [CrossRef]
70. Pereyra-Castro, S.C.; Pérez-Pérez, V.; Hernández-Sánchez, H.; Jiménez-Aparicio, A.; Gutiérrez-López, G.F.; Alamilla-Beltrán, L. Effect of composition and homogenization pressure of chia oil emulsions elaborated by microfluidization. *Rev. Mex. Ing. Química* **2018**, *19*, 69–81. [CrossRef]
71. Rajasekaran, B.; Singh, A.; Nilsuwan, K.; Ma, L.; Nazeer, R.A.; Benjakul, S. Shrimp oil nanoemulsions prepared by microfluidization and ultrasonication: Characteristics and stability. *Rouyal Soc. Chem. Adv.* **2024**, *14*, 6135–6145. [CrossRef] [PubMed]
72. Mahdi Jafari, S.; He, Y.; Bhandari, B. Nano-emulsion production by sonication and microfluidization—A comparison. *Int. J. Food Prop.* **2006**, *9*, 475–485. [CrossRef]

73. Oliete, B.; Potin, F.; Cases, E.; Saurel, R. Microfluidization as homogenization technique in pea globulin-based emulsions. *Food Bioprocess Technol.* **2019**, *12*, 877–882. [CrossRef]
74. Olson, D.W.; White, C.H.; Richter, R.L. Effect of pressure and fat content on particle sizes in microfluidized milk. *J. Dairy Sci.* **2004**, *87*, 3217–3223. [CrossRef]
75. Kumar, A.; Badgujar, P.C.; Mishra, V.; Sehrawat, R.; Babar, O.A.; Upadhyay, A. Effect of microfluidization on cholesterol, thermal properties and in vitro and in vivo protein digestibility of milk. *LWT* **2019**, *116*, 108523. [CrossRef]

gels MDPI

Review

Surimi and Low-Salt Surimi Gelation: Key Components to Enhance the Physicochemical Properties of Gels

Noman Walayat [1], María Blanch [2,*] and Helena M. Moreno [2,*]

[1] College of Tea Science and Tea Culture, Zhejiang Agriculture and Forestry University, Hangzhou 311300, China; 20232004@zafu.edu.cn
[2] Department Section of Galenic Pharmacy and Food Technology, Veterinary Faculty, Complutense University of Madrid, 28040 Madrid, Spain
* Correspondence: mblanchr@ucm.es (M.B.); helena.moreno@ucm.es (H.M.M.)

Abstract: Surimi-based products are nutritionally valuable due to their essential amino acid composition, their content of high-quality proteins with excellent digestibility, and their low fat content. However, to achieve the desired texture, a significant amount of salt (1–3%) must be added, which could compromise their health benefits. This study provides an overview of surimi production, the gelation mechanism of myosin, and the most relevant gelation enhancers that could be used in manufacturing low-salt surimi-based products. Reducing the salt content in surimi-based products presents a significant challenge for the industry, not only from technological and sensory perspectives but also in response to the growing demand of consumers for healthier food options. So, this manuscript highlights several strategies for achieving optimal quality characteristics in relation to functional properties for the surimi products industry. In addition, surimi as a raw material is often misunderstood by consumers, who may question its nutritional value and, consequently, its consumption. Therefore, it is crucial to thoroughly explain the processing of this raw material and emphasize the importance of proper myofibrillar protein gelation to develop high-value surimi-based products.

Keywords: surimi; low salt; gelation challenges; physicochemical properties; technological additives

1. Introduction

The nutritional quality of fishery products, especially those made from surimi, will depend mainly on the species from which the surimi is made and the ingredients added to the product. Surimi is mainly composed of water and myofibrillar proteins, so the higher the percentage of surimi used in the production of surimi products (kamaboko, chikuwa, fish balls, sausages, etc.), the higher the nutritional quality of the product. Surimi contains proteins of high biological value and good digestibility (due to the composition of essential amino acids) and has a low fat content [1]. In addition, its possible bioactive effects have been studied, and it was found that its protein could be effective in the control of dementia [2], could prevent colon cancer [3], or could inhibit the absorption of fat and sugar [4], among other benefits.

However, the preparation of surimi gels requires the addition of a significant amount of salt (1–3%) to achieve the desired texture, which may compromise their health benefits and, consequently, their consumption. As a result, the surimi products industry is working to reduce the salt content due to growing health concerns in surimi products by replacing it with alternative compounds, such as seaweed extract, potassium chloride, protein hydrolysates, and amino acids, or by employing different processes that enable the production of low-salt surimi-based products [5].

Common salt (NaCl) is the main source of sodium in the diet. Sodium is necessary for the proper functioning of the body; however, epidemiological studies worldwide suggest that salt consumption is an inducer of increased blood pressure [6]. As reported by the World Health Organization [7], arterial hypertension is one of the leading causes of premature death worldwide. For that reason, one of the global goals related to non-communicable diseases is to reduce the prevalence of hypertension by 25% by 2030 compared to the 2010 baseline levels.

To face the increasing concerns about sodium intake, various countries and regions have enacted sodium labelling laws to indicate the salt content in food products. These labels generally classify sodium levels as no sodium (<5 mg), very low sodium (<35 mg), low sodium (<140 mg), reduced sodium (<25%), and light sodium (<50%) [8]; however, the criteria for sodium labelling vary across countries. The European Union adopted "Regulation N° 1924/2006 (2006) of the European Parliament and of the Council on nutrition and health claims made on foods", which applies to all nutrition and health claims made in commercial communications [9]. As a result, the food industry, particularly the surimi products industry, is facing a significant challenge in reducing the salt content of surimi gel products while maintaining their quality [10]. Although significant efforts have been made to develop innovative salt reduction processing methods, researchers have faced numerous challenges, such as compromised gel quality, increased costs, and safety concerns [11].

The global production of surimi has been estimated at approximately 800,000 tons in 2023. This volume reflects a robust global market, led by countries such as Japan, the United States, and Russia, which has shown an increasing interest in expanding its production capacity, exemplified by its plan to triple surimi production in the coming years. The industry is supported by strong demand due to the versatile use of surimi in processed food products and its popularity in gastronomy [12]. The social changes that have taken place in recent years have led to an increase in the consumption of fishery products and, more specifically, surimi-based products. This increase in the consumption of these products is mainly due to the fact that they have a competitive price compared to other fishery products [13,14].

In the processing of surimi-based products, the final texture is crucial, as these products are often analogues that must replicate the characteristic texture of the original product they aim to imitate. Approximately 20% to 25% of the world's produced surimi is used in the manufacture of crab substitute [13,15]. To achieve an adequate texture in these products, a significant amount of salt (commonly 1–3% NaCl) is required to promote the thermal gelation of myofibrillar proteins [16–18]. As stated before, this fact makes it difficult to comply with nutritional requirements that call for a reduction in salt in the diet and seek to promote the consumption of healthy foods. Meanwhile, considering the role of salt in the techno-functional characteristics of surimi-based products, mainly texture, its reduction represents an important technological challenge for industry. According to this, different technological additives have been studied as potential enhancers of the gelation of low-salt surimi-based products [5,19–23]. Therefore, the objective of this work is to provide an in-depth descriptions of surimi as a raw material and the process of surimi gel formation, as well as to explore various alternatives that the surimi products industry can employ to compensate for the reduction in salt in terms of physicochemical properties, particularly texture. This includes optimizing the use of technological additives that promote surimi gelation without compromising the final physicochemical properties of the final products.

2. Surimi

The origin of the word "surimi" is Japanese and means "minced fish". Basically, it refers to the frozen block of fish myofibrillar protein concentrate that is mixed with cryoprotectants and finally undergoes freezing. This product is sold to food processors, who blend it with other ingredients to impart texture, taste, and colour [24]. In Japan, surimi has traditionally been used to produce a product called kamaboko (a mildly flavoured fish gel that is nearly tasteless and widely used in Japanese cuisine). However, it is now also employed in the production of substitutes for high-value seafood products, aiming to replicate their texture, flavour, and appearance [25]. Additionally, particularly in Japan, but also in the United States and Europe, various surimi-based products, either alone or combined with other ingredients, are becoming increasingly popular as snacks, representing a growing sector in the food industry. Surimi commercialization was initially limited to Japan and some Asian countries, but its worldwide expansion took place in the 1960s, when the use of low-molecular-weight carbohydrates (sucrose and sorbitol) was optimized as efficient cryoprotectants of fish protein, facilitating its commercialization in the freezing stage [15].

Surimi was mainly made from Alaska pollock (*Theragra chalcogramma*), a very abundant but undervalued fish due to its soft texture. However, the surimi obtained from this fish species has excellent gelling properties and is the base of the technological development of surimi-based products [15,26]. The increase in surimi production to meet consumers' needs, the raw material costs, market fluctuations, and the use of new technologies have boosted its production from other species of greater availability and/or lower economic value, such as hake (*Merluccius gayi; Merluccius productus*), horse mackerel (*Trachurus trachurus*), sardine (Sardine sp.), Pacific halibut (*Atheresthes stomias*), southern blue whiting (*Micromesistius australis*), blue-tailed hake (*Macruronus novaezelandiae*), Japanese bogue (Nemimpterus sp.), etc. These species are underutilized in many parts of the world and have sufficient properties for the manufacture of quality surimi [14,27,28].

2.1. Surimi Processing

The surimi manufacturing process begins with the capture of the fish and ends with the storage of the surimi in a frozen state (Figure 1).

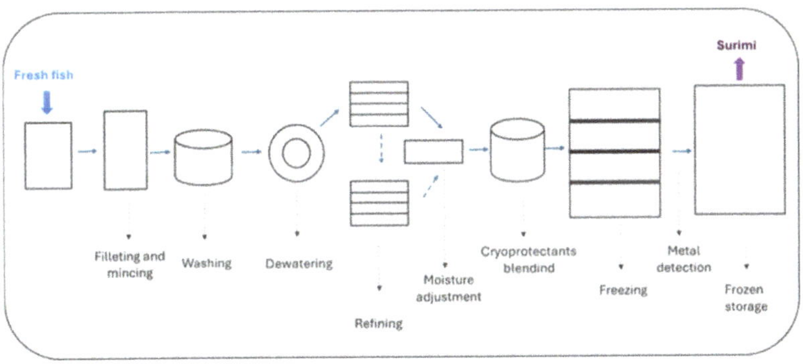

Figure 1. Flow chart of surimi manufacturing.

This process has been extensively described by several authors [15,25] and can be synthesized as follows: the fish is quickly beheaded and gutted to prevent intestinal enzymes from migrating to the muscle and damaging myofibrillar proteins. It is then washed thoroughly with water, followed by the removal of skin, bones, scraps, cartilage, and as many impurities as possible before being minced. The minced fish is washed several times with water or saline solutions (at temperatures between 5 and 10 °C) to remove soluble proteins, mainly sarcoplasmic proteins, and other impurities that could reduce the gelation capacity of the surimi. The next step is the refining process, which eliminates additional impurities before moisture adjustment (to around 80%). In addition to that, cryoprotectants are then incorporated to ensure that the proteins maintain maximum functionality after freezing, resulting in high-quality surimi. Commonly used cryoprotectants include sucrose and sorbitol (4% of each), while polyphosphates (0.3%) are also added to facilitate protein solubilization and enhance gel elasticity. Depending on the species, other additives may be used to adjust the pH, inhibit enzymes, and/or chelate metals, helping to preserve protein functionality. Surimi is typically frozen in contact plate freezers (to a thermal centre temperature of −20 °C) in standard 10 kg blocks. It is then packaged in cardboard boxes containing two 10 kg blocks of surimi. According to surimi processing, factors such as the solubility of myofibrillar proteins as affected by ionic strength, pH, washing cycles, the wash/meat (W/M) ratio, and proteolysis are important considerations taken into account by surimi processing industries before selecting the fish species used to make surimi.

2.2. Surimi Quality

The key factor that determines surimi quality and is closely linked to its freshness is the functionality of surimi myofibrillar protein [29]. This functionality primarily depends on whether processing was carried out properly and under optimal conditions, as well as the type of fish being used.

There is not a standardized method for raw surimi quality evaluation. In the case of Alaska pollock (*Theragra chalcogramma*) surimi, its quality is assessed by its gel-forming ability (measured through texture tests) although gel properties change greatly depending on the cooking process especially preheating procedures for surimi gelation [15]. During surimi production, the use of large machinery (Figure 1) presents a significant risk of contamination or cross-contamination. One of the relevant hazards that might occur during raw surimi processing is the potential inclusion of metal fragments [15]. Nevertheless, current industrial processing includes a metal detector before frozen storage (Figure 1).

On the other hand, Surimi is the main ingredient in surimi seafood products, which are cooked and/or pasteurized during production, therefore the presence of microorganisms in raw surimi is more of a quality concern than a safety issue [15].

Surimi is usually classified according to the properties exhibited by Alaska pollock (*Theragra chalcogramma*) surimi. Its quality is assessed by its gel-forming ability (measured through texture tests), colour (the whiter the colour, the higher the quality), purity (the complete absence of blood, bits of skin, and dark meat indicates greater purity), homogeneity, and microbiological and enzymatic quality [24]. These characteristics result in a three-grade grading system: the primary grade (SA, FA, A), which refers to the surimi that comes from the first refining process; the secondary grade (KA), which comes from a second refining process; and the recovered grade (KB, RA), which refers to the surimi that has undergone two refining processes and comes from discards.

3. Surimi Gelation

Surimi-based products primarily rely on thermal gelation, which is influenced by several factors. Therefore, it is important to describe the factors that contribute to the proper gelation process.

3.1. Main Factors Affecting Gelation Process

3.1.1. Protein Content

The gelling capacity of surimi is mainly attributed to the presence of myofibrillar proteins, especially myosin, that account for about 55–60% of the total protein content [17]. In the gelation phenomenon, myosin must be dissociated by the addition of salts, and then myosin heads predominantly aggregate through disulfide bonds, while the hydrophobic effect drives tail aggregation [30–33]. During the washing process in the production of surimi, the non-functional components present in the fish muscle (blood, pigments, impurities, and sarcoplasmic proteins, which are largely enzymes) are removed, and the functional myofibrillar proteins are concentrated, which enhances the gelation capacity of the surimi.

3.1.2. Salt Content

The addition of salt, usually 1–3%, is necessary to solubilize and disperse the myofibrillar proteins [17,34]. In fact, gelation does not occur or does so defectively in the absence of salt because myofibrillar proteins are insoluble at a low ionic strength, and to facilitate the formation of a well-organized and stabilized protein network, the protein must be solubilized. At low salt concentrations (<0.3 mol/L), myosin tends to form a discontinuous network that is a porous structure [35], while high-salt conditions (≥0.3 mol/L) enable the formation of a three-dimensional network gel [17]. During the homogenization process with NaCl, the salt ions (Cl^- and Na^+) bind to the oppositely charged groups on the protein surface, resulting in a disruption of the intermolecular ionic bonds of the myofibrillar proteins. This increases the affinity of proteins (mainly myosin) for water molecules and allows for their solubilization by partially unfolding the structure of the myosin molecule, favouring gelation [17,18].

3.1.3. pH

Changes in pH can alter molecular interactions and the spatial conformations of proteins. When the pH is close to the isoelectric point, the proteins present zero net charge; therefore, the repulsion between proteins is lower and they tend to precipitate due to minimal solubility, giving rise to less hydrated and less firm gels [30]. When the pH is located above or below the isoelectric point, the proteins are negatively or positively charged so that the chains of the protein molecules repel each other and are therefore more susceptible to binding with the water molecules present in the medium, increasing the water retention capacity of the protein gels [22,25].

3.1.4. Effect of Endogenous Transglutaminase (TGe)

The endogenous transglutaminase is an enzyme naturally found in fish muscle that catalyses the formation of non-disulfide covalent bonds (ε-(γ-glutamyl) lysine), resulting in the formation of myosin polymers, which is correlated with increased gel strength [16,36]. TGe activity is calcium ion-dependent, so it can be modulated depending on the presence of calcium ions in the medium. TGe acts at low temperatures (<40 °C); therefore, to favour its action, the gelation process starts with a "setting" period, which consists of starting with a 5–40 °C treatment for a certain period of time to obtain highly deformable gels called suwari gel [37,38].

3.2. Thermal Gelation of Myofibrillar Proteins

Focusing on the modification that occurs in proteins during the gelation process, these can be described in two phases: denaturation or unfolding of the proteins, followed by intermolecular aggregation of the proteins (Figure 2). The gel formation process begins with the mixing process of surimi and salt, which solubilizes the proteins, resulting in the formation of a viscous mass. The progressive increase in temperature results in the formation of a matrix that finally transforms into a gel with viscoelastic solid properties [22,39].

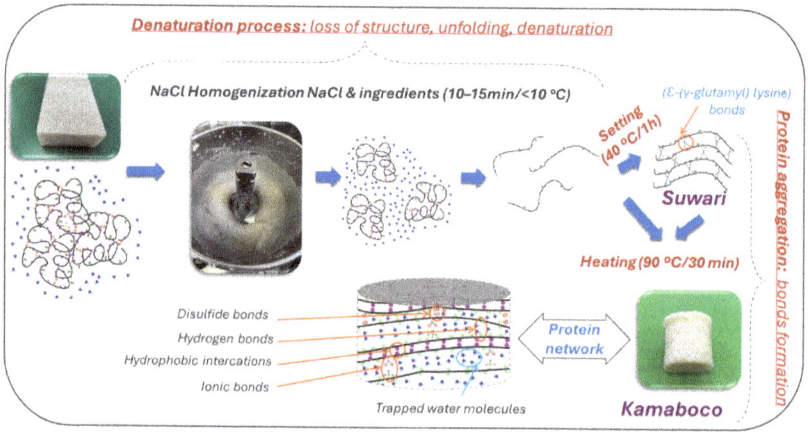

Figure 2. Denaturation or unfolding of proteins, followed by intermolecular aggregation of myofibrillar proteins during thermal gelation of surimi.

3.2.1. Denaturation Process of Myofibrillar Proteins

The gelation process starts with the solubilization of the proteins by the addition of salt (commonly NaCl). During this process, myosin loses its quaternary, tertiary, and secondary structures, so the α-helix structure stabilized by non-covalent bonds, mainly hydrogen bonds, unfolds. The loss of this structure gives rise to the formation of a β-sheet structure stabilized by non-covalent bonds [16]. This denaturation results in the exposure of reactive groups, resulting in a high hydration of the proteins upon interaction with the water molecules present in the medium. These reactive groups will be involved in the formation of bonds of different types during the aggregation stage (Figure 2). Denaturation proceeds continuously and can occur to varying degrees, and during this process, a large number of hydrogen bonds are broken between the carboxyl and amino groups of the polypeptide chain of proteins responsible for keeping the native structure of the protein folded [17,40,41].

3.2.2. Myofibrillar Protein Aggregation Process

Aggregation refers to the formation of protein–protein interactions that lead to the creation of large, high-molecular-weight protein complexes. Gelation is the process where previously unfolded proteins aggregate into a three-dimensional structure, with protein–protein and protein–solvent interactions forming an organized protein matrix that can retain a substantial amount of water molecules [42]. As the thermal denaturation temperature of proteins varies among different fish species, it is necessary to study the most appropriate time and temperature according to the specific properties of fish species [10].

During the initial phase of the heating process, myosin molecules undergo denaturation, which causes the helical structure of the tail to gradually unfold and reveal the previously hidden active groups. When the molecule is in its folded state, the activity of disulfide bonds is low; however, when proteins are unfolded by any agent, the reactivity and formation of these bonds increase [43]. Consequently, these proteins connect through intermolecular interactions. As the temperature continues to rise, hydrophobic interactions and disulfide bonds between proteins are strengthened, leading to the formation of an organized three-dimensional reticular gel structure [44,45] (Figure 2).

The main types of bonds involved in the gelation of myofibrillar proteins are hydrogen bonding, ionic interactions or salt bridges, hydrophobic interactions, and covalent bonds [16]. Ionic interactions are those formed between positive and negative charges on the protein surface. At a neutral pH, the carboxyl groups (COO^-) are negatively charged, while the amino groups (NH_2^+) are positively charged. These groups attract each other, forming such interactions. The addition of salt to surimi breaks the ionic bonds and aids in the dispersion of myofibrillar proteins [41]. On the other hand, hydrophobic interactions, like disulfide bonds, are formed by the action of heat or high-pressure (>300 MPa) treatments [46]. The formation of these bonds also occurs as a consequence of the unfolding of the protein structure, which exposes the hydrophobic residues and reactive groups inside the protein molecule. The association of the hydrophobic zones decreases the entropy of the system, resulting in balanced bonding, which concludes with the formation of an ordered three-dimensional structure [17]. Finally, disulfide (S-S) bonds are formed by the oxidation of sulfhydryl groups present on cysteine residues. Their formation is considered to be thermo-irreversible [44,45] (Figure 2). Additionally, the addition of oxidants can accelerate the formation of intermolecular disulfide bonds [47,48].

The gelation process of surimi is significantly affected by the heating method and conditions applied. Research has shown that the breaking force achieved with two-stage heating (firstly at 40 °C/60 min and secondly at 90 °C/30 min) is considerably stronger than that obtained with a single heating treatment [49]. This is because a single high-temperature heating treatment (121 °C) leads to myosin degradation due to the disintegration of the protein's secondary structure and the weakening of binding forces (hydrophobic interactions and ionic bonds) responsible for maintaining the stability of the gel. The resulting gel has a highly porous and discontinuous microstructure. In contrast, low-temperature pre-treatment (40 °C) assists TGe action, advocating the formation of covalent cross-linking bonds of ε-(γ-glutamyl) lysine. TGe efficiently binds protein molecules together, avoiding the thermal breakdown of myosin at high temperatures. The result is a dense three-dimensional network. For this reason, performing surimi gelation in two steps is the most convenient heating gelation method to obtain a firm and elastic surimi gel [20,50,51].

4. Using Gelation Enhancers as a Strategy for Surimi Gelation with a Low Salt Content

Salt plays a crucial role in the production of surimi-based products, making its reduction a significant challenge for the food industry in the manufacturing of these products.

4.1. Substitution of NaCl with Other Salts

This is the most used method; however, modifications in the presence of metal salt and in their amount can alter the forces and spatial conformation relating to myosin molecules, thereby affecting the properties of the resulting gel [52]. In particular, substitution with potassium chloride has certain disadvantages, such as adding strange flavours to the product; in fact, the substitution of 50% NaCl with 50% KCl has been described to significantly increase bitterness and to reduce the salt appreciation [53], although substitution with 25% KCl resulted in a gel with very appropriate physicochemical properties, which can be attributed to the great exposure of amino acid residues and hydrophobic groups in combination with the decrease in the alpha helix presence and the growth in the beta sheet presence of myofibrillar proteins that, at the end, enhance the water-holding capacity, maintaining the desired taste and improving the gel properties. Other salts, such as $CaCl_2$ and $MgCl_2$, have been tested without very positive results since the solubilization of myofibrillar proteins is facilitated, resulting in an organized and dense gel network with significant water-holding capacity [54]. Therefore, the type of salt and the amount added must be carefully tested in each case.

However, compared to salt, using the ingredients mentioned above could considerably increase the cost of the products, add extra calories, and reduce the nutritional value of the low-salt surimi gel.

4.2. *Incorporation of Different Gelation Enhancers in Low-Salt Surimi Gels*

The addition of small quantities of ingredients and/or adjuvants that improve or modify certain characteristics of foods is a common practice in the food industry. Therefore, many exogenous additives have been studied in the manufacture of surimi-based products and low-salt surimi-based products to improve the gel properties [22]. All of these ingredients serve a technological function; however, they also have several drawbacks from a nutritional standpoint. They could contribute extra empty calories to the final product and may contain allergens (such as milk protein and egg white), among other concerns.

4.2.1. Addition of Microbial Transglutaminase

The use of microbial transglutaminase (MTGase) to enhance the texture of myobibrillar protein gels from both fish and meat has been extensively studied [23,55]. However, MTGase might diminish gel strength due to excessive cross-linking, which could reduce the interaction between proteins and water as it constantly enhances cross-linking among proteins [56]. This could result in water loss within the system, causing the gel to become overly firm and less flexible, ultimately leading to an inelastic and fragile gel [23,57]. So, the use of MTGase should be optimized according to the surimi fish species, gelation process, and the presence of other technological adjuvants or salts [20,55].

4.2.2. Addition of Polyphenols

Polyphenols have been studied as gelation enhancers since they can function as protein cross-linkers. These compounds possess multiple hydroxyl groups that can trigger the formation of hydrogen bonds and interact with hydrophobic amino acids via their non-polar aromatic ring [58]. Additionally, phenols can oxidize to quinones, which may react with sulfhydryl and amino groups in the protein to create a covalent bond (C–S or C–N) [59]. Polyphenols from apple [60], tea [61], and olive leaf powder [62], among others, have demonstrated a positive effect on the gel strength of heat-induced surimi gels, although their effect on other physicochemical properties, such as colour and water-holding capacity, has to be considered, as well as the effect of the interaction with other technological adjuvants [23].

4.2.3. Addition of Phosphates

Phosphates are commonly added to surimi as cryoprotectants, usually in the form of sodium tripolyphosphate or tetrasodium pyrophosphate. These compounds prevent actin and myosin binding and thus myosin aggregation in frozen storage, resulting in improved gelling ability in different surimi types [25,63,64]. However, although they are sodium salts and, in percentage terms, provide a sodium content similar to that of NaCl (31.24% compared to the 39.34% that NaCl would provide), their advantage lies in the fact that they are added in lower percentages than NaCl to achieve similar effects [65]. Sodium pyrophosphate can enhance the water-holding capacity and strength of surimi gel [63,66]. On the other hand, a correlation has been described between the effect of sodium phosphate and the pH of the medium, indicating that an increase in the phosphate levels at an alkaline pH in rainbow trout surimi resulted in better emulsification properties and poorer gelation ability [67]. A positive effect in enhancing surimi gelation in low-salt surimi gels has also been described for tetra-sodium pyrophosphate at very low concentrations

(0.05%) [5]. However, although phosphate compounds have been proven as promising processing agents, they most likely have a detrimental effect on gel properties as they may chelate the Ca^{2+} ion. This could hinder the gelling of surimi induced by endogenous transglutaminase [63], which indicates the need to carefully study the range at which they are added to surimi.

4.2.4. Addition of Hydrocolloids

Polysaccharides (e.g., agar gum, K-carrageenan, curdlan, and fucoidan) and some proteins, such as gelatine, are hydrocolloids that significantly impact the texture and functional characteristics of restructured and surimi gels [33,68–70].

Hydrocolloids can absorb water and expand during heating, which fills in the network of surimi gels and exerts pressure on the protein network, thus increasing surimi gel strength [33,71–73]. In addition, some hydrocolloids can form a thermo-irreversible gel once the temperature rises to 80 °C, enhancing the gel strength of surimi [74]. Recently, yeast β-glucan, konjac glucomannan, and deacetylated konjac glucomannan, as functional polysaccharides, have been used as additives to improve surimi gel properties [75]. In general, hydrocolloids can interact with myofibrillar proteins through non-covalent bonds and electrostatic interactions. These interactions help stabilize the gel network and improve its texture, enhancing both gel strength and elasticity. As a result, they form a more cohesive and stable gel matrix, which improves the overall mechanical properties. However, the best effect on the mechanical properties of low-salt surimi gels was observed when treatment such as high-pressure processing was applied to aid the gelation process [21].

On the other hand, the addition of a small amount of gelatine has been reported to increase surimi gel's strength, which could be attributed to its water binding ability [76]. In addition, the effect of hydroxypropyl-methylcellulose (HPMC), a polysaccharide derived from cellulose able to form a thermo-reversible gel, has been checked as a gelation enhancer. Although it did not result in a well-built combinative gel, the rheological and textural properties of horse mackerel surimi gel were enhanced [77]. Nevertheless, the addition of a higher concentration of hydrocolloids might inhibit the cross-linking of surimi proteins and disrupt the formation of an organized matrix, leading to a reduction in gel strength due to the dilution effect of myofibrillar proteins in surimi because of hydrocolloid addition [73].

4.2.5. Addition of Amino Acids

A few recent studies have stated the positive effect of some amino acids on surimi gelation because these amino acids are amphoteric above their isoelectric point [5,78]. They can supply ionic strength to the medium, which is particularly interesting because, under low-salt conditions, myosin remains as insoluble fibres that are difficult to fully solubilize and disperse [79]. However, the charged amino acids in the dissolution of myofibrillar protein may promote the orderly interaction of protein molecules during heating. Firstly, basic amino acids such as Arg, Lys, and His can cooperate electrostatically with exposed negatively charged amino acid residues in the protein chain, modifying protein solubility. Additionally, basic amino acids can attach with aromatic amino acid residues through π-cation interactions, disrupting hydrogen bonds in the myosin backbone and partially preventing myosin aggregation. Lys can form complexes with metal ions, leading to the dissociation of actomyosin in proteins, thereby improving surimi gel strength. Glu and Asp can bind with amino acid residues like Ser, Thr, and Tyr on protein chains via ion–dipole interactions or interact with amino acid residues on myosin through hydrogen bonding during heating [80]. L-histidine, L-arginine, L-glutamine, and L-lysine have been research hotspots as additives for low-salt surimi gels [81–83]. Studies have reported that L-arginine (L-Arg) stabilizes proteins, prevents aggregation, and induces hydrophobic amino acid groups to engage with one another, enhancing the solubility of myofibrillar proteins. Compared to L-histidine, L-Arg has shown a more evident effect on the myofibrillar protein gel [84]. Additionally, the effect of cystine and L-lysine in low-salt surimi gels has also been described in [78]. Cystine is a weak oxidant; it causes the oxidation of SH groups on the surface of proteins, leading to the formation of S-S bonds [27], and lysine enhances the formation of cross-links mediated by endogenous transglutaminase that are established between lysine and glutamine amino acids. These amino acids improve the gel strength and thus the techno-functional properties of surimi gels by initially promoting protein denaturation or the unfolding of myofibrillar proteins, enabling the establishment of different types of bonds and resulting in compact and well-structured network gels. The combination of some of these amino acids with other components, such as oxidized caffeic acid, has also shown a strong potential for enhancing the gel properties of low-salt surimi gel [85].

4.2.6. Addition of Proteins

The most frequently utilized proteins are egg white, whey protein concentrate, chicken or porcine plasma protein, etc. [86]. Nevertheless, due to an allergy to different animal proteins, mainly from cow's milk, lactose intolerance problems, or religion, certain people cannot consume specific animal proteins. Therefore, plant proteins have gained relevance as effective additives to enhance surimi products' gelation since they can interact with myofibrillar proteins through hydrophobic interactions and disulfide bonds, thus modifying the gel structure and functionality [19,87–89].

Concerning plant proteins, the most popular and reasonably priced plant proteins are soy isolate protein (SPI) [87], low-lectin bean protein (LLBP) [88], wheat gluten (WG) [89], and pea protein (PP) [18]. The effect of the incorporation of soy protein isolate has been reported to be effective in improving the mechanical properties in silver carp surimi gels in a relatively low concentration (6 g/100 g) [87]. The effect of the incorporation of LLBP in a range from 14 to 17 g/100 g in low-salt Alaska pollock surimi gels indicated a high correlation with the salt content. Better gels were achieved when the salt content was decreased, likely due to the presence of saline ions that modify the electrostatic equilibrium of the charges that stabilize the gel network in the gels with a reduced salt content [18]. WB and hydrolysed WB have also shown a positive effect on enhancing surimi gel properties [89]. The effect of these proteins has demonstrated their ability as gelation enhancers by enhancing conformational flexibility and structural stabilizing ability [86,90]. Although a limited number of existing studies have initially assessed the potential use of these plant proteins as gel enhancers in surimi gels within a range of less than 10 g/100 g, not all of them have been studied with a low salt content in the medium, and the mechanisms underpinning the reinforcement effects are still unclear, suggesting that the gel-enhancing effectiveness of plant proteins depends on their concentration and could be related to the presence of different salt levels in the system.

5. Techno-Functional Properties of Low-Salt Surimi Gels

Surimi-based products have a series of techno-functional properties, which depend fundamentally on the quality of the raw material (protein functionality and protein concentration), the additives incorporated, and the gelation process carried out. According to [15], the three most important functional characteristics in surimi-based products, related to or derived from the protein's functionality, are colour, flavour, texture, and water holding capacity.

5.1. Texture

Texture is one of the most important sensory properties of surimi-based gels since it determines the degree of consumer acceptance.

Instrumental analysis is very useful for its study since there are several parameters that correlate well with its sensory analysis. One of the most used methods is the Puncture Test, which mimics the deformation that takes place during chewing of the food until the food breaks [55] and determines the breaking force (N) and breaking deformation (mm) (Figure 2).

Another method commonly used is the Texture Profile Analysis (TPA), which consists of a mechanical compression that aims to mimic the chewing process and determines the hardness (N), adhesiveness (N.s), springiness (mm), and cohesiveness [55].

At reduced salt levels, the challenges in myosin unfolding and assembly often result in structural disorder and weak gels [82], which are typically reflected in a reduced gel strength (Figure 3). However, not all additives used as gelation enhancers for low-salt surimi gels have the same effect on the mechanical properties. Therefore, regulating the addition levels is essential to optimize the quality of surimi gels and prevent adverse effects on the formation of the gel network [22].

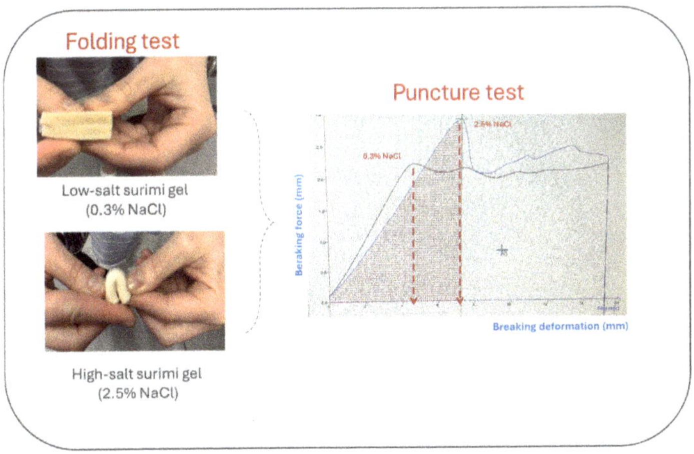

Figure 3. Effect of salt reduction on texture of surimi gels.

5.2. Water-Holding Capacity (WHC)

This is an essential characteristic of surimi gels that significantly affects their texture, mouthfeel, and overall quality. This property can be described as the ability of a food to retain its own moisture or added water (bound water, immobile water, and free water) when subjected to pressure, centrifugation, or heating forces [91]. These water molecules are firmly held within the gel matrix through hydrogen bonds and protein–water interactions [92]. The reduction in salt in surimi gels generally implies a reduction in the WHC, but the addition of different technological additives that act as gelation enhancers could increase the WHC. Depending on the additive, the enhancement is attributed to different processes: the exposure of hydrophilic groups, which bind more effectively with water molecules [93]; the increase in the contact area between protein and water molecules [87]; or the increase in the repulsion of protein groups due to the predominance of negative charges on the protein groups, which disorganizes the structure of the proteins and enhances the sites available to bind to water molecules [5].

5.3. Colour

Colour is one of the parameters that is modified during the gelation process due to protein aggregation. It is usually measured in the CIE L*a*b* colour space. In this space, colour is defined in a three-dimensional space, and each parameter corresponds to an axis, where L* indicates lightness (light to dark), a* is a colour coordinate from red to green, and b* is a colour coordinate from yellow to blue [15]. In surimi gels, special interest is usually paid to the brightness since it gives an idea of the degree of aggregation, and a lighter colour is associated with a higher quality. Due to protein aggregation, the amount of protein–protein bonds increases considerably, forming a compact protein structure that reflects light to a greater extent and therefore increases the luminosity [94]. According to this, all additives that enhance protein aggregation in low-salt surimi gels would also increase the lightness of the gel, which is not necessarily higher than the gels with a higher level of salt.

5.4. Flavour

The decrease in salt content in surimi-based products can considerably weaken their flavour, especially the perception of saltiness. The salty flavour of surimi gels has been related to the presence of adenosine triphosphate and its degradation products and to the formation of free amino acids, mainly alanine, a sweet amino acid, during thermal treatment [95,96]. In addition, it has been described that the addition of a small amount of yeast extract to reduced-salt surimi gels can enhance the salty and umami flavours of the product and, at the same time, reduce the sour and bitter flavours linked to the umami amino acids [80].

Achieving the optimal values of all of these parameters in low-salt surimi gels represents a major technological challenge for the surimi industry, which indicates that further research is required to achieve the desired properties.

6. Conclusions and Future Directions

In this paper, a comprehensive review of the most relevant low-salt surimi gelation enhancers has been provided, considering their feasibility for use by surimi product manufacturers or industries. Undoubtedly, salt reduction in surimi-based products poses a significant challenge for the industry, not only from technological and sensory perspectives but also in response to the growing consumer demand for healthier food options. In this context, surimi quality as a raw material is often poorly understood by consumers, who may question the nutritional value of this raw material and, by extension, the resulting surimi-based products. Therefore, it is essential to thoroughly describe the processing of this raw material and the importance of proper myofibrillar protein gelation to achieve products that are appealing to consumers. However, it is essential to inform consumers about the boundaries of surimi-based products by transmitting scientific information related to surimi processing, gelation, the effect of salt on the technological properties of the final products, and their nutritional value. To achieve this, education programs in schools, high schools, and universities, as well as well-designed marketing campaigns on TV, radio, and social media, would be effective options. As scientists, the best we can do is conduct research to provide society with reliable information and help disseminate it.

While surimi-based products typically require the addition of high levels of salt (1–3%) to obtain an adequate texture, such formulations may compromise their health benefits. In response, numerous countries and regions have introduced sodium-labelling regulations to indicate the level of salt in food products. Despite considerable efforts to develop low-salt surimi-based products, challenges remain, including issues with gel quality, increased costs, and a higher calorie content, all of which require further optimization. Moreover, it is important to recognize that both surimi-based and low-salt surimi-based products can be enriched with various ingredients (such as fibres, proteins, vitamins, and omega-3 fatty acids) that enhance their nutritional value and improve their final texture. Additionally, surimi-based products are fishery-derived, bone-free items that can serve as a valuable protein source for specific groups, such as children, young adults, and the elderly.

In this regard, substantial efforts must be directed toward the selection of surimi enhancers, either alone or in combination with different gelation methods (e.g., 3D printing, microwave processing, ultrasound, and high-pressure processing) that can enhance the physicochemical and sensory properties of low-salt surimi-based products. The most significant challenge for future research is the development of specific surimi gelation methods (additives and gelation techniques) that are affordable for the surimi-based products industry, which, together with consumers, remain the ultimate beneficiary of these advancements. Accordingly, the interaction between surimi production industries and researchers is fundamental to making progress in low-salt surimi gelation.

Author Contributions: N.W.: Conceptualization, data curation, writing—original draft, and writing—review and editing. M.B.: Conceptualization, data curation, writing—review and editing, and funding acquisition. H.M.M.: Conceptualization, supervision, validation, writing—original draft, writing—review and editing, and funding acquisition. All authors have read and agreed to the published version of the manuscript.

Funding: This research was supported by the Complutense University of Madrid (Spain) and UCM Project FEI-EU-24-09.

Institutional Review Board Statement: Not applicable.

Informed Consent Statement: Not applicable.

Data Availability Statement: No new data were created or analyzed in this study.

Conflicts of Interest: The authors declare no conflicts of interest.

References

1. Suzuki, T.; Tsuchiya, T. Nutrition and health benefits of surimi seafood. In *Surimi and Surimi Seafood*, 3rd ed.; Park, J.W., Ed.; CRC Press: Boca Raton, FL, USA, 2013; Section III; p. 603.
2. Ojima, F. *Effect of the Steamed Fish Paste Products on Dementia Prevention by Stimulating the Production of Nerve Growth Factor (NGF) Research Report on Health Benefit of Surimi Seafood*; ZENKAMA: Tokyo, Japan, 2010; pp. 66–67.
3. Fukunaga, K. *Colon Cancer Inhibition Derived from Eating to the Kamaboko and Identified the Inhibiting Component Research Report on Health Benefit of Surimi Seafood*; ZENKAMA: Tokyo, Japan, 2010; pp. 23–31.
4. Yazawa, K.; Yamaguchi, K. *Effect of Surimi Seafood on Life Style Related Disease in Mice Research Report on Health Benefit of Surimi Seafood*; ZENKAMA: Tokyo, Japan, 2010; pp. 44–49.
5. Cando, D.; Herranz, B.; Borderías, J.A.; Moreno, H.M. Different additives to enhance the gelation of surimi gel with reduced sodium content. *Food Chem.* **2016**, *196*, 791–799. [CrossRef] [PubMed]
6. Pérez, J.H.; Unanua, A.P. Arterial hypertension: Everest. 2002.
7. World Health Organization. Available online: https://www.who.int/news-room/fact-sheets/detail/hypertension (accessed on 16 December 2024).
8. Nurmilah, S.; Cahyana, Y.; Utama, G.L.; Aït-Kaddour, A. Strategies to reduce salt content and its effect on food characteristics and acceptance: A review. *Foods* **2022**, *11*, 3120. [CrossRef]
9. Regulation (EC) No 1924/2006. Regulation of the European Parliament and of the Council of 20 December 2006 on nutrition and health claims made on foods. *Off. J. Eur. Union* **2006**, *404*, 9–25.
10. Wang, X.; Luo, N.; Guo, C.; Wang, X.; Xia, S. Enhancing gel strength and saltiness perception of low-salt surimi gels: Synergistic effects of lysine assisted with water bath-microwave heating. *Food Biosci.* **2024**, *61*, 104827. [CrossRef]
11. Hu, Y.; Badar, I.H.; Liu, Y.; Zhu, Y.; Yang, L.; Kong, B.; Xu, B. Advancements in production, assessment, and food applications of salty and saltiness-enhancing peptides: A review. *Food Chem.* **2024**, *453*, 139664. [CrossRef] [PubMed]
12. Seafood News. 2023. Available online: https://www.seafoodnews.com/Story/1276788/Maruha-Nichiro-Estimates-Worldwide-Surimi-Production-in-2023-Reached-800000-Tons (accessed on 15 December 2024).

13. Surimi Market Analysis. Surimi Market Size, Share & Trends Analysis Report By Source (Tropical, Cold Water), By Form (Frozen, Fresh), By Distribution Channel (B2B, B2C), By Region, And Segment Forecasts, 2023–2030. 2023. Available online: https://www.grandviewresearch.com/industry-analysis/surimi-market-report (accessed on 15 December 2024).
14. Market Research Survey. Surimi Market Study by Fish Surimi and Meat Surimi for HoReCa, Food Processing, Households, Pahrmacutical, and Animal Feed, 2023–2033. 2023. Available online: https://www.factmr.com/report/5014/surimi-market (accessed on 15 December 2024).
15. Park, J.W.; Nozaki, H.; Suzuki, T.; Beliveau, J. Historical review of Surimi technology and market developments. In *Surimi and Surimi Seafood*, 3rd ed.; Park, J.W., Ed.; CRC Press: Boca Raton, FL, USA, 2013; Section I; p. 3.
16. Kim, Y.; Park, J. Negative roles of salt in gelation properties of fish protein isolate. *J. Food Sci.* **2008**, *73*, C585–C588. [CrossRef] [PubMed]
17. Lanier, T.C.; Carvajal-Rondanelli, P.; Vadlamudi, R.K. Surimi gelation chemistry. In *Surimi and Surimi Seafood*, 3rd ed.; Park, J.W., Ed.; CRC Press: Boca Raton, FL, USA, 2013; Section I; p. 101.
18. Shimada, M.; Takai, E.; Ejima, D.; Arakawa, T.; Shiraki, K. Heat-induced formation of myosin oligomer-soluble filament complex in high-salt solution. *Int. J. Biol. Macromol.* **2015**, *73*, 17–22. [CrossRef] [PubMed]
19. Borderías, A.J.; Tovar, C.; Domínguez-Timón, F.; Díaz, M.T.; Pedrosa, M.M.; Moreno, H.M. Characterization of healthier mixed surimi gels obtained through partial substitution of myofibrillar proteins by pea protein isolates. *Food Hydrocoll.* **2020**, *107*, 105976. [CrossRef]
20. Cando, D.; Herranz, B.; Borderías, A.J.; Moreno, H.M. Effect of high pressure on reduced sodium chloride surimi gels. *Food Hydrocoll.* **2015**, *51*, 176–187. [CrossRef]
21. Hongviangjan, W.; Sompongse, W. Combined effects of high-pressure processing and polysaccharides on the characteristics and microstructure of low-salt threadfin bream surimi gel. *Int. J. Food Sci. Technol.* **2024**, *59*, 8312–8320. [CrossRef]
22. Li, Q.; Jin, H.; Xia, M.; Sun, H.; Zeng, T.; Wang, Y.; Lu, L.; Cai, Z. Sucrose-phosphate osmotic system improves the quality characteristics of reduced-salt salted egg yolk: Profiling from protein structure and lipid distribution perspective. *Food Chem.* **2024**, *445*, 138750. [CrossRef] [PubMed]
23. Zhong, Q.; Wang, Y.; Tian, Y.; Zhuang, Y.; Yang, H. Effects of anthocyanins and microbial transglutaminase on the physicochemical properties of silver carp surimi gel. *J. Texture Stud.* **2023**, *54*, 541–549. [CrossRef]
24. Vidal-Giraud, B.; Chateau, D. World surimi market. In *GLOBEFISH Research Programme*; FAO: Rome, Italy, 2007; Volume 89, p. 125.
25. Park, J.W.; Graves, D.; Draves, R.; Yongsawatdigul, J. Manufacture of Surimi: Harvest to Frozen Block. In *Surimi and Surimi Seafood*, 3rd ed.; Park, J.W., Ed.; CRC Press: Boca Raton, FL, USA, 2013; Section I; p. 55.
26. Lanier, T.C. New tecnologies in surimi manufacture. In *Surimi Technology*; Lanier, T.C., Lee, C.M., Eds.; Marcel Dekker Inc: New York, NY, USA, 1992; pp. 167–207.
27. Chen, W.L.; Chow, C.J.; Ochiai, Y. Effects of some food additives on the gel-forming ability and color of milkfish meat paste. *Fish. Sci.* **1999**, *65*, 777–783. [CrossRef]
28. Sánchez-Alonso, I.; Haji-Maleki, R.; Borderias, A.J. Wheat fiber as a functional ingredient in restructured fish products. *Food Chem.* **2007**, *100*, 1037–1043. [CrossRef]
29. Kobayashi, Y.; Park, J.W. Biochemical and physical characterizations of fish protein isolate and surimi prepared from fresh and frozen whole fish. *LWT-Food Sci. Technol.* **2017**, *77*, 200–207. [CrossRef]
30. Zayas, J.F. Solubility of proteins. In *Functionality of Proteins in Food*; Zayas, J.F., Ed.; Springer: Berlin/Heidelberg, Germany, 1997; Chapter 2; pp. 6–75.
31. Zhang, C.; Lu, M.; Ai, C.; Cao, H.; Xiao, J.; Imran, M.; Chen, L.; Teng, H. Ultrasonic treatment combined with curdlan improves the gelation properties of low-salt *Nemipterus virgatus* surimi. *Int. J. Biol. Macromol.* **2023**, *248*, 125899. [CrossRef] [PubMed]
32. Wei, Q.J.; Zhang, W.W.; Wang, J.J.; Thakur, K.; Hu, F.; Khan, M.R.; Zhang, J.G.; Wei, Z.J. Effect of κ-carrageenan on the quality of crayfish surimi gels. *Food Chem. X* **2024**, *22*, 101497. [CrossRef]
33. Zhang, L.; Li, Q.; Shi, J.; Zhu, B.; Luo, Y. Changes in chemical interactions and gel properties of heat-induced surimi gels from silver carp (*Hypophthalmichthys molitrix*) fillets during setting and heating: Effects of different washing solutions. *Food Hydrocoll.* **2018**, *75*, 116–124. [CrossRef]
34. Wang, G.; Liu, M.; Cao, L.; Yongsawatdigul, J.; Xiong, S.; Liu, R. Effects of different NaCl concentrations on self-assembly of silver carp myosin. *Food Biosci.* **2018**, *24*, 1–8. [CrossRef]
35. Ye, Y.; Liu, X.; Bai, W.; Zhao, W.; Zhang, Y.; Dong, H.; Pan, Z. Effect of microwave ultrasonic combination treatment on heating-induced gel properties of low-sodium tilapia surimi during gel setting stage and comparative analysis. *LWT-Food Sci. Technol.* **2022**, *161*, 113386. [CrossRef]
36. Zhu, J.; Cheng, Y.; Ouyang, Z.; Yang, Y.; Ma, L.; Wang, H.; Zhang, Y. 3D printing surimi enhanced by surface crosslinking based on dry-spraying transglutaminase, and its application in dysphagia diets. *Food Hydrocoll.* **2023**, *140*, 108600. [CrossRef]
37. Yongsawatdigul, J.; Worratao, A.; Park, J. Effect of endogenous transglutaminase on threadfin bream surimi gelation. *J. Food Sci.* **2002**, *67*, 3258–3263. [CrossRef]

38. Núñez-Flores, R.; Cando, D.; Borderías, A.J.; Moreno, H.M. Importance of salt and temperature in myosin polymerization during surimi gelation. *Food Chem.* **2018**, *239*, 1226–1234. [CrossRef] [PubMed]
39. Kristinsson, H.G.; Hultin, H.O. Changes in conformation and subunit assembly of cod myosin at low and high pH and after subsequent refolding. *J. Agric. Food Chem.* **2003**, *24*, 7187–7196. [CrossRef] [PubMed]
40. Gao, X.; Yin, T.; Hu, Y.; You, J.; Xiong, S.; Liu, R. Effect of high intensity ultrasound on gelation properties of silver carp surimi with different salt contents. *Ultrason. Sonochem.* **2021**, *70*, 105326. [CrossRef] [PubMed]
41. Zayas, J.F. Gelling properties of proteins. In *Functionality of Proteins in Food*; Zayas, J.F., Ed.; Springer: Berlin/Heidelberg, Germany, 1997; Chapter 7; pp. 310–366.
42. Messens, W.; Van Camp, J.; Huyghebaert, A. The use of high pressure to modify the functionality of food proteins. *Trends Food Sci. Technol.* **1997**, *8*, 107–112. [CrossRef]
43. Visschers, R.W.; de Jongh, H.H.J. Disulphide bond formation in food protein aggregation and gelation. *Biotechnol. Adv.* **2005**, *23*, 75–80. [CrossRef] [PubMed]
44. Feng, M.; Pan, L.; Yang, X.; Sun, J.; Xu, X.; Zhou, G. Thermal gelling properties and mechanism of porcine myofibrillar protein containing flaxseed gum at different NaCl concentrations. *LWT-Food Sci. Technol.* **2018**, *87*, 361–367. [CrossRef]
45. Zhang, Y.; Bai, G.; Jin, G.; Wang, Y.; Wang, J.; Puolanne, E.; Cao, J. Role of low molecular additives in the myofibrillar protein gelation: Underlying mechanisms and recent applications. *Crit. Rev. Food Sci. Nutr.* **2022**, *64*, 3604–3622. [CrossRef]
46. Gilleland, G.; Lanier, T.C.; Hamann, D. Covalent bonding in pressure induced fish protein gels. *J. Food Sci.* **1997**, *62*, 713–733. [CrossRef]
47. Balange, A.K. *Enhancement of Gel Strength of Surimi Using Oxidized Phenolic Compound*; Food Science and Technology Prince of Songkla University: Tambon Ruesamilae, Thailand, 2009.
48. Walayat, N.; Liu, J.; Nawaz, A.; Aadil, R.M.; López-Pedrouso, M.; Lorenzo, J.M. Role of Food Hydrocolloids as Antioxidants along with Modern Processing Techniques on the Surimi Protein Gel Textural Properties, Developments, Limitation and Future Perspectives. *Antioxidants* **2022**, *11*, 486. [CrossRef] [PubMed]
49. Guo, X.; Shi, L.; Xiong, S.; Hu, Y.; You, J.; Huang, Q.; Yin, T. Gelling properties of vacuum-freeze dried surimi powder as influenced by heating method and microbial transglutaminase. *LWT-Food Sci. Technol.* **2019**, *99*, 105–111. [CrossRef]
50. Wang, Y.; Liu, C.; Lang, H.; Hu, Z.; Wang, X.; Guo, Z.; Jiang, L. Effects of microwave on the structural and emulsifying properties and interfacial properties of oxidized soybean protein aggregates. *Food Chem. X* **2023**, *19*, 100861. [CrossRef]
51. Xie, Y.; Zhao, K.; Yang, F.; Shu, W.; Ma, J.; Huang, Y.; Cao, X.; Liu, Q.; Yuan, Y. Modification of myofibrillar protein structural characteristics: Effect of ultrasound-assisted first stage thermal treatment on unwashed Silver Carp surimi gel. *Ultrason. Sonochem.* **2024**, *107*, 106911. [CrossRef] [PubMed]
52. Qian, X.; Lin, S.; Chen, T.; Li, S.; Wang, S.; Li, C.; Wang, R.; Sun, N. Evaluation of the texture characteristics and taste of shrimp surimi with partial replacement of NaCl by non-sodium metal salts. *Food Chem.* **2024**, *459*, 140403. [CrossRef]
53. Desmond, E. Reducing salt: A challenge for the meat industry. *Meat Sci.* **2006**, *74*, 188–196. [CrossRef] [PubMed]
54. Zheng, J.; Han, Y.; Ge, G.; Zhao, M.; Sun, W. Partial substitution of NaCl with chloride salt mixtures: Impact on oxidative characteristics of meat myofibrillar protein and their rheological properties. *Food Hydrocoll.* **2019**, *96*, 36–42. [CrossRef]
55. Moreno, H.M.; Carballo, J.; Borderías, A.J. Influence of alginate and microbial transglutaminase as binding ingredients on restructured fish muscle processed at low temperature. *J. Sci. Food Agric.* **2008**, *88*, 1529–1536. [CrossRef]
56. Dong, X.; Pan, Y.; Zhao, W.; Huang, Y.; Qu, W.; Pan, J.; Qui, H.; Prakash, S. Impact of microbial transglutaminase on 3D printing quality of *Scomberomorus niphonius* surimi. *LWT-Food Sci. Technol.* **2020**, *124*, 109123. [CrossRef]
57. Seighalani, F.Z.B.; Bakar, J.; Saari, N.; Khoddami, A. Thermal and physicochemical properties of red tilapia (*Oreochromis niloticus*) surimi gel as affected by microbial transglutaminase. *Anim. Prod. Sci.* **2017**, *57*, 993. [CrossRef]
58. Sharma, S.; Majumdar, R.K.; Mehta, N.K. Gelling properties and microstructure of the silver carp surimi treated with pomegranate (*Punica granatum* L.) peel extract. *J. Food Sci. Technol.* **2022**, *59*, 4210–4220. [CrossRef] [PubMed]
59. Quan, T.H.; Benjakul, S.; Sae-leaw, T.; Balange, A.K.; Maqsood, S. Protein–polyphenol conjugates: Antioxidant property, functionalities and their applications. *Trends Food Sci. Technol.* **2019**, *91*, 507–517. [CrossRef]
60. Sun, L.; Sun, J.; Thavaraj, P.; Yang, X.; Guo, Y. Effects of thinned young apple polyphenols on the quality of grass carp (*Ctenopharyngodon idellus*) surimi during cold storage. *Food Chem.* **2017**, *224*, 372–381. [CrossRef] [PubMed]
61. Wu, J.; Li, C.; Li Li Laihao Yang, X.; Wang, Y.; Zhou, W. Improved physicochemical properties and product characteristics of tilapia surimi by tea polyphenols during chilled storage. *LWT-Food Sci. Technol.* **2022**, *167*, 113822. [CrossRef]
62. Arsyad, M.A.; Akazawa, T.; Ogawa, M. Effects of Olive Leaf Powder on Mechanical Properties of Heat-Induced Surimi Gel. *J. Aquat. Food Prod.* **2018**, *28*, 2–13. [CrossRef]
63. Julavittayanukul, O.; Benjakul, S.; Visessanguan, W. Effect of phosphate compounds on gel-forming ability of surimi from bigeye snapper (*Priacanthus tayenus*). *Food Hydrocoll.* **2006**, *20*, 1153–1163. [CrossRef]
64. Zheng, M.; Hong, J.; Chuai, P.; Chen, Y.; Ni, H.; Li, Q.; Jiang, Z. Impacts of agar gum and fucoidan on gel properties of surimi products without phosphate. *Food Sci. Nutr.* **2022**, *10*, 3759–3771. [CrossRef] [PubMed]

65. Ruusunen, M.; Puolanne, E. Reducing sodium intake from meat products. *Meat Sci.* **2005**, *70*, 531–541. [CrossRef]
66. Lee, J.; Yuan, P.; Heidolph, B.B.; Park, J.W. Physicochemical properties of frozen Alaska Pollock fillets and surimi as affected by various sodium phosphates. *J. Food Process. Preserv.* **2018**, *42*, e13530. [CrossRef]
67. Díaz-Vela, J.; Pérez-Chabela, M.D.L.; Totosaus, A. Efecto del pH y de la adición de fosfatos de sodio sobre las propiedades de gelificación y emulsión de surimi de trucha arco-iris (*Oncorhynchus mykiss*). *Food Sci. Technol.* **2008**, *28*, 691–695. [CrossRef]
68. Iglesias-Otero, M.A.; Borderias, J.; Tovar, C.A. Use of konjac glucomannan as additive to reinforce the gels from low-quality squid surimi. *J. Food Eng.* **2010**, *101*, 281–288. [CrossRef]
69. Salehi, B.; Sharifi-Rad, J.; Seca, A.M.L.; Pinto, D.C.G.A.; Michalak, I.; Trincone, A.; Mishra, A.P.; Nigam, M.; Zam, W.; Martins, N. Current trends on seaweeds: Looking at chemical composition, phytopharmacology, and cosmetic applications. *Molecules* **2019**, *24*, 4182. [CrossRef]
70. Zhong, H.; Gao, X.; Cheng, C.; Liu, C.; Wang, Q.; Han, X. The structural characteristics of seaweed polysaccharides and their application in gel drug delivery systems. *Mar. Drugs* **2020**, *18*, 658. [CrossRef]
71. Chen, J.; Deng, T.; Wang, C.; Mi, H.; Yi, S.; Li, X.; Li, J. Effect of hydrocolloids on gel properties and protein secondary structure of silver carp surimi. *J. Sci. Food Agric.* **2020**, *100*, 2252–2260. [CrossRef] [PubMed]
72. Chen, H.H.; Xue, C. Effect of hydrocolloids on the gel properties of horse-mackerel surimi. *Trans. Chin. Soc. Agric.* **2009**, *40*, 119–125. [CrossRef]
73. Wu, C.; Yuan, C.; Chen, S.; Liu, D.; Ye, X.; Hu, Y. The effect of curdlan on the rheological properties of restructured ribbonfish (*Trichiurus* spp.) meat gel. *Food Chem.* **2015**, *179*, 222–231. [CrossRef] [PubMed]
74. Hu, Y.; Liu, W.; Yuan, C.; Morioka, K.; Chen, S.; Liu, D.; Ye, X. Enhancement of the gelation properties of hairtail (*Trichiurus haumela*) muscle protein with curdlan and transglutaminase. *Food Chem.* **2015**, *176*, 115–122. [CrossRef]
75. He, X.; Lv, Y.; Li, X.; Yi, S.; Zhao, H.; Xu, Y.; Li, J. Effect of oat β-glucan on gel properties and protein conformation of silver carp surimi. *J. Sci. Food Agric.* **2023**, *103*, 3367–3375. [CrossRef]
76. Kaewudom, P.; Benjakul, S.; Kijroongrojana, K. Properties of surimi gel as influenced by fish gelatin and microbial transglutaminase. *Food Biosci.* **2013**, *1*, 39–47. [CrossRef]
77. Chen, H.H. Thermal gelation behaviors of surimi protein mixed with Hydroxypropylmethylcellulose. *Fish. Sci.* **2006**, *72*, 679–685. [CrossRef]
78. Cando, D.; Borderías, A.J.; Moreno, H.M. Combined effect of aminoacids and microbial transglutaminase on gelation of low salt surimi content under high pressure processing. *Innov. Food Sci. Emerg. Technol.* **2016**, *36*, 10–17. [CrossRef]
79. Dai, H.; Chen, X.; Peng, L.; Ma, L.; Sun, Y.; Li, L.; Wan, Q.; Zhan, Y. The mechanism of improved myosin gel properties by low dose rosmarinic acid addition during gel formation. *Food Hydrocoll.* **2020**, *106*, 105869. [CrossRef]
80. Wu, D.; Xiong, J.; Li, P.; Zhang, Y.; Li, F.; Yin, T.; Huang, Q. Dual enhancement effects of different yeast extract on gel properties and saltiness perception of low-salt surimi gel from silver carp. *Food Hydrocoll.* **2024**, *152*, 109925. [CrossRef]
81. Jiang, Q.; Chen, N.; Gao, P.; Yu, D.; Yang, F.; Xu, Y.; Xia, W. Influence of L-arginine addition on the gel properties of reduced-salt white leg shrimp (*Litopenaeus vannamei*) surimi gel treated with microbial transglutaminase. *LWT-Food Sci. Technol.* **2023**, *173*, 114310. [CrossRef]
82. Wang, J.; Huang, X.; Zhang, Y.; Li, S.; Dong, X.; Qin, L. Effect of sodium salt on meat products and reduction sodium strategies—A review. *Meat Sci.* **2023**, *205*, 109296. [CrossRef] [PubMed]
83. Takai, E.; Yoshizawa, S.; Ejima, D.; Arakawa, T.; Shiraki, K. Synergistic solubilization of porcine myosin in physiological salt solution by arginine. *Int. J. Biol. Macromol.* **2013**, *62*, 647–651. [CrossRef] [PubMed]
84. Gao, R.; Shi, T.; Sun, Q.; Li, X.; McClements, D.J.; Yuan, L. Effects of L-arginine and L-histidine on heat-induced aggregation of fish myosin: Bighead carp (*Aristichthys nobilis*). *Food Chem.* **2019**, *295*, 320–326. [CrossRef] [PubMed]
85. Xiong, Z.; Shi, T.; Zhang, W.; Kong, Y.; Yuan, L.; Gao, R. Improvement of gel properties of low salt surimi using low-dose l-arginine combined with oxidized caffeic acid. *LWT-Food Sci. Technol.* **2021**, *145*, 111303. [CrossRef]
86. Zhao, Y.; Wei, K.; Chen, J.; Wei, G.; Li, J.; Zheng, B.; Song, Y.; Gao, P.; Zhou, R. Enhancement of myofibrillar protein gelation by plant proteins for improved surimi gel characteristics: Mechanisms and performance. *LWT-Food Sci. Technol.* **2024**, *198*, 116045. [CrossRef]
87. Zhang, X.; Guo, Q.; Shi, W. Ultrasound-assisted processing: Changes in gel properties, water-holding capacity, and protein aggregation of low-salt *Hypophthalmichthys molitrix* surimi by soy protein isolate. *Ultrason. Sonochem.* **2023**, *92*, 106258. [CrossRef]
88. Moreno, H.M.; Díaz, M.T.; Borderías, A.J.; Domínguez-Timón, F.; Varela, A.; Tovar, C.A.; Pedrosa, M.M. Effect of Different Technological Factors on the Gelation of a Low-Lectin Bean Protein Isolate. *Plant Foods Hum. Nutr.* **2022**, *77*, 141–149. [CrossRef] [PubMed]
89. Zhang, L.; Zhang, F.; Wang, X. Effects of hydrolyzed wheat gluten on the properties of high-temperature (≥100 °C) treated surimi gels. *Food Hydrocoll.* **2015**, *45*, 196–202. [CrossRef]
90. Moreno, H.M.; Tovar, C.A.; Domínguez-Timón, F.; Cano-Báez, J.; Díaz, M.T.; Pedrosa, M.M.; Borderías, A.J. Gelation of commercial pea protein isolate: Effect of microbial transglutaminase and thermal processing. *Food Sci. Technol.* **2020**, *40*, 800–809. [CrossRef]

91. Zayas, J.F. Water holding capacity of proteins. In *Functionality of Proteins in Food*; Zayas, J.F., Ed.; Springer: Berlin/Heidelberg, Germany, 1997; Chapter 3; p. 76.
92. Liu, C.; Li, W.; Lin, B.; Yi, S.; Ye, B.; Mi, H.; Li, J.; Wang, J.; Li, X. Comprehensive analysis of ozone water rinsing on the water-holding capacity of grass carp surimi gel. *LWT-Food Sci. Technol.* **2021**, *150*, 111919. [CrossRef]
93. Zhang, C.; Chen, L.; Lu, M.; Ai, C.; Cao, H.; Xiao, J.; Zhong, S.; Teng, H. Effect of cellulose on gel properties of heat-induced low-salt surimi gels: Physicochemical characteristics, water distribution and microstructure. *Food Chem. X* **2023**, *19*, 100820. [CrossRef] [PubMed]
94. Cando, D.; Borderías, A.J.; Moreno, H.M. Influence of amino acid addition during the storage life of high pressure processed low salt surimi gels. *LWT-Food Sci. Technol.* **2017**, *75*, 599–607. [CrossRef]
95. Luo, X.; Xiao, S.; Ruan, Q.; Gao, Q.; An, Y.; Hu, Y.; Xiong, S. Differences in flavor characteristics of frozen surimi products reheated by microwave, water boiling, steaming, and frying. *Food Chem.* **2022**, *372*, 131260. [CrossRef]
96. Zhao, C.J.; Schieber, A.; Gänzle, M.G. Formation of taste-active amino acids, amino acid derivatives and peptides in food fermentations–A review. *Food Res. Int.* **2016**, *89*, 39–47. [CrossRef] [PubMed]

gels

MDPI

Article

Physicochemical and Functional Properties of Yanbian Cattle Bone Gelatin Extracted Using Acid, Alkaline, and Enzymatic Hydrolysis Methods

Song Zhang [1], Duanduan Zhao [1], Lu Yin [1], Ruixuan Wang [1], Zhiyan Jin [1], Hongyan Xu [1,*] and Guangjun Xia [2,*]

1 Department of Food Science and Engineering, College of Agriculture, Yanbian University, Yanji 133000, China; 2022010679@ybu.edu.cn (S.Z.); 18943362092@163.com (D.Z.); 18743675027@163.com (L.Y.); 2024010885@ybu.edu.cn (R.W.); 15526722180@163.com (Z.J.)

2 Department of Animal Science, College of Agriculture, Yanbian University, Yanji 133000, China

* Correspondence: xuhongyan@ybu.edu.cn (H.X.); ybuac@ybu.edu.cn (G.X.); Tel.: +86-13944709169 (H.X.); Fax: +86-0433-2435583 (H.X.)

Abstract: Yanbian cattle, a high-quality indigenous breed in China, were selected due to their unique biological characteristics, underutilized bone byproducts, and potential as a halal-compliant gelatin source, addressing the growing demand for alternatives to conventional mammalian gelatin in Muslim-majority regions. This study investigates the physicochemical and functional properties of gelatin extracted from Yanbian cattle bones using three different methods: acid, alkaline, and papain enzymatic hydrolysis. The extraction yields and quality of gelatin were evaluated based on hydroxyproline content, gel strength, viscosity, amino acid composition, molecular weight distribution, and structural integrity. Specifically, A gelatin, prepared using 0.075 mol/L hydrochloric acid, achieved the highest yield (18.64%) among the acid-extraction methods. B gelatin, extracted with 0.1 mol/L sodium hydroxide, achieved the highest yield (21.06%) among the alkaline-extraction methods. E gelatin, obtained through papain hydrolysis, exhibited the highest yield (25.25%) among the enzymatic methods. Gelatin extracted via papain enzymatic hydrolysis not only retained better protein structure but also exhibited higher hydroxyproline content (19.13 g/100 g), gel strength (259 g), viscosity (521.67 cP), and superior thermal stability. Structural analyses conducted using SDS-PAGE, GPC, FTIR, XRD, and CD spectroscopy confirmed that papain extraction more effectively preserved the natural structure of collagen. Furthermore, amino acid composition analysis revealed that gelatin extracted via papain hydrolysis contained higher levels of essential residues, such as glycine, proline, and hydroxyproline, emphasizing the mild and efficient nature of enzymatic treatment. These findings suggest that, compared with acid and alkaline extraction methods, enzymatic hydrolysis has potential advantages in gelatin production. Yanbian cattle bone gelatin shows promise as an alternative source for halal gelatin production. This study also provides insights into optimizing gelatin production to enhance its functionality and sustainability.

Keywords: Yanbian cattle; bone; gelatin; enzymatic hydrolysis; physicochemical properties

1. Introduction

Gelatin, a biopolymer derived from the partial hydrolysis of collagen, is widely utilized in the food, pharmaceutical, and biomedical industries due to its biocompatibility, biodegradability, and excellent film-forming and gelling properties [1,2]. For example, a recent study explored the influence of gelatin/guar gum mixtures on the rheological and textural properties of restructured ricotta cheese. The combination of gelatin and guar gum significantly improved the texture and viscosity of the cheese, making it a promising candidate for food products requiring improved mouthfeel and stability [3]. Zhao et al. used citric acid as a cross-linking agent to form a film matrix through the condensation reaction between bovine bone gelatin and chitosan. Using the blueberry anthocyanin–iron ion complex as a pH-sensitive indicator, they developed a highly sensitive indicator film to effectively monitor the freshness of pork [4]. Traditional gelatin extraction from mammalian sources (e.g., bovine bones) faces limitations in meeting halal certification requirements, driving demand for alternative processing methods and underutilized raw materials [5]. Yanbian cattle, a high-quality Chinese breed with unique collagen composition due to genetic and environmental factors, offer a promising yet underexplored gelatin source [6]. Comparative studies on extraction methods for this specific collagen source remain scarce, particularly regarding structural preservation and functional performance.

Currently, three main methods are used for gelatin production: the alkaline method, the acid method, and the enzymatic method. Ismail and Abdullah treated fish skin with hydrochloric acid (HCl) at different concentrations (0.05–0.2 M), followed by thermal extraction of gelatin in a water bath at 45 °C. The results showed that the gelatin yields were 18.86% and 20.95% when using 0.05 M and 0.2 M HCl, respectively [7]. Similarly, Amertaningtyas et al. compared acid (0.25 M HCl) and alkaline (0.25 M NaOH) pretreatments for cattle hide-derived gelatin, revealing that HCl pretreatment yielded superior outcomes in terms of extraction efficiency, protein content, and gel strength. Alkaline methods, while advantageous for mature collagen sources (e.g., cattle bones and ligaments) due to their ability to preserve cross-linked structures and minimize yield loss, may inadvertently reduce yields if excessively prolonged, as prolonged alkali exposure increases collagen solubility in aqueous phases [8,9]. For instance, Mafazah et al. identified 0.01 M NaOH with a 12 h treatment as optimal for fish skin gelatin extraction, noting that alkaline conditions induce deamidation of glutamine and asparagine residues, shifting the isoelectric point from 7–9 to 5 [10,11]. However, alkaline-derived gelatin typically exhibits lower yields and higher lipid residues compared to acid-based methods [12]. In contrast, enzymatic strategies utilizing proteases (e.g., papain) offer a sustainable alternative by hydrolyzing collagen from animal byproducts into gelatin with enhanced efficiency. This biocatalytic approach not only shortens processing times but also improves yield and reduces waste generation [13]. Ma et al. selected a protease with specific catalytic specificity, pepsin, which remains stable and effectively cleaves peptide bonds under low pH conditions. They then innovatively combined the demineralization process (using hydrochloric acid to dissolve hydroxyapatite in bones) with pepsin hydrolysis in a single-step reaction. By eliminating the lime treatment and demineralization steps found in traditional methods, this approach reduces environmental pollution. As a result, a high-quality gelatin product with improved physicochemical properties was obtained compared to the gelatin extracted by traditional methods [14]. Zaitsev et al. have demonstrated that the molecular weight distribution (MWD) of gelatin significantly influences its properties [15]. To preserve the structural integrity of collagen, it is crucial to select enzymes that cleave collagen chains gently while minimizing damage to critical structural domains. Enzymes such as papain exhibit advantages in this regard, as they specifically target and cleave certain peptide bonds rather than randomly disrupting the three-dimensional structure of collagen. This selective cleavage can help maintain or enhance the content of key amino acids such as hydroxyproline, which plays a vital role in determining the properties of gelatin [16,17]. However, comparative studies on extraction methods for specific collagen sources remain scarce, particularly regarding structural preservation and functional performance.

This study aimed to comprehensively evaluate and compare the properties of gelatin extracted from Yanbian cattle bone using acid, alkaline, and enzymatic hydrolysis. Specifically, the study focused on two primary objectives: (1) Structural analysis: Investigating the molecular structural characteristics and molecular weight distribution of gelatin using spectroscopic and chromatographic techniques to understand the degree of collagen degradation and preservation of the triple-helix structure in each extraction method. (2) Functional property assessment: Examining the functional properties of gelatin, including moisture content, crude protein content, ash content, gel strength, and viscosity, which are critical for its application in the food industry. The findings provide valuable insights into how different extraction methods affect the characteristics and performance of Yanbian cattle bone gelatin, thereby contributing to the development of high-value-added products from sustainable and underutilized animal byproducts.

2. Results and Discussion

2.1. Yield of Extraction

Figure 1a shows the effect of different hydrochloric acid concentrations on gelatin yield. The yield of A 0.075 gelatin was significantly higher than that of other concentration groups ($p < 0.05$). As the concentration of hydrochloric acid increased, the yield initially increased, reaching its maximum at 0.075 mol/L, and then decreased. This phenomenon may be attributed to the fact that an appropriate acid concentration can effectively break cross-linked structures, producing more gelatin, while excessively high concentrations may lead to excessive degradation of collagen, thereby reducing the yield. A certain concentration of acid disrupts the cross-linking between collagen fibers, leading to the degradation of collagen chains into smaller peptides. This degradation not only reduces the molecular weight of collagen but also results in the loss of functional amino acids, such as hydroxyproline, which are essential for maintaining the structural integrity of collagen and the yield of gelatin [18]. Figure 1b shows the effect of different sodium hydroxide concentrations on gelatin yield. The yield of B 0.1 gelatin was significantly higher than that of other samples ($p < 0.05$). The results indicate that an increase in sodium hydroxide concentration facilitates the removal of non-collagenous proteins and lipids, thereby promoting the release of gelatin. However, when the concentration exceeded 0.1 mol/L, the yield slightly decreased, which may be attributed to the strong corrosiveness of the alkali causing excessive structural damage and affecting the extraction efficiency. Compared to acid treatment, the hydrolysis process of collagen under lower concentrations of NaOH is relatively milder. While alkaline conditions can also degrade the collagen structure, the effect is slower, allowing for better preservation of the collagen's molecular structure. Hydroxide ions (OH-) primarily act on non-collagenous proteins and lipids, rather than directly attacking the collagen backbone, thus the hydrolysis process is relatively slower, and the degree of degradation is milder. Under lower concentrations of sodium hydroxide, the alkaline hydrolysis generally occurs at a slower and less pronounced rate than acidic hydrolysis. This difference is supported by findings in the literature, such as Gómez-Guillén et al., who reported that acid treatments at higher concentrations lead to a more extensive breakdown of collagen's molecular structure compared to alkaline treatments [19]. Additionally, Cao et al. also observed that acid pretreatment causes more severe molecular weight reduction in collagen, whereas alkaline treatment preserves the collagen structure to a greater extent [20]. Figure 1c illustrates the effect of different enzyme treatments on gelatin yield. E papain gelatin showed a significantly higher yield compared to other enzyme-treated samples ($p < 0.05$). The catalytic mechanism of papain is determined by the cysteine-25 triad located at its active site, which operates effectively when the pH is maintained between 3.0 and 9.0 and the temperature ranges from 60 to 70 °C.

During this process, asparagine-175 and histidine-159 work synergistically to cleave peptide bonds by attracting the carbonyl carbon of the peptide backbone, thereby releasing the amino-terminal end [16]. This explains the strong catalytic activity of papain, making it suitable for extracting bone gelatin. The high degree of cross-linking within collagen molecules is primarily determined by the presence of certain amino acids, such as proline and hydroxyproline, which are essential for the formation and stabilization of the collagen triple helix. The cross-linking is further influenced by environmental factors such as pH, temperature, and the concentration of specific ions or chemicals used during extraction. In particular, acidic or alkaline conditions can disrupt these cross-links, causing either partial or complete degradation of the collagen structure. The degree of cross-linking in collagen is essential for its functional properties, such as gel strength, viscosity, and thermal stability [21]. On the other hand, enzymatic hydrolysis using proteases offers a milder approach that specifically targets collagen without causing excessive degradation, thus enabling the more efficient release of high-quality gelatin. Therefore, the release conditions, such as pH, enzyme concentration, and reaction time, need to be carefully controlled to achieve an optimal balance between cross-link disruption and the preservation of collagen structure, leading to high yields and high-quality gelatin [22].

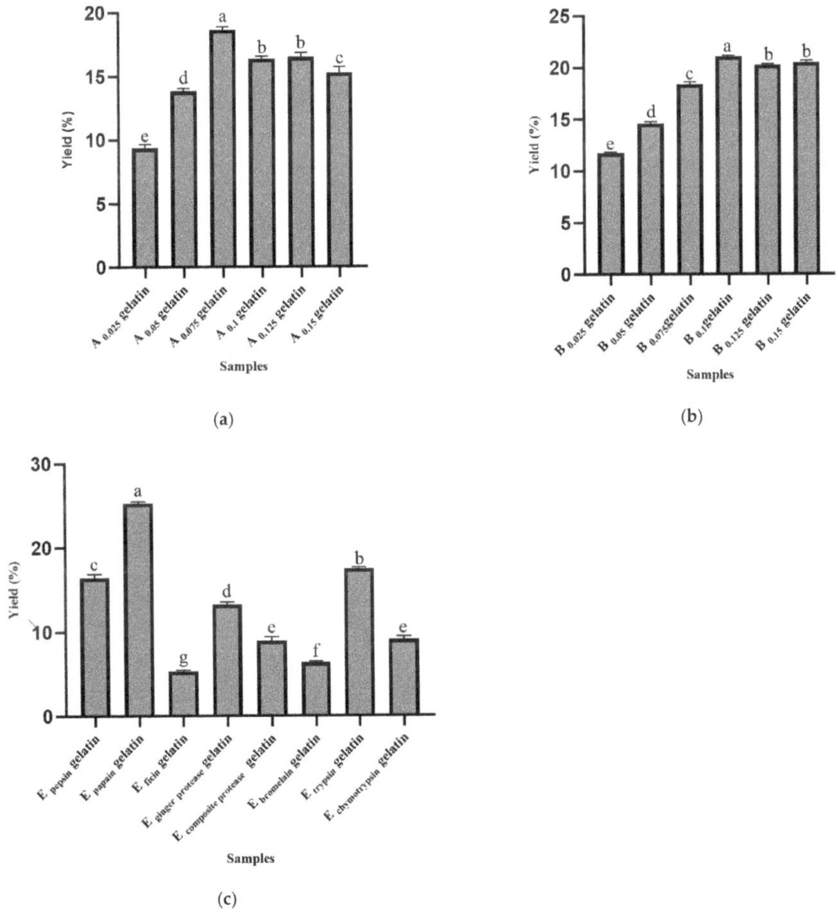

Figure 1. Yields (%) extracted from Yanbian cattle bone using different methods. (**a**) $A_{0.025}$ gelatin, $A_{0.05}$ gelatin, $A_{0.075}$ gelatin, $A_{0.1}$ gelatin, $A_{0.125}$ gelatin, and $A_{0.15}$ gelatin denote the gelatins extracted

with different hydrochloric acid concentrations. (**b**) $B_{0.025}$ gelatin, $B_{0.05}$ gelatin, $B_{0.075}$ gelatin, $B_{0.1}$ gelatin, $B_{0.125}$ gelatin, and $B_{0.15}$ gelatin denote the gelatins extracted with different sodium hydroxide concentrations. (**c**) E_{pepsin} gelatin, E_{papain} gelatin, E_{ficin} gelatin, $E_{ginger\ protease}$ gelatin, $E_{composite\ protease}$ gelatin, $E_{bromelain}$ gelatin, $E_{trypsin}$ gelatin, and $E_{chymotrypsin}$ gelatin denote the gelatins extracted with different enzymes. Results are presented as mean ± SD ($n = 3$). Statistical analysis was performed between samples for the same property. Different superscript letters (a–g) within the same color of bar chart indicate statistically significant differences as determined by one-way ANOVA followed by Duncan's post hoc test ($p < 0.05$).

2.2. Physicochemical Properties

2.2.1. Moisture Content

As shown in Table 1, moisture content was consistent across the three methods: enzymatic, alkaline, and acid. This similarity suggests that the final drying processes were effectively controlled and that the extraction method had minimal impact on moisture retention in the gelatin samples. The lack of significant differences in moisture content across the different extraction methods (acid, alkaline, and enzymatic) can be attributed to the effectiveness of the final drying process in all methods. After the gelatin was extracted, it underwent a similar drying procedure, which controlled the moisture content and likely minimized the impact of the extraction method on the final moisture levels. Moisture content in gelatin is primarily influenced by the drying conditions rather than the extraction process itself [23]. In gelatin production, stable moisture content is important, as it affects the shelf life, handling, and susceptibility to microbial growth [24]. Since there were no significant differences in moisture content among the methods, this parameter does not distinguish product quality. Therefore, the primary factors influencing gelatin quality are more likely related to protein integrity and yield rather than moisture content.

Table 1. Hydroxyproline and chemical properties of Yanbian cattle bone gelatin.

Type of Gelatin	Hydroxyproline Content (g/100 g)	Moisture (%)	Ash (%)	Crude Protein (%)
A gelatin	11.51 ± 0.20^{c}	7.67 ± 0.29^{a}	0.94 ± 0.07^{a}	77.57 ± 0.46^{c}
B gelatin	13.48 ± 0.35^{b}	7.47 ± 0.26^{a}	1.00 ± 0.06^{a}	84.16 ± 0.42^{b}
E gelatin	19.13 ± 0.13^{a}	7.40 ± 0.11^{a}	1.00 ± 0.09^{a}	87.54 ± 0.40^{a}

Results are presented as mean ± SD ($n = 3$). Statistical analyses were performed on samples with the same properties. Different superscript letters (a–c) within the same column indicate statistically significant differences, as determined by one-way ANOVA followed by Duncan's post hoc test ($p < 0.05$).

2.2.2. Ash Content

Ash content, an important indicator of gelatin quality, reflects the residual inorganic matter in the product. As shown in Table 1, the ash content of gelatin obtained from different extraction methods showed no significant differences. These ash values indicated that the three methods were similarly effective in removing inorganic substances. Ash content is closely related to the purity of gelatin, with lower ash contents indicating higher purity, which is particularly important for applications in food and pharmaceuticals. The low ash content observed in all gelatin samples suggests high purity, making these gelatins suitable for applications requiring high quality [25]. Furthermore, low ash content contributes to an improved transparency and sensory quality of gelatin, preventing color darkening and enhancing transparency. It also helps to maintain desirable gel strength and solubility.

2.2.3. Crude Protein Content

As shown in Table 1, enzymatic extraction yielded the highest crude protein content, followed by alkaline and the acid methods. A higher protein content typically indicates a greater proportion of collagen in the final product, as impurities and non-collagen proteins are minimized [26]. The reason why enzymatic extraction results in a higher crude protein content is that during enzymatic hydrolysis, the enzyme specifically targets collagen molecules, selectively cleaving peptide bonds and avoiding excessive degradation of the collagen. This gentle treatment preserves more collagen molecules, resulting in a higher proportion of collagen in the final extracted gelatin. Therefore, enzymatic extraction better preserves the structure and protein content of collagen. In contrast, acid or alkaline extraction methods may cause partial degradation of collagen, leading to a reduction in protein content [15]. The selective cleavage of papain likely reduces non-collagen protein contaminants, thereby enhancing the purity and functionality [27]. In contrast, acid extraction may partially hydrolyze the collagen backbone, resulting in reduced protein content and molecular breakdown [28]. The high protein content of the enzymatically extracted gelatin is particularly valuable for applications requiring high gel strength and purity.

2.2.4. Hydroxyproline Content

As shown in Table 1, hydroxyproline content, a key marker of collagen integrity, was highest in enzymatically extracted gelatin, followed by the alkaline and the acid methods. The higher hydroxyproline content in enzymatic gelatin reflects better preservation of the triple-helical structure of collagen, as hydroxyproline stabilizes collagen fibers through hydrogen bonding and supports gel strength [29]. The selective action of papain likely prevents the extensive molecular breakdown observed with hydrochloric acid and sodium hydroxide treatments, which can cause excessive denaturation or decarboxylation of hydroxyproline residues.

In acidic environments, hydrogen ions (H+) can promote the removal of a hydrogen ion from the carboxyl group (-COOH) of hydroxyproline residues, forming a carboxylate salt, which may lead to a decarboxylation reaction. Under stronger acidic conditions, acids not only protonate functional groups such as amino and carboxyl groups but may also cause collagen degradation. The decarboxylation of hydroxyproline alters its chemical structure, thereby affecting the stability of collagen. Excessive acidic treatment can lead to the decarboxylation of hydroxyproline, which in turn reduces the quality of gelatin, particularly affecting its structural stability. In alkaline conditions, hydroxide ions (OH-) can deprotonate the carboxyl group, causing it to enter a deprotonated state. This process can disrupt the normal structure of hydroxyproline, leading to the loss of a hydrogen ion and subsequent decarboxylation. Deprotonation in alkaline environments may not only result in the decarboxylation of hydroxyproline but also further disrupt the collagen triple-helix structure, affecting the stability of collagen and the functionality of gelatin [30]. In contrast to these two treatments, papain, as a cysteine protease, exhibits strong selectivity. It specifically cleaves peptide bonds within collagen, primarily targeting non-collagenous components while having a minimal impact on the core collagen structure, thereby preserving the integrity of collagen. This selective action ensures that the hydroxyproline content remains higher, leading to improved gelatin quality [31]. This structural preservation enhances the functional properties of gelatin, particularly gel strength, elasticity, and thermal stability—critical for high-value applications such as pharmaceuticals, biomedical materials, and high-quality food gels [32].

2.2.5. Gel Strength

Gelatin gel strength is a critical parameter reflecting the structural integrity and functionality of extracted collagen. As shown in Table 2, E gelatin, prepared using the enzymatic method, exhibited the highest gel strength, significantly outperforming B gelatin from the alkaline method and A gelatin from the acid method. The superior gel strength of E gelatin indicates that the papain process preserves the triple-helical structure of collagen more effectively. The gentle and specific action of papain minimizes the breakdown of collagen chains compared to the more aggressive hydrolysis observed with hydrochloric acid and sodium hydroxide treatments. The triple-helical structure of collagen is formed by three tightly intertwined polypeptide chains (α-chains) stabilized through hydrogen bonds and van der Waals forces. The non-helical telopeptide regions at the termini of collagen molecules serve as critical sites for intermolecular cross-linking. Papain, a cysteine protease with substrate specificity, preferentially hydrolyzes the non-helical telopeptide regions of collagen molecules, which are characterized by loose conformations or specific amino acid sequences, while avoiding direct cleavage of the triple-helical core domain. This selective cleavage releases intact helical domains, thereby preserving the integrity of the triple-helical structure. The triple-helical regions, enriched with glycine, proline, and hydroxyproline, exhibit strong resistance to papain-mediated proteolysis due to their rigid conformation conferred by the closely packed arrangement of these amino acids [33]. Consequently, the core triple-helical structure remains undamaged. The retention of relatively intact triple-helical fragments enables the formation of a denser three-dimensional network during gelation through hydrogen bonding and hydrophobic interactions. Furthermore, the residual triple-helical structures act as "physical cross-linking junctions", enhancing intermolecular interactions within the gelatin matrix and ultimately improving gel strength. This makes E gelatin ideal for applications requiring robust gel formation, such as high-quality gelatin desserts, gummy candies, and pharmaceutical gel capsules, where mechanical strength and texture are essential [34].

Table 2. Gel strength, viscosity, pH, and isoelectric point of Yanbian cattle bone gelatin.

Type of Gelatin	Gel Strength (g)	Viscosity (cP)	pH	Isoelectric Point
A gelatin	197.33 ± 13.58^{c}	484.33 ± 5.13^{b}	5.31 ± 0.01^{c}	7.44 ± 0.22^{b}
B gelatin	207.67 ± 11.59^{b}	482.67 ± 6.81^{b}	7.41 ± 0.01^{a}	5.42 ± 0.06^{c}
E gelatin	259.00 ± 10.54^{a}	521.67 ± 7.37^{a}	6.44 ± 0.01^{b}	8.5 ± 0.14^{a}

Results are presented as mean ± SD (n = 3). Statistical analyses were performed on samples with the same properties. Different superscript letters (a–c) within the same column indicate statistically significant differences, as determined by one-way ANOVA followed by Duncan's post hoc test ($p < 0.05$).

2.2.6. Viscosity

Viscosity plays a crucial role in determining the texture and stability of gelatin solutions, thereby influencing their suitability as thickening agents in various food and pharmaceutical products [35]. As shown in Table 2, the papain-extracted gelatin (E gelatin) also showed the highest viscosity, compared to the acid and alkaline extractions. The high gel strength and high viscosity of E gelatin are interconnected, as both properties rely on the structural integrity of collagen molecules [24]. The triple-helix structure of collagen plays a crucial role in forming a strong gel network. E gelatin retains a higher content of hydroxyproline, effectively maintaining the collagen triple-helix structure, which allows the gelatin to form a stronger gel network upon cooling, thereby enhancing gel strength. At the same time, E gelatin better preserves the integrity of the collagen structure, resulting in a higher molecular weight, which increases its water-holding capacity and thus improves its viscosity. This means that larger and more intact collagen molecules enable the gelatin to form a more stable gel, increasing the viscosity of the solution, and ultimately resulting in higher gel strength and viscosity.

2.2.7. pH

As shown in Table 2, A gelatin exhibits significant differences in pH compared to B gelatin and E gelatin. This indicates that gelatin produced using the acidic method retains a certain level of acidity, it is likely due to the use of hydrochloric acid during the extraction process. During this process, some of the acid may remain in the final gelatin product, causing it to be acidic [36]. whereas gelatin produced using the alkaline method shows a neutral pH, and the enzymatic method yields a slightly acidic pH. The near-neutral to slightly acidic pH value of the enzymatic method may help maintain the natural properties of gelatin and improve its stability in certain food systems. The higher pH value of the alkaline method may be disadvantageous for food applications sensitive to pH changes, whereas the lower pH of acidic gelatin may limit its use in certain neutral food systems. However, acidic gelatin could be advantageous for acidic food formulations, as a neutral pH minimizes undesirable interactions with other food systems, ensuring consistent texture and taste [37]. Additionally, compared to gelatin A, gelatin B and gelatin E are more suitable for incorporation into food systems. The higher pH value of the alkaline method may be disadvantageous for food applications sensitive to pH changes, whereas the lower pH of acidic gelatin may limit its use in certain neutral food systems. However, acidic gelatin could be advantageous for acidic food formulations, as a neutral pH minimizes undesirable interactions with other food systems, ensuring consistent texture and taste [37].

2.2.8. Isoelectric Point

The isoelectric point represents the pH at which gelatin carries no net charge. As shown in Table 2. The high isoelectric point of E gelatin may result from the selective hydrolysis of specific amino acid residues by papain, yielding a structure closer to its native state. Differences in the isoelectric points directly affect the solubility and stability of gelatin under different pH conditions. The high isoelectric point of E gelatin suggests better solubility and thermal stability in neutral and alkaline conditions. In contrast, the lower isoelectric point of B gelatin may render it more prone to precipitation in acidic environments, potentially affecting its dispersion and emulsification properties in some food applications [38].

2.3. *Rheological Properties of Gelatin*

The gelation and melting temperatures are key parameters for evaluating the gelling characteristics of gelatin [20]. Figure 2 shows the dynamic viscoelastic curves of A, B, and E gelatins during the cooling and heating processes. The storage modulus (G') and loss modulus (G') intersect, defining the gelation and melting temperatures, respectively [39]. Figure 2a shows the rheological behavior of A gelatin. During cooling, the G' (storage modulus) of A gelatin significantly increases at around 24.1 °C, indicating the onset of gelation. As the temperature decreases further, the G' value rises rapidly, reflecting the progressive formation and stabilization of the gelatin network. During the cooling process of A gelatin, a phase transition occurs from a liquid to a gel state. This process is driven by molecular interactions between collagen molecules, particularly hydrogen bonding, which promotes the aggregation and network formation of gelatin molecules. As the gel structure gradually forms, the G' value increases [40]. The triple-helix structure of collagen may be partially disrupted during the extraction process, but as the temperature decreases, the triple-helix structure begins to reassemble and stabilize, strengthening the gel network and causing a significant increase in G' [41]. Conversely, G'' (loss modulus) decreases at lower temperatures, indicating the dominance of elasticity in the gelatin structure. A gelatin exhibits a melting temperature of approximately 33.0 °C, indicating relatively low thermal stability. This suggests that acid-extracted gelatin forms elastic gels at lower temperatures, making it suitable for applications requiring quick setting and low melting points. Figure 2b shows the rheological characteristics of gelatin B. Compared to A gelatin, B gelatin has a slightly higher gelation point (24.3 °C) and a melting temperature (33.6 °C). The G′ and G″ modulus curves were relatively smooth, indicating that alkaline treatment had a milder impact on the collagen structure. This results in stable rheological properties over a broader temperature range. Figure 2c shows the rheological analysis of gelatin E. Compared to A and B gelatins, E gelatin exhibits a significantly higher gelation point (approximately 28.9 °C) and the highest melting temperature (36.7 °C). The enzymatic treatment with papain significantly enhances the thermal stability of the gelatin. Additionally, the G′ modulus of E gelatin remains higher than that of A and B samples across all temperature ranges, demonstrating superior gel-forming capability and structural stability. Ma et al. and Sarbon et al. reported that the intrinsic differences in the gelatin structure and the processing methods used in gelatin production can lead to variations in gelling ability [14,36]. Abedinia et al. also reported that the source of gelatin, breed and age of the animal, molecular weight, peptide chain cleaving position, and the concentration of amino acid residue in gelatin are the key contributors to the rheological properties of gelatin [36]. This is consistent with the conclusions of this study.

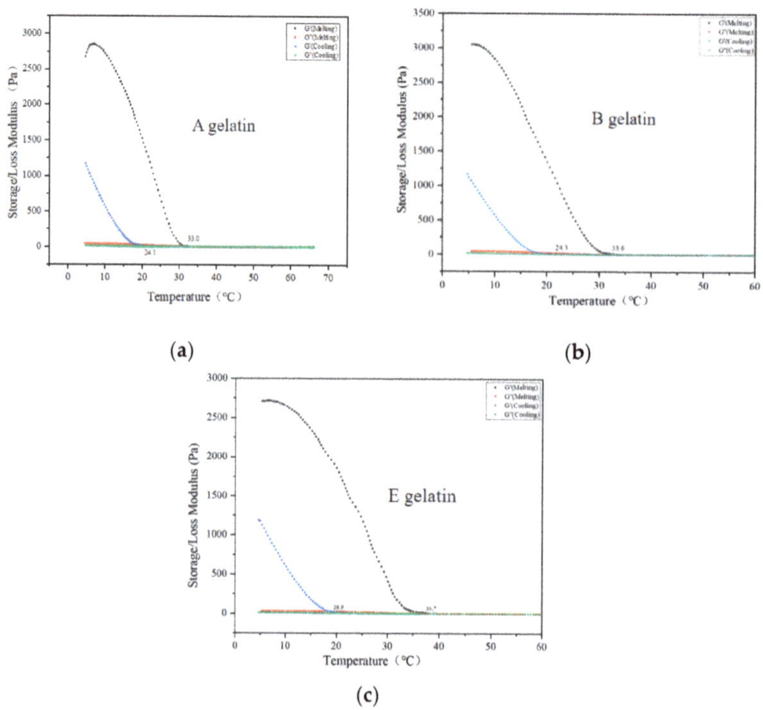

Figure 2. Viscoelastic properties (**a**–**c**) of A gelatin, B gelatin, and E gelatin. A gelatin denotes the gelatins extracted with 0.075 mol/L hydrochloric acid. B gelatin denotes the gelatins extracted with 0.1 mol/L sodium hydroxide. E gelatins denote the gelatins extracted with papain. hey are the highest-yielding samples from the three methods (acid method, alkaline method, and enzymatic method), respectively.

2.4. The Protein Structure of Gelatin

2.4.1. SDS-PAGE

SDS-PAGE analysis of gelatin extracted from Yanbian cattle bones using different methods—acidic (A gelatin), alkaline (B gelatin), and enzymatic (E gelatin)—revealed notable differences in the preservation of collagen protein subunits (Figure 3a). All three extraction methods displayed distinct bands at around 140 kDa, corresponding to the α1 and α2 chains, characteristic of collagen proteins. This result is consistent with previous studies on gelatin [14,41–43]. However, the intensity and clarity of these bands varied, reflecting the degree of preservation or degradation of the collagen structure during extraction. For acid-extracted gelatin (A), the α1 and α2 bands were relatively weak, indicating that the acidic environment caused significant structural breakdown, resulting in partial hydrolysis of collagen. This finding correlates with the prior FTIR analysis, which suggested that acid extraction reduces the integrity of the protein structure, resulting in a lower molecular weight and potentially reduced gel strength and viscosity. Hydrochloric acid protonates the amino and carboxyl functional groups in collagen, disrupting the hydrogen bonds and other non-covalent interactions between collagen molecules, leading to their unraveling and degradation. As the structure is damaged, the collagen molecules become shorter and irregular, resulting in a significant decrease in molecular weight. The smaller molecular weight prevents the gelatin from forming a strong gel network upon cooling, which causes a reduction in gel strength and viscosity [44]. In contrast, alkaline-extracted gelatin (B) exhibits clearer and stronger bands at α1 and α2, suggesting better retention of collagen subunits. However, the use of a strong alkali, such as sodium hydroxide, causes some degradation, as evidenced by slightly diffuse bands. This intermediate retention of the collagen structure positions alkali-extracted gelatin between acidic and enzymatic methods in terms of gel strength and functional properties, making it suitable for general food additive applications. Notably, enzymatic extraction (E) shows the most prominent and well-defined α1 and α2 bands, indicating minimal structural damage during the extraction process. Selective hydrolysis by papain enabled efficient extraction while preserving high-molecular-weight components. These results align with previous FTIR findings, which demonstrated superior retention of the protein's secondary structure and functional properties (Figure 4a). Consequently, enzymatically extracted gelatin offers a higher gel strength and viscosity, making it ideal for high-value applications in pharmaceuticals and premium food formulations [45]. This underscores the significance of the molecular weight distribution, structure, and subunit composition of gelatin (including α chains and low-molecular-weight protein fragments) in determining the functional and physical properties of gelatin.

2.4.2. GPC

Gel Permeation Chromatography (GPC) analysis of gelatin extracted from Yanbian cattle bones using different methods—acid extraction (A gelatin), alkaline extraction (B gelatin), and enzymatic extraction (E gelatin)—demonstrated significant differences in molecular weight distribution among the samples (Figure 3b,c). The GPC chromatogram revealed that gelatin A (acid-extracted) exhibited a major peak at approximately 26 min, indicating a lower molecular weight due to extensive hydrolysis caused by acidic conditions, which generated a higher proportion of small molecular fragments. In contrast, gelatin B (alkaline extract) showed a broader distribution, with peaks between 24 and 28 min, reflecting the presence of both high- and low-molecular-weight components. This suggests that, while alkaline treatment partially degrades collagen structure, it retains a moderate level of molecular integrity compared to the acid method. Notably, gelatin E (enzymatically extracted) showed a prominent peak at an earlier retention time (15–20 min), indicative of higher-molecular-weight fractions. Selective hydrolysis by papain preserved larger collagen chains, resulting in shorter retention times in the chromatogram. Papain selectively hydrolyzes collagen through its specific and mild enzymatic action, and the mechanism by which it protects larger collagen chains is primarily reflected in its selective hydrolysis characteristics. Papain is a cysteine protease, and its mechanism involves a catalytic triad consisting of cysteine, histidine, and asparagine residues. The enzyme selectively targets peptide bonds that are more susceptible to hydrolysis, while preserving those bonds that are crucial for the structural integrity of the collagen molecule. The protease cleaves peptide bonds at specific sites within the collagen molecule, thereby removing non-collagenous impurities, while retaining key structural components of collagen, particularly the larger collagen chains [16]. The molecular weight distribution chart further supports these observations. For gelatin A, 40% of the content was in the 10–100 kDa range, 32% was below 10 kDa, and only 13% exceeded 300 kDa, indicating a high proportion of low-molecular-weight fragments. Gelatin B displayed a relatively balanced distribution, with 34% in the <10 kDa range, 38% in the 10–100 kDa range, and smaller fractions in the 100–300 kDa (14%) and >300 kDa (13%) ranges, suggesting moderate degradation and a diverse molecular weight profile suitable for general food additive applications. In contrast, gelatin E showed a significant shift toward higher molecular weights, with 39% of its content exceeding 300 kDa and only 3% below 10 kDa, reflecting minimal structural breakdown during enzymatic extraction.

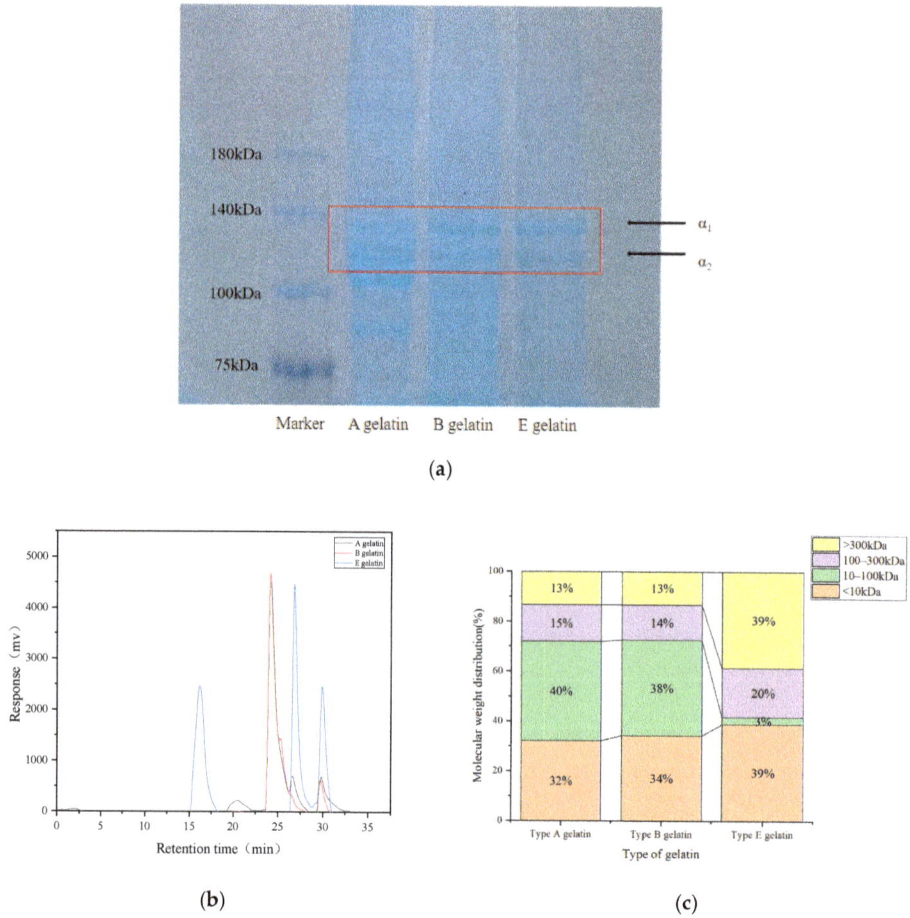

Figure 3. SDS-PAGE patterns: (**a**) molecular weight distribution and (**b**) molecular weight ratio (**c**) of A gelatin, B gelatin, and E gelatin.

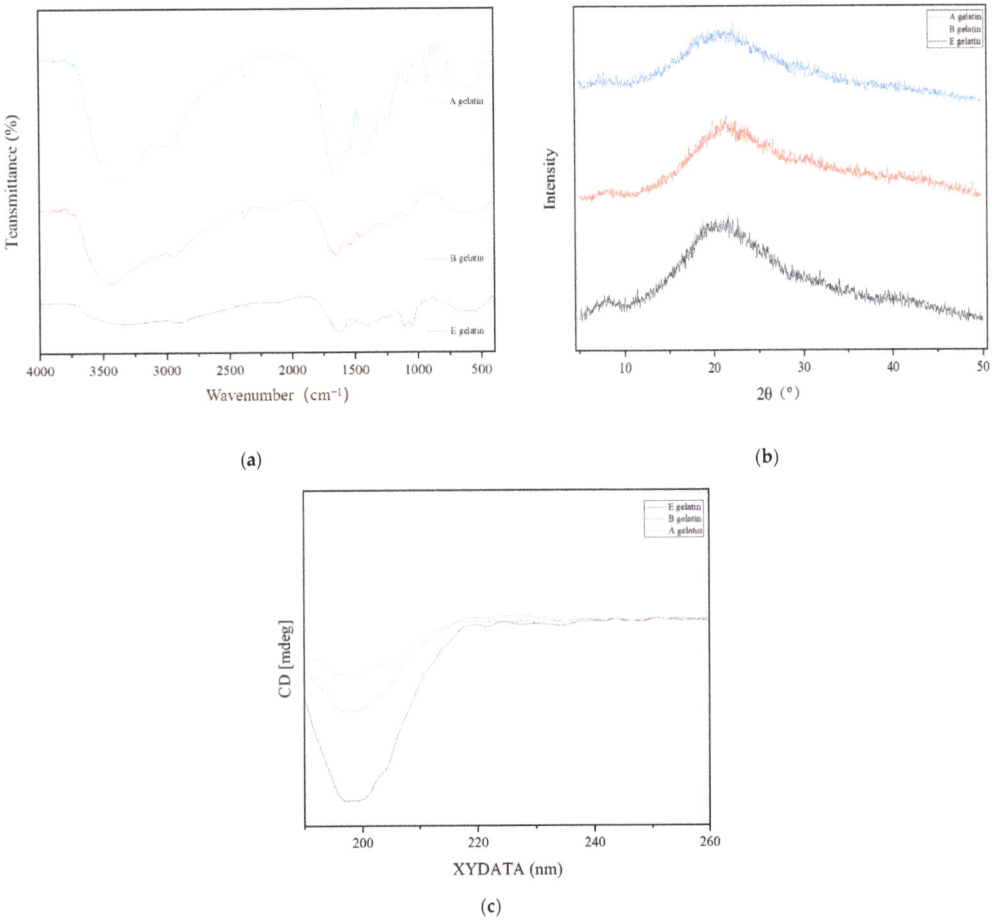

Figure 4. FTIR spectra: (**a**) XRD patterns and (**b**) CD spectra (**c**) of A gelatin, B gelatin, and E gelatin.

2.5. Functional Groups of Gelatin

Table 3 and Figure 4a present the FTIR spectra, wavenumber ranges, and assignments of the gelatin extracted from Yanbian cattle bone. The Amide A band for the three types of gelatin (A gelatin, B gelatin, and E gelatin) exhibited wavenumber ranges of 3348.42–3435.22 cm^{-1}. Gelatin A and B exhibited higher wavenumbers, attributed to hydrolysis by hydrochloric acid and sodium hydroxide, which altered the hydrogen-bonding environment of the protein [46]. This process weakened hydrogen bonding and increased the freedom of N–H stretching vibrations [47]. Similar results were reported by Ahmad et al. when extracting gelatin using different enzymes [48,49]. The Amide B band appeared within 2931.8–3064.89 cm^{-1}, with gelatin A showing the highest wavenumber (3064.89 cm^{-1}). This increase was attributed to acid hydrolysis, which reduces molecular cross-linking and enhances molecular vibrational freedom [50]. The Amide I band, primarily arising from C=O stretching vibrations, serves as a critical fingerprint region for protein secondary structure [51]. The results indicated that all three types of gelatin retained a certain degree of protein secondary structure. This result ties well with previous studies by Ahmad et al., Samatra et al., and Ma et al. who found a similar wavenumber for gelatin pre-treated with different types of enzymes [14,27,48,52]. The wavenumber is generally indicative of the hydrogen-bonding strength involving the carbonyl group (C=O), with higher wavenumbers indicating weaker hydrogen bonds [53]. Gelatin E displayed the lowest wavenumber (1629.85 cm^{-1}), suggesting that papain hydrolysis led to stronger hydrogen bonding involving carbonyl groups compared to the other two hydrolysis methods. The amide II band, mainly consisting of N–H bending and C–N stretching vibrations, is influenced by hydrogen bonding and the degree of protein folding [54]. Gelatin A showed the lowest wavenumber (1462.04 cm^{-1}), indicating that papain hydrolysis has a lesser impact on molecular cross-linking and hydrogen bonding compared to hydrolysis by hydrochloric acid or sodium hydroxide. The amide III band, generated by a combination of C–N stretching and N–H bending vibrations, is another important region for studying protein secondary structures [55]. Gelatin E exhibited the highest wavenumber (1247.94 cm^{-1}), indicating that enzymatic hydrolysis has the least effect on the stability of C–N bonds. Meanwhile, a lower amplitude indicates a significant shift to a disordered coiled structure due to molecular disorder [48].

Table 3. FTIR spectra wavenumber and assignments of Yanbian cattle bone gelatin.

Type of Gelatin	Wavenumber (cm^{-1})				
	Amide A	**Amide B**	**Amide I**	**Amide II**	**Amide III**
A gelatin	3390.86	3064.89	1643.35	1550.77	1238.30
B gelatin	3435.22	2935.66	1641.42	1529.55	1242.16
E gelatin	3348.42	2931.80	1629.85	1462.04	1247.94
Assignment	Corresponds to the N–H stretching vibration	Corresponds to another mode of N–H stretching vibration	Primarily arises from the C=O stretching vibration	Mainly composed of N–H bending vibration and C–N stretching vibration	Results from a combination of C–N stretching and N–H bending vibrations

Overall, among the three types of gelatins, E gelatin had a more compact molecular structure, lower vibration wavenumbers, and the most intact protein secondary structure. Gelatin B exhibited moderate structural damage with intermediate wavenumbers, while gelatin A showed the greatest disruption of hydrogen bonding and secondary structures, resulting in higher wavenumbers for most vibration bands and increased freedom from molecular vibrations.

2.6. XRD Analysis

XRD analysis is commonly used to analyze the crystallinity and triple-helical structure of gelatin [56]. Based on the XRD analysis of gelatin extracted from Yanbian cattle bones using acid (A gelatin), alkaline (B gelatin), and enzymatic (E gelatin) methods, distinct differences in crystallinity and structural characteristics were observed (Figure 4b). All three types of gelatin exhibited broad diffraction peaks, reflecting primarily an amorphous structure with some degree of crystallinity. Gelatin, a denatured form of collagen, generally lacks a highly ordered structure but retains minor crystalline regions due to residual sequences from the native triple helix [57]. The XRD pattern of A gelatin (acid-extracted) shows a broad, low-intensity peak centered around 21° (2θ), suggesting significant disruption of the collagen structure by acid treatment. This led to a more amorphous gelatin with minimal crystalline regions, a result of extensive hydrolysis under acidic conditions that degraded ordered structures. In contrast, B gelatin (alkaline-extracted) displayed a moderately sharper and higher-intensity peak around 21° (2θ), indicating partial retention of the crystalline regions of the original collagen. Alkaline extraction caused less aggressive degradation compared to acid treatment, preserving a moderate degree of structural order. Notably, E gelatin (enzymatically extracted) showed the highest peak intensity at around 20° (2θ), indicating that enzymatic extraction best preserved the ordered regions within the gelatin structure. The sharper and more intense peaks reflect a higher degree of crystallinity, likely due to selective hydrolysis by enzymes such as papain, which minimizes structural disruption while effectively extracting gelatin [58].

2.7. Amino Acid Composition of Gelatin

The concentration of amino acids in gelatin varies depending on the source and extraction method used [59,60]. Table 4 presents the amino acid composition of Yanbian cattle bone gelatin prepared using acidic (A gelatin), alkaline (B gelatin), and enzymatic (E gelatin) methods, each showing distinct variations. Hydrochloric acid extraction under acidic conditions tends to hydrolyze and degrade collagen chains. Under acidic conditions, some essential amino acids, such as glycine, glutamic acid, and leucine, are subject to varying degrees of damage and loss. Glutamic acid and leucine are crucial for collagen properties; however, the strong hydrolytic ability of the acid extraction method disrupts the triple-helix structure of collagen, resulting in lower levels of these amino acids compared to B gelatin and E gelatin. Furthermore, basic amino acids, such as lysine, are prone to partial degradation under acidic conditions, leading to reduced lysine content. Consequently, gelatin extracted using the acid method exhibited lower amounts of glycine, glutamic acid, and lysine, as well as a lower overall amino acid composition. In contrast, extraction using sodium hydroxide (the alkaline method) retains certain amino acids due to the higher pH, although it still leads to collagen degradation. Amino acids such as glycine, glutamic acid, and leucine demonstrate higher stability in alkaline environments. This suggests that, compared to gelatin A, gelatin B retains more of its original collagen structure. Studies have shown that alkaline treatment helps remove non-collagenous substances, resulting in purer collagen. When properly controlled, this method preserves more of the native collagen structure [61].

Table 4. Amino acid compositions (residues per 100%) of A gelatin, B gelatin, and E gelatin.

Amino Acid	A Gelatin	B Gelatin	E Gelatin
Asp	3.06 ± 0.37	5.01 ± 0.41	5.06 ± 0.33
Thr	1.03 ± 0.58	1.90 ± 0.83	1.95 ± 0.78
Ser	2.46 ± 0.26	3.43 ± 0.11	3.36 ± 0.16
Glu	6.23 ± 0.27	8.65 ± 0.22	8.27 ± 0.17
Gly	33.25 ± 0.48	37.40 ± 0.51	36.55 ± 0.41
Ala	13.19 ± 0.35	14.08 ± 0.21	14.09 ± 0.28
Cys	0.00 ± 0.00	0.00 ± 0.00	0.00 ± 0.00
Val	1.96 ± 0.80	2.46 ± 0.74	2.42 ± 0.79
Met	0.29 ± 0.55	0.75 ± 0.32	0.54 ± 0.27
Ile	1.06 ± 0.58	1.37 ± 0.62	1.36 ± 0.63
Leu	2.17 ± 0.21	2.97 ± 0.12	3.01 ± 0.11
Tyr	0.14 ± 0.05	0.29 ± 0.42	0.31 ± 0.35
Phe	1.37 ± 0.34	1.49 ± 0.51	1.54 ± 0.44
Lys	2.46 ± 0.15	3.03 ± 0.11	3.16 ± 0.25
NH3	3.15 ± 0.08	3.83 ± 0.22	3.78 ± 0.18
His	0.47 ± 0.21	0.59 ± 0.31	0.61 ± 0.23
Arg	4.83 ± 0.18	5.54 ± 0.25	5.52 ± 0.22
Imino Acid (Pro + Hyp)	12.26 ± 0.11	16.44 ± 0.12	22.31 ± 0.07

Enzymatic extraction (using papain) causes the least damage to collagen due to its high specificity and mild reaction conditions. This method effectively preserves the natural structure of collagen, resulting in higher levels of key amino acids such as glycine, glutamic acid, lysine, proline, and hydroxyproline. These amino acids are essential for maintaining the triple-helical structure of gelatin, as well as its physical and chemical properties, such as gel strength and thermal stability. Moreover, enzymatic extraction conducted at lower temperatures and under neutral to mildly acidic conditions minimizes the risk of amino acid degradation, particularly for basic amino acids such as lysine. As a result, gelatin extracted using the enzymatic method exhibited a more complete amino acid composition and optimal physical and functional properties [15].

2.8. Circular Dichroism Spectrum of Gelatin

CD analysis of gelatin extracted from Yanbian cattle bones using acidic (A gelatin), alkaline (B gelatin), and enzymatic (E gelatin) methods revealed significant differences in the secondary structure of gelatin (Figure 4c and Table 5). E gelatin exhibited a relatively more ordered structure, characterized by prominent peaks around 210 nm, indicating a higher content of α-helix and β-sheet [62,63]. The secondary structure analysis showed that E gelatin contained 14.4% α-helix and 27.1% β-sheet, suggesting that enzymatic extraction preserved the collagen triple helix and retained more ordered structures. B gelatin, Similar results were reported by Ma et al. when extracting gelatin using different enzymes [14], on the other hand, showed a more balanced secondary structure with a significant proportion of β-turn (35.5%) and unordered coil (35.6%), while its α-helix content was 8.7% and β-sheet content was 20.2%. The CD spectrum for B gelatin exhibited a more gradual curve, indicating a less ordered structure compared to E gelatin but still more stable than A gelatin. A gelatin exhibited a more disordered structure, as indicated by the strong absorption peaks and higher content of unordered coil (44.9%) and β-turn (28.2%). The secondary structure analysis revealed that A gelatin had only 1.4% α-helix and 25.5% β-sheet, with a predominance of disordered regions likely due to the harsh conditions of acid extraction. These findings suggest that enzymatic extraction (E gelatin) better preserves the collagen triple helix, resulting in more ordered structures, while acid extraction (A gelatin) leads to excessive degradation and a greater proportion of disordered structures.

Table 5. Content of the secondary structure in three varieties of Gelatin (%).

Type of Gelatin	α-Helix	β-Sheet	β-Turn	Unordered Coil
A gelatin	1.4 ± 0.2	25.5 ± 0.1	28.2 ± 0.1	44.9 ± 0.2
B gelatin	8.7 ± 0.2	20.2 ± 0.3	35.5 ± 0.3	35.6 ± 0.1
E gelatin	14.4 ± 0.1	27.1 ± 0.2	35.2 ± 0.2	23.3 ± 0.1

3. Conclusions

This study comprehensively evaluated the extraction efficiency and physicochemical properties of gelatin obtained from Yanbian cattle bones using acid, alkaline, and enzymatic methods. The results demonstrated that enzymatic hydrolysis with papain produced the highest yield and preserved collagen structure, as evidenced by superior hydroxyproline content, gel strength, viscosity, and molecular weight distribution. Compared to acid and alkaline methods, enzymatic extraction minimized structural damage, retained functional groups, and maintained higher crystallinity, as confirmed by FTIR and XRD analyses. These findings suggest that papain-assisted enzymatic hydrolysis is an effective method for preparing bone gelatin from Yanbian cattle. Yanbian cattle bone gelatin shows promise as a viable halal gelatin alternative to traditional mammalian gelatin, with broad application prospects in fields such as food and pharmaceuticals.

4. Material and Methods

4.1. Materials

Yanbian cattle bones were purchased from the Yanji market and stored at −20 °C before use. Pepsin (15,600 U/g), papain (800 U/mg), ficin (40,000 U/g), ginger protease (400 U/mg), compound protease (120 U/mg), bromelain (50,000 U/g), pancreatin (100 U/mg), and trypsin (250 U/g) were purchased from Shanghai Yuanye Biotechnology Co., Ltd. (Shanghai, China). Analytical-grade chemicals were obtained from Tianjin Kemiou Chemical Reagent Co., Ltd. (Tianjin, China).

4.2. Pretreatment

Yanbian cattle bone powder was prepared based on the method described by Cao et al. [20], with minor modifications. The Yanbian cattle bone was thoroughly washed with distilled water and soaked in distilled water for 3 h to remove blood. The bone was then dried in an oven at 40 °C for 24 h. Subsequently, the dried bone was pulverized using a high-speed grinder and sieved through a 60-mesh sieve to obtain Yanbian cattle bone powder. The bone powder was mixed with 0.25 M EDTA (1:10, *v*/*v*) at 20 °C to remove calcium salts. The EDTA solution was replaced every 12 h for three consecutive days. After the demineralization process, the slurry was centrifuged at 12,000 rpm for 10 min at 4 °C to remove the supernatant. The residue was concentrated using a rotary evaporator, (Buchi, Flawil, St. Gallen, Switzerland) followed by freeze-drying to obtain demineralized Yanbian cattle bone powder. The resulting powder was stored at −80 °C for further use.

4.3. Gelatin Extraction

4.3.1. Enzymatic Extraction and Enzyme Selection

A total of 4.00 g of pre-treated Yanbian cattle bone powder was weighed, and enzymatic hydrolysis was performed under the optimal conditions (temperature and pH) for different enzymes. The conditions for each enzyme were as follows: pepsin (37 °C, pH 3), papain (60 °C, pH 6.5), ficin (37 °C, pH 10), ginger protease (60 °C, pH 6.5), compound protease (50 °C, pH 7.5), bromelain (60 °C, pH 7), pancreatin (37 °C, pH 7.5), and trypsin (37 °C, pH 8). The enzyme dosage was 140 U/g, with a solid-to-liquid ratio of 1:7 (*w*/*v*). After 3 h of enzymatic hydrolysis, the enzymes were inactivated by boiling in a water bath at 100 °C for 10 min. The resulting mixture was then extracted with deionized water at 60 °C for 5 h. After extraction, based on the method of Ma et al., the mixture was centrifuged at 12,000 rpm for 10 min, and the supernatant was collected. The supernatant was filtered through Whatman No. 4 filter paper to obtain a gelatin solution. The solution was subsequently frozen at −80 °C for 24 h and freeze-dried to obtain enzymatically treated bovine bone gelatin samples [14]. The most suitable enzyme was determined by comparing the gelatin yield from different treatments.

4.3.2. Acid Extraction and Determination of HCl Concentration

Based on the method of Cao et al. and Samatra et al. [13,20], with slight modifications, we aimed to maximize gelatin yield while avoiding excessive degradation of bovine bone collagen due to hydrochloric acid treatment; 4.00 g of pre-treated bovine bone powder was weighed and added to 1:7 (*w*/*v*) mixtures of hydrochloric acid solutions with concentrations of 0.025, 0.05, 0.075, 0.1, 0.125, and 0.15 mol/L. The mixtures were gently stirred at 4 °C for 24 h, with the solution replaced every 6 h. After treatment, the bovine bone residue was washed repeatedly with distilled water until the pH of the wash water became neutral. The washed bone powder was air-dried and weighed. The treated samples were mixed with deionized water at a 1:7 solid-to-liquid ratio (*w*/*v*) and extracted in a water bath at 60 °C for 5 h. After extraction, the mixture was centrifuged at 12,000 rpm for 10 min, and the supernatant was collected. The supernatant was filtered through Whatman No. 4 filter paper (Whatman, Maidstone, Kent, UK) to obtain a gelatin solution. The solution was then frozen at −80 °C for 24 h and freeze-dried to obtain acid-treated gelatin samples. The most suitable HCl concentration was determined by comparing the gelatin yield from different treatments.

4.3.3. Alkaline Extraction and Determination of NaOH Concentration

Slight modifications were made based on the experimental method of Ahmad et al. [48]. A total of 4.00 g of pre-treated bovine bone powder was weighed and added to 1:7 (*w*/*v*) mixtures of sodium hydroxide solutions with concentrations of 0.025, 0.05, 0.075, 0.1, 0.125, and 0.15 mol/L. The mixtures were gently stirred at 4 °C for 24 h, with the solution replaced every 6 h. After treatment, the bovine bone residue was washed repeatedly with distilled water until the pH of the wash water became neutral. The washed bone powder was air-dried and weighed. The treated samples were mixed with deionized water at a 1:7 solid-to-liquid ratio (*w*/*v*) and extracted in a water bath at 60 °C for 5 h. After extraction, the mixture was centrifuged at 12,000 rpm for 10 min, and the supernatant was collected. The supernatant was filtered through Whatman No. 4 filter paper to obtain a gelatin solution. The solution was then frozen at −80 °C for 24 h and freeze-dried to obtain alkaline-treated gelatin samples. The most suitable NaOH concentration was determined by comparing the gelatin yield from different treatments.

4.4. Extraction Yield of Yanbian Cattle Bone Gelatin

The yield of gelatin was calculated using the following Equation (1):

$$\text{Yield of Yanbian cattle bone gelatin } (\%) = \frac{W_g}{W_b} \times 100, \quad (1)$$

where W_g is the weight of dry gelatin and W_b is the wet weight of Yanbian cattle bone.

4.5. Characterization of Yanbian Cattle Bone Gelatin

In this experiment, the gelatin samples with the highest extraction yields from the three preparation methods were selected for further analysis. They were named as A gelatin, B gelatin, and E gelatin. The quality of bone gelatin was evaluated based on key indicators, including chemical properties, hydroxyproline content, gel strength, and viscosity. Additionally, supplementary parameters such as amino acid composition, molecular weight distribution and structure, rheological properties, crystal structure, and protein structure were also analyzed.

4.5.1. Chemical Properties

The moisture, ash, and crude protein contents were determined following the methods outlined in (AOAC), 2000 [64]. Each measurement was conducted in triplicate.

4.5.2. Hydroxyproline Content

Hydroxyproline content was determined as described by Ristaniemi et al. [38]. A 20 mg sample of freeze-dried gelatin was weighed and mixed with 3 mL of hydrochloric acid (6 mol/L). The mixture was hydrolyzed at 130 °C for 4 h. After cooling, the solution was transferred to a small beaker, and one drop of methyl red reagent was added. Subsequently, 1 mL of citrate buffer and 1 mL of chloramine-T solution were added, and the mixture was allowed to react at room temperature for 10 min. Next, 1 mL of perchloric acid (3.5 mol/L) was added, and the reaction was continued at room temperature for another 10 min. Afterward, 1 mL of color reagent was introduced, and the reaction was incubated in a water bath at 65 °C for 20 min. The mixture was then cooled to room temperature, and the absorbance was measured at 558 nm. The hydroxyproline content was quantified using a standard curve derived from hydroxyproline as a reference.

4.5.3. Gel Strength

The gelatin solution, consisting of 120 mL at a concentration of 6.67% (*w*/*v*), was transferred into a Bloom jar and matured at 10 °C for 17 h. Gel strength was assessed using the TA. XT Plus Texture Analyzer (Stable Micro Systems, England, UK) equipped with a flat-ended cylindrical probe with a diameter of 4.12 mm. The probe was set to descend at a constant rate of 0.5 mm/s and was allowed to penetrate the gel to a depth of 4 mm. The test was performed once, and the peak force (measured in g) encountered during the compression was recorded as the gel strength. The compression test was carried out at room temperature (approximately 20 °C) [19].

4.5.4. Viscosity

Gelatin (1.34 g) was dissolved in distilled water (20 mL) to prepare a 6.67% (*w*/*v*) gelatin solution, which was then heated to 60 °C. The viscosity of the solution was measured using a RheolabQC viscometer (Anton Paar, Graz, Austria). Triplicate measurements were performed to ensure accuracy, and the mean value was calculated [27].

4.5.5. Measurement of ζ -Potential and pH

The pH of Yanbian cattle gelatin was measured according to the method described by Ahmad et al. [65]. A 1% (*w*/*v*) gelatin solution was prepared by dissolving 0.2 g of gelatin in 20 mL of distilled water and cooling it at 25 °C. The pH meter (Thermo Fisher Scientific, Waltham, MA, USA) was calibrated with buffer solutions of pH 4.0 and 7.0, and measurements were repeated three times.

The ζ-potential analyzer (Zeta PALS, Malvern, UK) was used to measure the ζ-potential at room temperature (25 °C). Gelatin solution (0.5 mg/mL) dissolved in deionized water was adjusted to different pH levels (2, 3, 4, 5, 6, 7, 8, 9, 10, 11) using 0.1 M hydrochloric acid or 0.1 M sodium hydroxide. The isoelectric point (pI) was determined as the pH at which the ζ-potential reached zero.

4.5.6. Rheological Properties

The rheological behavior of the 6.67% (*w*/*v*) gelatin solution was investigated using an AR G2 rheometer (TA Instruments Inc., New Castle, DE, USA) equipped with a parallel-plate geometry with a diameter of 40 mm. Temperature sweeps were conducted from 5 °C to 60 °C and back to 5 °C at a rate of 2 °C/min, with a strain of 1% and a frequency of 1 Hz. The storage modulus (G′) and loss modulus (G′) were recorded during the temperature scans. All measurements were performed within the linear viscoelastic region of the rheometer, and each experiment was repeated at least three times.

4.5.7. Electrophoresis Analysis of Gelatin (SDS-PAGE)

Electrophoretic analysis was employed to evaluate the molecular weight distribution of the gelatin samples, as previously reported [66]. For the SDS-PAGE gel electrophoresis (6% stacking gel, 8% separating gel), gelatin samples (0.1 mg/mL) were mixed with 5× sample loading buffer in a 1:4 ratio, followed by heating in a boiling water bath for 5 min. A 10 μL aliquot of the sample was then loaded onto the gel. Electrophoresis was initially performed at 80 V, and once the bands entered the separating gel, the voltage was increased to 120 V until the electrophoresis was complete. After electrophoresis, the polyacrylamide gel was carefully removed and stained with Coomassie Brilliant Blue R-250 solution at room temperature for 30 min. The gel was then washed with distilled water to remove excess stain and destained with a decolorizing solution until the bands were clearly visible. Finally, the destained gel was photographed using a gel imaging system.

4.5.8. Molecular Weight Distribution (GPC)

The relative molecular weight distribution of gelatin proteins was determined using an Agilent 1260 High-Performance Liquid Chromatography (HPLC) system (Agilent Technologies, Santa Clara, CA, USA) coupled with three Waters Ultrahydrogel columns (TM120, TM250, and TM500 (Waters Corporation, Milford, MA, USA); dimensions: 7.8 mm × 300 mm). The experimental conditions included a flow rate of 1 mL/min, column and detector temperatures of 40 °C, and detection with a G1362A differential refractive index detector. The mobile phase consisted of a 0.1 M aqueous solution of $NaNO_3$.

For sample preparation, 0.020 g of gelatin was dissolved in 4 mL of the mobile phase using ultrasonic dissolution and filtered through a 0.45 μm filter membrane. A 40 μL sample was injected into the system, and the analysis runtime was 35 min. The relative molecular weight distribution of the gelatin proteins was determined by analyzing the obtained chromatograms.

4.5.9. Fourier Transform Infrared Spectroscopy (FT-IR)

The Fourier transform infrared (FTIR) spectra of gelatin were obtained using an FTIR spectrometer (Shimadzu, IRTRACER-100, Kyoto, Japan). Dried gelatin powder was analyzed at 25 °C, and transmittance (%) data were recorded from 400 cm^{-1} to 4000 cm^{-1}. The results were plotted as a function of the wavenumber (cm^{-1}).

4.5.10. X-Ray Diffraction (XRD)

XRD patterns of the gelatin samples were acquired using an X-ray diffractometer (Bruker AXS, Rheinfelden, Germany). Gelatin powder was evenly dispersed on a sample plate and scanned over a 2θ range from 5° to 50° in 0.05° increments. Cu Kα radiation was employed as the X-ray source, operating at 30 kV and 10 mA. The resulting XRD data were exported and visualized using OriginPro2021 graph-plotting software.

4.5.11. Amino Acid Analysis

Gelatin samples were hydrolyzed with 6 mol/L HCl in a hydrolysis tube and processed as previously described. After purging with nitrogen for 1 min to exclude air, the samples were heated at 110 °C for 24 h. The hydrolysates were filtered through a 0.45 μm filter membrane, and the filtrates were analyzed using an LC-MS/MS system (Shimadzu, LCMS-8050, Kyoto, Japan) equipped with a C18 chromatographic column (Discovery HS F5-3, 15 cm × 2.1 mm, 3 μm, Supelco, Bellefonte, Pennsylvania, USA). The column temperature was maintained at 35 °C, and the flow rate was 0.6 mL/min. Data were processed using LabSolutions Ver. 5.99 SP3 software and LabSolutions Insight™ Ver. 3.5 (Shimadzu Corporation, Kyoto, Japan) [67].

4.5.12. Circular Dichroism (CD) Spectroscopy

The circular dichroism (CD) spectrum of gelatin was measured at 25 °C using a J-815 CD spectropolarimeter (Jasco, Tokyo, Japan). Approximately 100 mL of gelatin solution (0.5 mg/mL, dissolved in deionized water) was pipetted into a cuvette and scanned from 190 to 260 nm at a scanning speed of 50 nm/min.

4.5.13. Statistical Analysis

All experiments were conducted in triplicate, and the results are expressed as the mean ± standard deviation (SD). Statistical analysis of the data was performed using analysis of variance (ANOVA). To further analyze the test data, a multiple-range test was conducted using SPSS Statistics 23.0. Graphical representation of the data was achieved using Origin 2021 software. The means were compared using Duncan's test at a significance level of $p < 0.05$.

Author Contributions: Conceptualization, S.Z.; data curation, S.Z., D.Z. and L.Y.; formal analysis, R.W.; funding acquisition, H.X. and G.X.; investigation, D.Z. and Z.J.; methodology, S.Z., D.Z. and L.Y.; project administration, H.X. and G.X.; resources, H.X.; software, L.Y.; supervision, H.X. and G.X.; validation, R.W.; visualization, L.Y.; writing—original draft, S.Z.; writing—review and editing, S.Z., D.Z., L.Y., R.W., H.X. and G.X. All authors have read and agreed to the published version of the manuscript.

Funding: This work was supported by the Science and Technology Development Plan Projects of Jilin Province of China, (20230202056NC, YDZJ202402039CXJD).

Institutional Review Board Statement: Not applicable.

Informed Consent Statement: Not applicable.

Data Availability Statement: The raw data supporting the conclusions of this article will be made available by the authors on request.

Conflicts of Interest: The authors declare no conflicts of interest regarding the work described in this manuscript.

References

1. Alipal, J.; Pu'Ad, N.M.; Lee, T.; Nayan, N.; Sahari, N.; Basri, H.; Idris, M.; Abdullah, H. A review of gelatin: Properties, sources, process, applications, and commercialisation. *Mater. Today Proc.* **2021**, *42*, 240–250. [CrossRef]
2. Sani, M.A.; Tavassoli, M.; Salim, S.A.; Azizi-Lalabadi, M.; McClements, D.J. Development of green halochromic smart and active packaging materials: TiO_2 nanoparticle- and anthocyanin-loaded gelatin/κ-carrageenan films. *Food Hydrocoll.* **2022**, *124*, 107324. [CrossRef]
3. Hesarinejad, M.A.; Lorenzo, J.M.; Rafe, A. Influence of gelatin/guar gum mixture on the rheological and textural properties of restructured ricotta cheese. *Carbohydr. Polym. Technol. Appl.* **2021**, *2*, 100162. [CrossRef]
4. Zhao, H.; Liu, W.; Min, C.; Qi, Y.; Chen, X.; Zhang, H. pH-responsive color indicator film based on gelatin/chitosan cross-linking with anthocyanin-Fe^{2+} chelate for pork freshness monitoring. *Food Hydrocoll.* **2025**, *162*, 110895. [CrossRef]
5. Hosseini-Parvar, S.H.; Keramat, J.; Kadivar, M.; Khanipour, E.; Motamedzadegan, A. Optimising conditions for enzymatic extraction of edible gelatin from the cattle bones using response surface methodology. *Int. J. Food Sci. Technol.* **2009**, *44*, 467–475. [CrossRef]
6. Lv, J.; Feng, J.; Zhong, H.; Lou, Y.; Wang, Y.; Liu, S.; Xu, H.; Xia, G. Preparation, characterization and antioxidant effect of polypeptide mineral-chelate from Yanbian cattle bone. *LWT* **2023**, *187*, 115353. [CrossRef]
7. Ismail, N.; Abdullah, H.Z. The extraction of gelatin from black tilapia fish skins with different acid concentration. *J. Phys. Conf. Ser.* **2019**, *1150*, 012041. [CrossRef]
8. Ramli, R.A.; Razali, U.H.M.; Noor, N.Q.I.M. Optimization of extraction conditions of gelatin from buffalo (*Bubalus bubalis*) skins using response surface methodology. *Heliyon* **2023**, *9*, e14367. [CrossRef]
9. Liu, D.; Wei, G.; Li, T.; Hu, J.; Lu, N.; Regenstein, J.M.; Zhou, P. Effects of alkaline pretreatments and acid extraction conditions on the acid-soluble collagen from grass carp (*Ctenopharyngodon idella*) skin. *Food Chem.* **2015**, *172*, 836–843. [CrossRef]
10. Mafazah, E.M.; Pranoto, Y.; Rohman, A. Extracting of yellowfin tuna (*Thunnus albacares*) fish skin gelatin as influenced by alkaline concentration and extraction times. *IOP Conf. Ser. Earth Environ. Sci.* **2018**, *139*, 012047. [CrossRef]
11. Hattrem, M.N.; Molnes, S.; Haug, I.J.; Draget, K.I. Interfacial and rheological properties of gelatin based solid emulsions prepared with acid or alkali pretreated gelatins. *Food Hydrocoll.* **2015**, *43*, 700–707. [CrossRef]
12. Zhou, Y.; Yang, H. Effects of calcium ion on gel properties and gelation of tilapia (*Oreochromis niloticus*) protein isolates processed with pH shift method. *Food Chem.* **2019**, *277*, 327–335. [CrossRef]
13. Samatra, M.Y.; Noor, N.Q.I.M.; Razali, U.H.M.; Bakar, J.; Shaarani, S.M. Bovidae-based gelatin: Extractions method, physicochemical and functional properties, applications, and future trends. *Compr. Rev. Food Sci. Food Saf.* **2022**, *21*, 3153–3176. [CrossRef]
14. Ma, Y.; Zeng, X.; Ma, X.; Yang, R.; Zhao, W. A simple and eco-friendly method of gelatin production from bone: One-step biocatalysis. *J. Clean. Prod.* **2019**, *209*, 916–926. [CrossRef]
15. Zaitsev, S.Y. Changes in the Amino Acid Composition of Gelatin After Treatment of Bovine Collagen with Enzyme Preparations. *Mosc. Univ. Chem. Bull.* **2023**, *78*, 292–298. [CrossRef]
16. Amri, E.; Mamboya, F. Papain, a plant enzyme of biological importance: A review. *Am. J. Biochem. Biotechnol.* **2012**, *8*, 99–104. [CrossRef]

17. Samatra, M.Y.; Azmi, A.; Shaarani, S.M.; Hartina, U.; Razali, U.H.M. Characterisation of gelatin extracted from buffalo (*Bubalus bubalis*) bone using papain pre-treatment. *J. Agric. Food Eng.* **2020**, *1*, 1–5. [CrossRef]
18. Ahmed, M.; Verma, A.K.; Patel, R. Collagen extraction and recent biological activities of collagen peptides derived from sea-food waste: A review. *Sustain. Chem. Pharm.* **2020**, *18*, 100315. [CrossRef]
19. Gómez-Guillén, M.C.; Giménez, B.; López-Caballero, M.E.; Montero, M.P. Functional and bioactive properties of collagen and gelatin from alternative sources: A review. *Food Hydrocoll.* **2011**, *25*, 1813–1827. [CrossRef]
20. Cao, S.; Wang, Y.; Xing, L.; Zhang, W.; Zhou, G. Structure and physical properties of gelatin from bovine bone collagen influenced by acid pretreatment and pepsin. *Food Bioprod. Process.* **2020**, *121*, 213–223. [CrossRef]
21. Kanwate, B.W.; Kudre, T.G. Effect of various acids on physicochemical and functional characteristics of gelatin from swim bladder of rohu (*Labeo rohita*). *J. Food Sci. Technol.* **2017**, *54*, 2540–2550. [CrossRef]
22. Milovanovic, I.; Hayes, M. Marine Gelatine from Rest Raw Materials. *Appl. Sci.* **2018**, *8*, 2407. [CrossRef]
23. Kanwate, B.W.; Ballari, R.V.; Kudre, T.G. Influence of spray-drying, freeze-drying and vacuum-drying on physicochemical and functional properties of gelatin from Labeo rohita swim bladder. *Int. J. Biol. Macromol.* **2019**, *121*, 135–141. [CrossRef]
24. Duthen, S.; Levasseur-Garcia, C.; Kleiber, D.; Violleau, F.; Vaca-Garcia, C.; Tsuchikawa, S.; Raynaud, C.D.; Daydé, J. Using near-infrared spectroscopy to determine moisture content, gel strength, and viscosity of gelatin. *Food Hydrocoll.* **2021**, *115*, 106627. [CrossRef]
25. Wang, Y.; Regenstein, J.M. Effect of EDTA, HCl, and Citric Acid on Ca Salt Removal from Asian (Silver) Carp Scales Prior to Gelatin Extraction. *J. Food Sci.* **2009**, *74*, C426–C431. [CrossRef]
26. Kusumaningrum, I.; Pranoto, Y.; Hadiwiyoto, S. Extraction optimization and characterization of gelatine from fish dry skin of Spanish mackerel (*Scomberromorus commersoni*). *IOP Conf. Ser. Earth Environ. Sci.* **2018**, *144*, 012036. [CrossRef]
27. Ahmad, T.; Ismail, A.; Ahmad, S.A.; Khalil, K.A.; Kee, L.T.; Awad, E.A.; Sazili, A.Q. Physicochemical characteristics and molecular structures of gelatin extracted from bovine skin: Effects of actinidin and papain enzymes pretreatment. *Int. J. Food Prop.* **2019**, *22*, 138–153. [CrossRef]
28. Mulyani, S.; Setyabudi, F.S.; Pranoto, Y.; Santoso, U. Physicochemical Properties of Gelatin Extracted from Buffalo Hide Pretreated with Different Acids. *Korean J. Food Sci. Anim. Resour.* **2017**, *37*, 708–715. [CrossRef]
29. Luo, T.; Kiick, K.L. Collagen-like peptides and peptide–polymer conjugates in the design of assembled materials. *Eur. Polym. J.* **2013**, *49*, 2998–3009. [CrossRef]
30. Matinong, A.M.E.; Chisti, Y.; Pickering, K.L.; Haverkamp, R.G. Collagen Extraction from Animal Skin. *Biology* **2022**, *11*, 905. [CrossRef]
31. Li, P.; Wu, G. Roles of dietary glycine, proline, and hydroxyproline in collagen synthesis and animal growth. *Amino Acids* **2018**, *50*, 29–38. [CrossRef] [PubMed]
32. Musso, Y.S.; Salgado, P.R.; Mauri, A.N. Smart edible films based on gelatin and curcumin. *Food Hydrocoll.* **2017**, *66*, 8–15. [CrossRef]
33. Usman, M.; Ishaq, A.; Mac Regenstein, J.; Sahar, A.; Aadil, R.M.; Sameen, A.; Khan, M.I.; Alam, A. Valorization of animal by-products for gelatin extraction using conventional and green technologies: A comprehensive review. *Biomass Convers. Biorefin.* **2023**, *11*, 1–13. [CrossRef]
34. Singh, P.; Benjakul, S.; Maqsood, S.; Kishimura, H. Isolation and characterisation of collagen extracted from the skin of striped catfish (*Pangasianodon hypophthalmus*). *Food Chem.* **2011**, *124*, 97–105. [CrossRef]
35. Zhou, P.; Mulvaney, S.J.; Regenstein, J.M. Properties of Alaska Pollock Skin Gelatin: A Comparison with Tilapia and Pork Skin Gelatins. *J. Food Sci.* **2006**, *71*, C313–C321. [CrossRef]
36. Abedinia, A.; Ariffin, F.; Huda, N.; Nafchi, A.M. Extraction and characterization of gelatin from the feet of Pekin duck (*Anas platyrhynchos domestica*) as affected by acid, alkaline, and enzyme pretreatment. *Int. J. Biol. Macromol.* **2017**, *98*, 586–594. [CrossRef]
37. Khalesi, H.; Emadzadeh, B.; Kadkhodaee, R.; Fang, Y. Whey protein isolate-Persian gum interaction at neutral pH. *Food Hydrocoll.* **2016**, *59*, 45–49. [CrossRef]
38. Ristaniemi, A.; Torniainen, J.; Stenroth, L.; Finnilä, M.; Paakkonen, T.; Töyräs, J.; Korhonen, R. Comparison of water, hydroxyproline, uronic acid and elastin contents of bovine knee ligaments and patellar tendon and their relationships with biomechanical properties. *J. Mech. Behav. Biomed. Mater.* **2020**, *104*, 103639. [CrossRef]
39. Park, H.E.; Gasek, N.; Hwang, J.; Weiss, D.J.; Lee, P.C. Effect of temperature on gelation and cross-linking of gelatin methacryloyl for biomedical applications. *Phys. Fluids* **2020**, *32*, 033102. [CrossRef]
40. Feng, J.; Tian, H.; Chen, X.; Cai, X.; Shi, X.; Wang, S. Interaction between fish gelatin and tremella polysaccharides from aqueous solutions to complex coacervates: Structure and rheological properties. *Food Hydrocoll.* **2023**, *138*, 108439. [CrossRef]
41. Samatra, M.Y.; Razali, U.H.M.; Shaarani, S.M.; Roslan, J.; Ramli, R.A.; Izzreen, M.N.N.Q. Physicochemical and functional properties of buffalo (*Bubalus bubalis*) bone gelatin extracted using acid pre-treatment. *Futur. Foods* **2024**, *10*, 100428. [CrossRef]
42. Ahammed, S.; Liu, F.; Wu, J. Effect of transglutaminase crosslinking on solubility property and mechanical strength of gelatin-zein composite films. *Food Hydrocoll.* **2021**, *116*, 106649. [CrossRef]

43. Casanova, F.; Mohammadifar, M.A.; Jahromi, M.; Petersen, H.O.; Sloth, J.J.; Eybye, K.L.; Kobbelgaard, S.; Jakobsen, G.; Jessen, F. Physico-chemical, structural and techno-functional properties of gelatin from saithe (*Pollachius virens*) skin. *Int. J. Biol. Macromol.* **2020**, *156*, 918–927. [CrossRef]
44. Shi, W.; Yin, Z.; Liu, Y.; Han, J.; Sun, J. Physiochemical property and structure of gelatin obtained from Chinese soft-shelled turtle carapace by three pretreatment methods. *J. Food Sci. Technol.* **2024**, *27*, 1–12. [CrossRef]
45. Balti, R.; Jridi, M.; Sila, A.; Souissi, N.; Nedjar-Arroume, N.; Guillochon, D.; Nasri, M. Extraction and functional properties of gelatin from the skin of cuttlefish (*Sepia officinalis*) using smooth hound crude acid protease-aided process. *Food Hydrocoll.* **2011**, *25*, 943–950. [CrossRef]
46. Matmaroh, K.; Benjakul, S.; Prodpran, T.; Encarnacion, A.B.; Kishimura, H. Characteristics of acid soluble collagen and pepsin soluble collagen from scale of spotted golden goatfish (*Parupeneus heptacanthus*). *Food Chem.* **2011**, *129*, 1179–1186. [CrossRef]
47. Tintor, Đ.; Ninković, K.; Milošević, J.; Polović, N.Đ. Gaining insight into protein structure via ATR-FTIR spectroscopy. *Vib. Spectrosc.* **2024**, *134*, 103726. [CrossRef]
48. Ahmad, T.; Ismail, A.; Ahmad, S.A.; Khalil, K.A.; Awad, E.A.; Akhtar, M.T.; Sazili, A.Q. Recovery of Gelatin from Bovine Skin with the Aid of Pepsin and Its Effects on the Characteristics of the Extracted Gelatin. *Polymers* **2021**, *13*, 1554. [CrossRef]
49. Abedinia, A.; Nafchi, A.M.; Sharifi, M.; Ghalambor, P.; Oladzadabbasabadi, N.; Ariffin, F.; Huda, N. Poultry gelatin: Characteristics, developments, challenges, and future outlooks as a sustainable alternative for mammalian gelatin. *Trends Food Sci. Technol.* **2020**, *104*, 14–26. [CrossRef]
50. Liu, Y.; Luo, C.; Wang, S.; Iglesia, E.; Liu, H. Acid Catalysis Mediated by Aqueous Hydronium Ions Formed by Contacting Zeolite Crystals with Liquid Water. *J. Am. Chem. Soc.* **2024**, *146*, 35185–35198. [CrossRef]
51. De Meutter, J.; Goormaghtigh, E. Evaluation of protein secondary structure from FTIR spectra improved after partial deuteration. *Eur. Biophys. J.* **2021**, *50*, 613–628. [CrossRef] [PubMed]
52. Ahmad, T.; Ismail, A.; Ahmad, S.A.; Khalil, K.A.; Kee, L.T.; Awad, E.A.; Sazili, A.Q. Extraction, characterization and molecular structure of bovine skin gelatin extracted with plant enzymes bromelain and zingibain. *J. Food Sci. Technol.* **2020**, *57*, 3772–3781. [CrossRef]
53. Emamian, S.; Lu, T.; Kruse, H.; Emamian, H. Exploring Nature and Predicting Strength of Hydrogen Bonds: A Correlation Analysis Between Atoms-in-Molecules Descriptors, Binding Energies, and Energy Components of Symmetry-Adapted Perturbation Theory. *J. Comput. Chem.* **2019**, *40*, 2868–2881. [CrossRef]
54. López-Lorente, Á.I.; Mizaikoff, B. Mid-infrared spectroscopy for protein analysis: Potential and challenges. *Anal. Bioanal. Chem.* **2016**, *408*, 2875–2889. [CrossRef] [PubMed]
55. Asher, S.A.; Ianoul, A.; Mix, G.; Boyden, M.N.; Karnoup, A.; Diem, M.; Schweitzer-Stenner, R. Dihedral ψ Angle Dependence of the Amide III Vibration: A Uniquely Sensitive UV Resonance Raman Secondary Structural Probe. *J. Am. Chem. Soc.* **2001**, *123*, 11775–11781. [CrossRef]
56. Sharma, N.; Vuppu, S. A sustainable approach for conversion of leather trimming wastes into non-edible gelatine and its physicochemical analysis, optimization, FTIR, XRD characterization, and statistical study. *Biomass Convers. Biorefin.* **2024**, *25*, 1–20. [CrossRef]
57. Tang, C.; Zhou, K.; Zhu, Y.; Zhang, W.; Xie, Y.; Wang, Z.; Zhou, H.; Yang, T.; Zhang, Q.; Xu, B. Collagen and its derivatives: From structure and properties to their applications in food industry. *Food Hydrocoll.* **2022**, *131*, 107748. [CrossRef]
58. Niu, H.; Piaggi, P.M.; Invernizzi, M.; Parrinello, M. Molecular dynamics simulations of liquid silica crystallization. *Proc. Natl. Acad. Sci. USA* **2018**, *115*, 5348–5352. [CrossRef] [PubMed]
59. Rather, J.A.; Akhter, N.; Ashraf, Q.S.; Mir, S.A.; Makroo, H.A.; Majid, D.; Barba, F.J.; Khaneghah, A.M.; Dar, B. A comprehensive review on gelatin: Understanding impact of the sources, extraction methods, and modifications on potential packaging applications. *Food Packag. Shelf Life* **2022**, *34*, 100945. [CrossRef]
60. Noor, N.Q.I.M.; Razali, R.S.; Ismail, N.K.; Ramli, R.A.; Razali, U.H.M.; Bahauddin, A.R.; Zaharudin, N.; Rozzamri, A.; Bakar, J.; Shaarani, S.M. Application of Green Technology in Gelatin Extraction: A Review. *Processes* **2021**, *9*, 2227. [CrossRef]
61. Venupriya, V.; Krishnaveni, V.; Ramya, M. Effect of acidic and alkaline pretreatment on functional, structural and thermal properties of gelatin from waste fish scales. *Polym. Bull.* **2023**, *80*, 10533–10567. [CrossRef]
62. Akita, M.; Nishikawa, Y.; Shigenobu, Y.; Ambe, D.; Morita, T.; Morioka, K.; Adachi, K. Correlation of proline, hydroxyproline and serine content, denaturation temperature and circular dichroism analysis of type I collagen with the physiological temperature of marine teleosts. *Food Chem.* **2020**, *329*, 126775. [CrossRef] [PubMed]
63. Arya, V.; Dutta, A.; Muthuswami, R. CD Spectroscopy to Study DNA-Protein Interactions. *J. Vis. Exp.* **2022**, *180*, e63147. [CrossRef]
64. Association of Official Analytical Chemists (AOAC). *Official Methods of Analysis of AOAC International*, 17th ed.; Association of Official Analytical Chemists: Gaithersburg, MD, USA, 2000.
65. Ahmad, T.; Ismail, A.; Ahmad, S.A.; Khalil, K.A.; Awad, E.A.; Leo, T.K.; Imlan, J.C.; Sazili, A.Q. Characterization of gelatin from bovine skin extracted using ultrasound subsequent to bromelain pretreatment. *Food Hydrocoll.* **2018**, *80*, 264–273. [CrossRef]

66. Sinthusamran, S.; Benjakul, S.; Kishimura, H. Characteristics and gel properties of gelatin from skin of seabass (*Lates calcarifer*) as influenced by extraction conditions. *Food Chem.* **2014**, *152*, 276–284. [CrossRef]
67. Amertaning, D.; Bachrudin, Z.; Chin, K.B.; Erwanto, Y. Characteristics of Gelatin Extracted from Indonesian Local Cattle Hides Using Acid and Base Curing. *Pak. J. Nutr.* **2019**, *18*, 443–454. [CrossRef]

gels

MDPI

Article

High-Quality Application of Crayfish Muscle in Surimi Gels: Fortification of Blended Gels by Transglutaminase

Hongyi Wang [1,2,†], Qiang Li [1,2,†], Mengru Yang [1,2], Hong Wang [1,2], Mengtao Wang [1,2], Lin Lin [1,2,*] and Jianfeng Lu [1,2,*]

1 Engineering Research Center of Bio-Process, Ministry of Education, School of Food and Biological Engineering, Hefei University of Technology, Hefei 230009, China; wanghongyi.hfut@gmail.com (H.W.); lq00742021@163.com (Q.L.); 2023111323@mail.hfut.edu.cn (M.Y.); 13505612025@163.com (H.W.); wmt13027660032@163.com (M.W.)

2 Anhui Province Key Laboratory for Agriculture Products Modern Processing, School of Food and Biological Engineering, Hefei University of Technology, Hefei 230009, China

* Correspondence: linlin@hfut.edu.cn (L.L.); lujf@sibs.ac.cn (J.L.)

† These authors have contributed equally to the work.

Abstract: The application of crayfish muscle in surimi products is a potential way to promote their processing and ensure that it is of a high value. In this study, a one-way completely randomized design was used to prepare mixed surimi gels with different proportions of crayfish muscle. The effect of transglutaminase (TGase) on the improvement in the structural properties, water-binding capacity, micromorphology and protein conformation of blended gels was explored using mass spectrometry, centrifugation, scanning electron microscopy, and Fourier transform infrared spectroscopy. The results of thus study were analyzed by one-way ANOVA showed that in the absence of TGase, crayfish muscle made the microstructure of the blended gel looser and rougher, with a reduction in the strength of the gel and a decrease in the water holding capacity. The addition of 0.6% TGase was able to ameliorate this negative effect by promoting the formation of key chemical bonds and changes in protein conformation, which ultimately led to the enhancement of the crayfish–surimi blended gel properties. Practically, this study provides a viable strategy for incorporating crayfish into surimi products, enabling the development of novel, high-quality seafood products with improved texture and moisture retention, thereby enhancing consumer appeal and reducing waste in crayfish processing.

Keywords: surimi; transglutaminase; crayfish muscle; blended gel; gel properties

1. Introduction

Crayfish (*Procambarus clarkii*) are native to northeastern Mexico and the South Central United States, and subsequently, their distribution has gradually expanded to almost the entire globe, with the exception of Antarctica and Oceania [1]. Crayfish have a strong ability to adapt to the environment and are an important economic freshwater product in China, with their aquaculture production reaching 3,161,000 tons in 2023 alone, an increase of 9.35% over that in the previous year [2]. In addition, crayfish also has a delicious flavor, nutritional value and high food value, and very good processing and utilization prospects [3]. In 2023, 1,402,300 tons of crayfish was processed in China, making up 44.36% of farmed production [2]. However, at present, processed crayfish are mainly distributed as fresh crayfish and single frozen crayfish (whole crayfish and crayfish tails). Serious homogenization of crayfish processing products limits the high-quality development of the crayfish processing industry.

The preparation of surimi gels shows high compatibility with a wide range of food-borne ingredients. The application of protein-based low-value ingredients to the development of surimi products to enhance their application is a widely used approach, and a large number of studies have been devoted to it. These proteins contain both plant and animal sources. Plant proteins, as a low-cost and widely available food processing ingredient, have received equal attention for their application in surimi gels. Different plant proteins can participate in hydrophobic interactions, disulfide bond formation, and protein structural rearrangements to couple with myofibrillar proteins in surimi to form blended gels [4]. Further enhancement of the addition of plant proteins could also lead to the development of softened surimi gels with higher in vitro digestibility [5]. Research related to low-value raw materials of animal origin in surimi gels is equally extensive. Chinese mitten crab meat was applied to the preparation of surimi gel in an amount of up to 10% of the gel [6]. Based on microwave puffing technology, surimi gels containing crab meat exhibiting degree of fibrillation of up to 2.23% have also been derived [7]. Animal-derived protein raw materials such as chicken breast, egg, and plasma are also thought to be able to participate in the formation of the gel matrix and are used in the preparation of surimi gels [8–10]. In addition, some protein hydrolysis products have strong functional properties due to them carrying a large number of surface active groups, and they often show good application in surimi gels [11,12]. In summary, the application of crayfish muscle in the preparation of surimi blended gels is a promising development strategy. It is not only conducive to solving the problem of the huge production of crayfish using a single form of processing, but also conducive to encouraging surimi processing enterprises to develop new high-value-added products. Therefore, the development of crayfish surimi blended gel is not only a technological breakthrough, but also an important strategy to promote industrial transformation and respond to the trend of efficient resource utilization and improvements in consumption.

The quality of surimi gels is mainly reflected by its gel properties. For this reason, rinsing is usually required during the preparation of surimi to remove components that are not conducive to gel formation, such as proteases and water-soluble proteins with poor gelation properties [13]. These components are also present in crayfish muscle, and the complexity of its composition also reduces the relative concentration of the main component in gel formation (myofibrillar protein). These factors could potentially severely limit the gel properties of surimi gels blended with crayfish. Currently, a large number of studies have been devoted to the enhancement of the gel properties of surimi. In this regard, some novel processing techniques have been introduced into the gelatinization process of surimi. Microwave heating can utilize the dielectric effect of surimi to increase the gel strength of low-salt surimi gels to 821.74 g·cm by promoting protein solubilization and cross-linking [14]. The mechanical effect and cavitation generated by ultrasonic treatment can promote the unfolding of structural proteins and expose more hydrophobic groups and reaction sites, thus enhancing cross-linking between protein molecules in surimi [15]. Appropriate radiofrequency heating induces the conversion of α-helixes into random curls and promotes hydrophobic interactions between and the cross-linking of proteins by disulfide bonds to improve the structural properties of grass carp protein gels [16]. In addition, ohmic heating is also an emerging method for celiac gel fortification [17]. The enhancement of surimi gel by exogenous additives is more stable and simple to implement compared to the physical field technique. Polysaccharide additives enhance the structural properties of surimi gels through the synergistic effects of matrix enhancement, water binding and encapsulation, and covalent–noncovalent interactions [18]. The phase behavior of polysaccharides during the gelatinization of surimi can also be used to judge their appropriate concentration to improve the physicochemical properties of blended gels [19]. TGase catalyzes the generation of ε-(γ-Gln)-Lys bonds from ε-amino group with the γ-carboxy amide group in glutamine and promotes intermyosin cross-linking [20]. Therefore, it is considered to be an effective fortifier for surimi gels. TGase could enhance the surimi blended gel properties of a variety of exogenous proteins intervening in surimi, including chickpea protein, sheep plasma protein, and crabmeat, to name a few [6,21,22]. Even in different gelation environments, including intrinsic environments such as salt ion concentration and extrinsic environments such as processing methods, TGase was able to promote cross-linking between proteins in the blended system [21,23]. The broad adaptability of TGase opens up possibilities for the use of crayfish muscle in surimi gels. However, it has not been clarified whether or not the lifting effect of TGase on surimi blended gels is affected by the type and concentration of ingredients other than surimi; it is also unclear what the coupling characteristics of crayfish muscle in this catalytic effect are. Related research will facilitate the high-quality application of crayfish muscle in surimi gels and promote the industrial innovation of crayfish and surimi products.

In this study, surimi-based blended gels containing crayfish muscle were investigated. The effects of different crayfish muscle additions (0%, 2.5%, 5%, 7.5%, and 10%) on the gel properties in the presence or absence of TGase were investigated by analyzing the textural features, water binding capacity, micro-morphology, and protein conformation of the blended gels. The results of this study may provide theoretical references for the high-value utilization of crayfish muscle in surimi products, to promote the utilization of crayfish in deep processing, and to alleviate the problem of homogenization of processed crayfish products.

2. Results and Discussion

2.1. Effect of TGase and Crayfish Meat on the Texture Properties of Blended Gels

2.1.1. Texture Properties

Figure 1A–C show the effect of different amounts of crayfish muscle on the breaking force, deformation and gel strength of surimi blended gels, and the intervention of TGase on this effect. A decreasing trend in gel strength and breaking force in gels without TGase was observed as the crayfish muscle content increased. Notably, with 5% inclusion of crayfish muscle, significant reductions in breaking force, deformation, and gel strength were documented, dropping from 580.28 g, 0.98 cm, and 566.89 g·cm to 539.64 g, 0.93 cm, and 503.72 g·cm, respectively ($p < 0.05$). These trends are similar to those found in previous studies for the addition of Chinese mitten crab meat to surimi gels; this may be due to the co-precipitation of myoplasmic proteins with myofibrillar proteins in the added crayfish, which weakened the structure of the gel [6,24]. Interestingly, the TGase-containing gels improved in all measured gel strength characteristics after the addition of moderate amounts of crayfish muscle, highlighting the positive impact of TGase on gel properties. Research indicates that TGase's role in forming ε-(γ-Glu)-Lys covalent bonds stabilizes the protein structure within the gel, substantially boosting gel strength [25]. The addition of microbial transglutaminase (MTG) has been found to improve gel properties notably [26]. As crayfish muscle content increased up to 7.5%, the differences in gel properties between TGase-added and non-TGase-added gels became more pronounced, though no further differences were seen between the 7.5% and 10% groups.

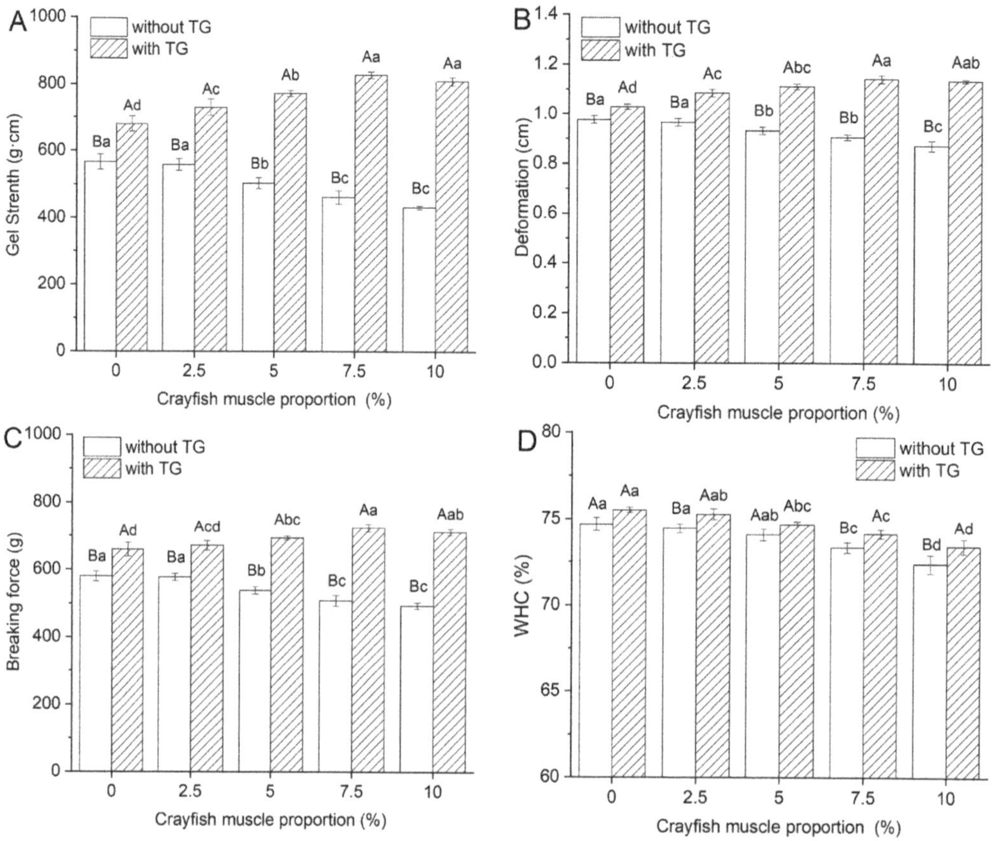

Figure 1. Changes in breaking force (**A**), deformation (**B**), gel strength (**C**), and WHC (**D**) of blended gels. Lowercase letters above each standard deviation bar reveal significant variations among gels with different crayfish muscle contents ($p < 0.05$). Uppercase letters highlight significant differences between samples with and without added TGase ($p < 0.05$).

2.1.2. Effect of TGase and Crayfish Muscle on the Water Holding Capacity of Blended Gels

Water holding capacity (WHC) is an indication of the ability of crayfish–surimi blended gels to bind to water, which can reflect the stability of the gel system laterally [27]. The effect of crayfish muscle and TGase addition on the WHC of the blended gel is shown in Figure 1D. The WHC of the blended gels decreased with increasing crayfish muscle content, with or without the addition of TGase. This decline was significant when the amount of crayfish muscle exceeded 7.5%. This reduction in WHC may be linked to the structural properties of crayfish proteins and their interactions within the gel matrix. However, at the same level of crayfish muscle addition, the blended gels containing TGase consistently showed higher WHC, suggesting that TGase improves the water holding capacity of these gels. This enhancement is attributed to the role of TGase in the promotion of myofibrillar proteins' cross-linking reactions, forming a denser gel network that effectively retains more water [28]. Therefore, TGase can effectively enhance the water retention capacity of crayfish–surimi blended gels and stabilize the gel system.

2.2. Effect of TGase and Crayfish Muscle on the Color Properties of Blended Gels

The magnitude of the whiteness value is often used as an indication of the tightness of the surimi gel structure; when the texture is tight, there is a greater ability to reflect light, which will exhibit a larger whiteness value. This is calculated from the red–green value (a*), the yellow–blue value (b*) and the luminance value (L*). Figure 2 illustrates the effect of crayfish muscle and TGase addition on each of the color characteristics of the surimi blended gels. The increase in the amount of crayfish muscle significantly affected the color properties. The increase in a* and b* may be attributed to the presence of blue and red carotenoproteins and the oxygen-carrying molecule, hemocyanin, found in crayfish [29]. On the other hand, the reduction in gel strength implies that the crayfish muscle may have disrupted the homogeneity of the texture of the surimi blended gels, resulting in reduced light reflectivity on the gel surface and poorer L* and whiteness [6]. The application of TGase did not notably affect the whiteness of surimi gels that lacked crayfish muscle ($p > 0.05$). However, at the same level of crayfish muscle addition, the surimi blended gel containing TGase showed lower a* and b* values as well as higher L* and whiteness ($p < 0.05$). This indicates that TGase contributes to the development of a denser gel network, which enhances the gel's ability to reflect visible light, thereby improving its whiteness. The enhancement of the whiteness of surimi gels by TGase is a more general phenomenon, even in different thermal gelation methods [23]. This phenomenon is consistent with the higher performance of gel strength and WHC. A comprehensive observation of the color characteristics of the crayfish–surimi blended gel (Figure 2E) showed that the effect of TGase on the overall color characteristics of the blended gel was greatest ($p < 0.05$) when the crayfish muscle addition was 7.5%. This was consistent with the fact that its gel strength was no longer significant ($p > 0.05$) after increasing to 827.79 g·mm. Overall, crayfish muscle may lead to a reduction in the color properties of surimi hybrid gels due to a number of reasons, including compositional characteristics, and TGase may attenuate this negative effect through cross-linking facilitation.

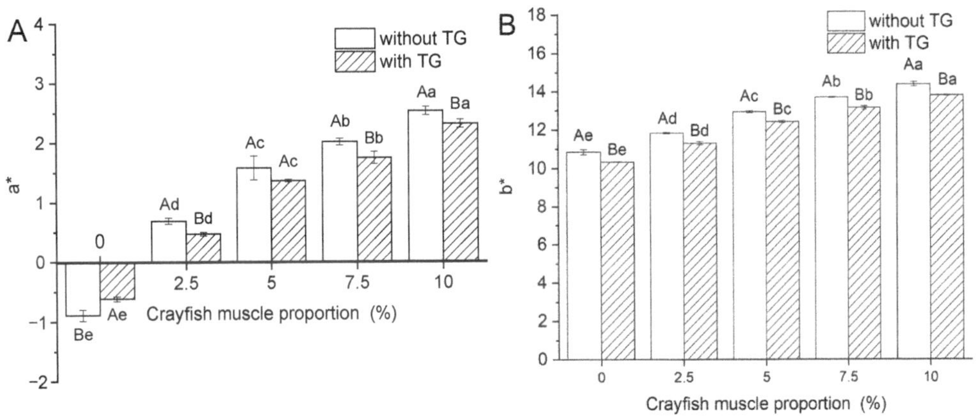

Figure 2. *Cont.*

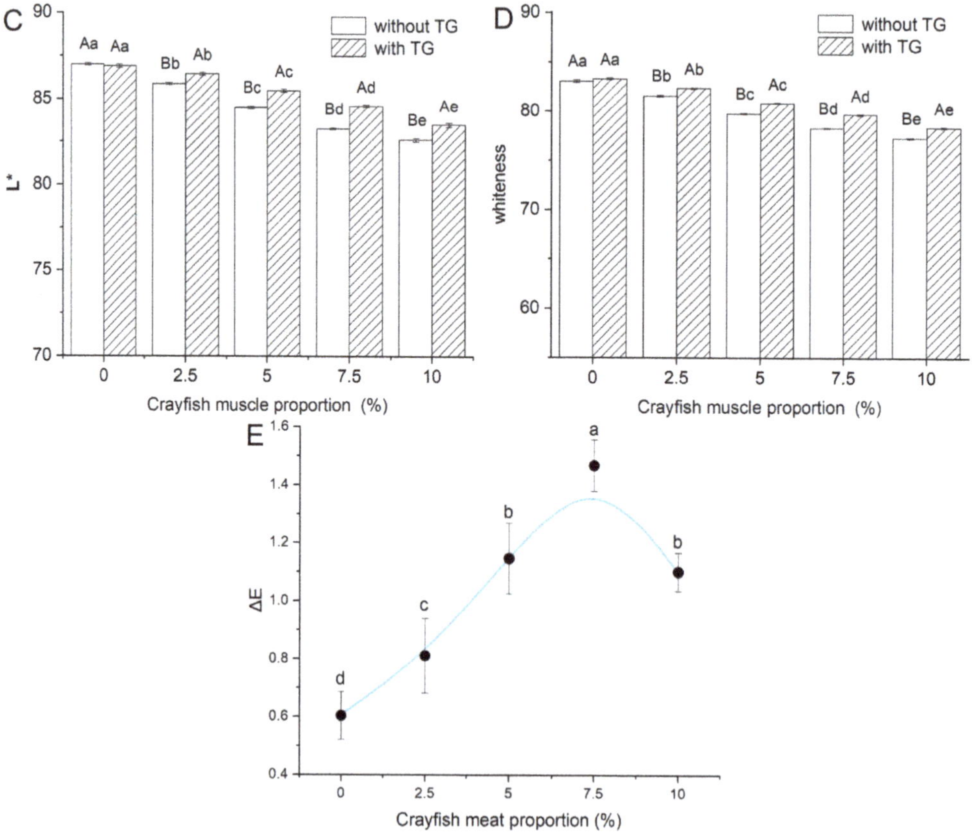

Figure 2. Changes in a* (**A**), b* (**B**), L* (**C**), whiteness (**D**) and ΔE (**E**) of blended gels. Lowercase letters above each standard deviation bar reveal significant variations among gels with different crayfish muscle contents ($p < 0.05$). Uppercase letters highlight significant differences between samples with and without added TGase ($p < 0.05$).

2.3. *Effect of TGase and Crayfish Muscle on the Water Distribution of Blended Gels*

The low-field nuclear magnetic resonance (LF-NMR) technique can be used to analyze the rate and proportion of water molecules migrating through a gel by measuring the relaxation properties of the hydrogen atoms in the gel [30]. LF-NMR was applied to investigate the effects of crayfish muscle and TGase addition on the co-water distribution characteristics of surimi blended gels, as shown in Figure 3A,B. The relaxation spectra show three separate peaks, each indicating a distinct water state. T_{2b} (0.1–1 ms) indicates water molecules tightly bound to the gel matrix. T_{22} (20–400 ms) indicates water molecules tightly encapsulated by the gel matrix. T_{23} (400–2000 ms) indicates free water molecules outside the gel matrix. The T_2 relaxation times directly reflect the mobility of water molecules; longer T_2 times suggest greater migration capabilities [31]. The T_{2b} relaxation time of the blended gel increased with increasing crayfish muscle addition, while the relaxation times of T_{22} and T_{23} decreased (Figure 3A). This suggests that the addition of crayfish muscle reduced the stability of bound water in the blended gel, while the stability of immobile and free water was improved. Among them, the peak area of T_{22} is clearly larger than the other peak areas, and its representation of water that does not flow easily is also the main observation in the related study. When TGase was added, the T_{2b} relaxation time and

T_{23} relaxation time of the blended gels were reduced to a degree that increased with the increase in the amount of crayfish muscle (Figure 3B). This suggests that the addition of TGase made the bound and free water more stable, while the relaxation time of the less mobile water did not change significantly.

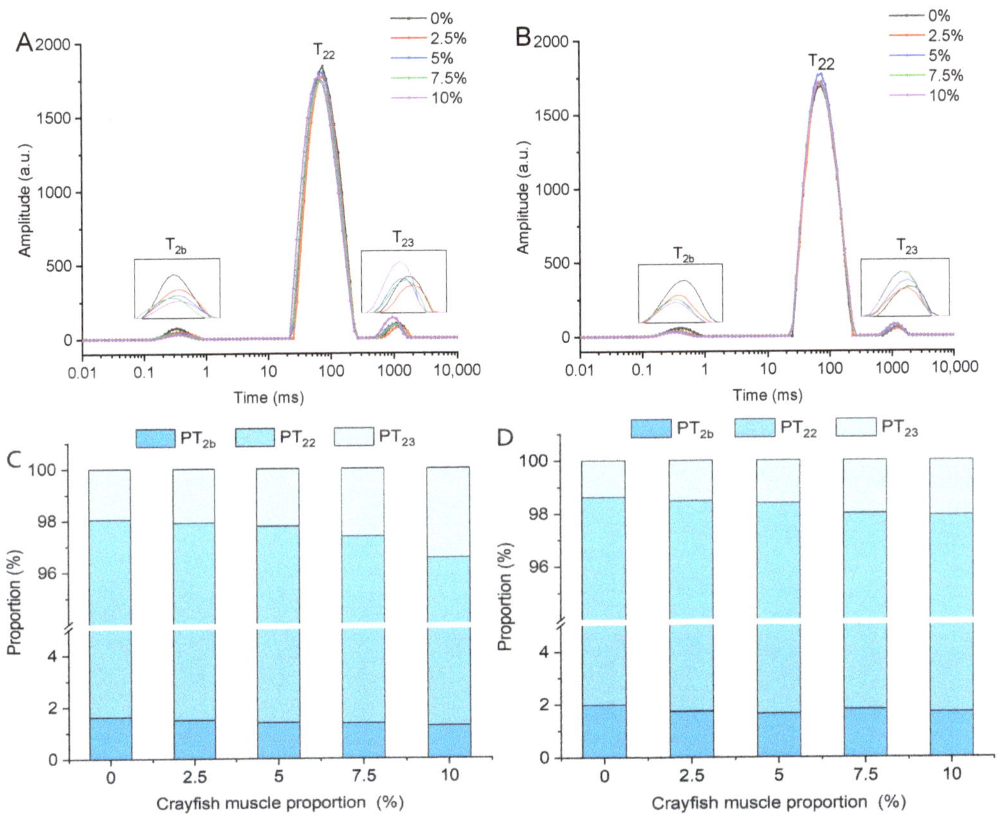

Figure 3. The effect of different treatments on the spin relaxation time (T_2) of blended gels treated (**A**) without TGase and (**B**) with TGase, as well as the percentage of relative area of peaks (**C**,**D**). Note: PT_{2b}, PT_{22}, and PT_{23} represent the peak area ratios of T_{2b}, T_{22}, and T_{23}, respectively.

The proportions of the three water distributions expressed by the peak area share are shown in Figure 3C,D. As the crayfish muscle content increased, the PT_{2b} of blended gel gradually decreased while PT_{23} gradually increased, suggesting that adding crayfish muscle enhanced the escape capability of bound water in the gel. Figure 3D shows that the addition of TGase altered the relaxation peak area percentages of bound water and free water. Compared to the group without TGase, the PT_{23} of the TGase-added group significantly decreased, indicating a decrease in free water content within the gel following the incorporation of TGase. However, as the content of crayfish muscle increased, PT_{23} in blended gels with TGase continued to show an upward trend. Additionally, after adding TGase, the proportion of bound water generally increased, while the proportion of immobile water exhibited no significant changes. Thus, the application of TGase resulted in an elevation in bound water content coupled with a reduction in the proportion of free water. This effect is due to TGase inhibiting the conversion of bound water to free water in blended gels, resulting in changes in water distribution and enhanced water retention.

Similarly, TGase-catalyzed cross-linking between soy protein and scallop muscle was found to be inhibited by the conversion of bound water [32].

2.4. Effect of TGase and Crayfish Muscle on the Magnetic Resonance Imaging of Blended Gels

Magnetic resonance imaging (MRI) can efficiently, accurately and non-destructively visualize the state of water distribution in a gel system. The darker-color (tending toward blue) portion of the MRI is meant to reflect a lower distribution of hydrogen protons, indicating the poorer water retention capacity of the gel, and vice versa (brighter color tending toward red) indicates a higher distribution of hydrogen protons and the better water retention capacity of the gel [33]. The effect of crayfish muscle addition and TGase on the hydrogen proton density in surimi blended gels is shown in Figure 4. Variations in these images reveal that with increasing additions of crayfish muscle, the samples exhibit fewer red areas and more yellow areas, indicating an increase in the uneven distribution of water. In addition, under the same level of crayfish muscle addition, a significant increase in red areas and a more homogeneous distribution was observed in the MRI of the blended gel after the introduction of TGase. This enhancement shows that adding TGase improved the bound water retention within the blended gel network, thereby improving the water holding capacity of the crayfish–surimi blended gel. This phenomenon is consistent with the state of moisture distribution and the results of the correlation analysis of WHC. This further validates the negative effect of crayfish muscle on surimi blended gels and the ameliorating effect of TGase on this negative effect.

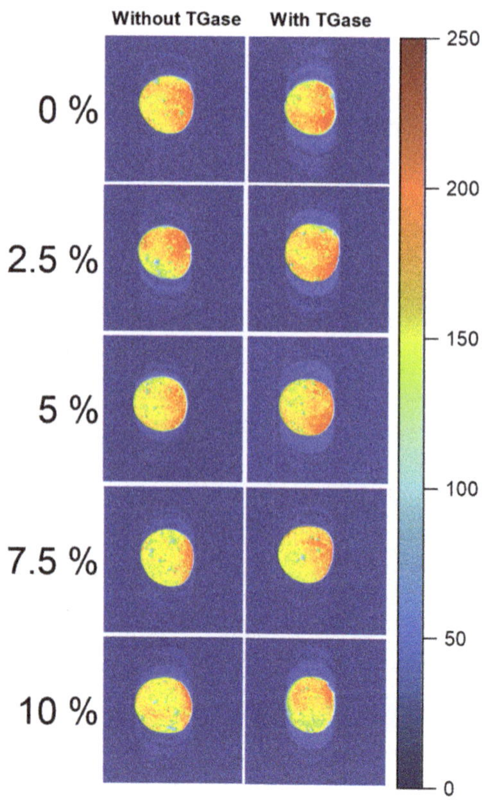

Figure 4. T2-weighted magnetic resonance images of surimi gel with different amounts of crayfish muscle (0%, 2.5%, 5%, 7.5% and 10%) in the samples with TGase and without TGase.

2.5. Effect of TGase and Crayfish Muscle on the Microstructural of Blended Gels

In Figure 5A, the pictures of each group on the left provide a visualization of the untreated crayfish–surimi blended gel. Visual observation cannot not reveal the effect of crayfish addition and TGase addition on the apparent differences in the surimi blended gels, both in terms of structural and color characteristics. Scanning electron microscopy (SEM) is a commonly used method to observe the microstructure of surimi gels [34]. It was able to visualize the extensive effect of crayfish muscle and TGase addition on the structure of the surimi blended gel. In Figure 5A, the images to the right of each group show the results of SEM observations. The results of porosity quantification in the images obtained using image analysis software are shown in Figure 5B,C. As the crayfish muscle content increased, the microstructure of the blended gel became rougher and more irregular, while the porosity increased. The structural damage induced by crayfish muscle inevitably led to a reduction in the water retention capacity of the blended gel, as well as a reduction in gel strength due to stress reduction. Thus, the microstructural observations are consistent with the results of correlation analyses such as that of WHC. The introduction of TGase was more effective in optimizing the microstructure of the blended gels. The addition of TGase reduced the porosity from 24.508% to 11.036% when the crayfish muscle content was 10%. This effect resulted from TGase influencing the protein molecule interactions and the covalent bonding between glutamine and lysine residues, allowing myofibrillar proteins to aggregate more effectively and form a dense protein gel network. According to recent research, reduced porosity enhances the gel's ability to trap excess moisture, thereby improving its textural properties [35]. The most significant effect of TGase on the reduction in the porosity (8.151%) of the blended gel was observed when crayfish muscle was added at 7.5%. This suggests that the molecular structure was most uniform at this point, with the most significant protein cross-linking, corresponding to the most compact three-dimensional gel network. This finding is consistent with the enhanced gel strength observed in the experiments.

2.6. Effect of TGase and Crayfish Muscle on the Chemical Interactions of Blended Gels

Noncovalent bonds, including hydrogen, ionic, and hydrophobic interactions, and disulfide bonds are critical in the formation of surimi gels [36]. The effect of crayfish muscle addition and TGase on the content of chemical interactions in the surimi blended gels is shown in Table 1. In myofibrillar proteins, interactions occur between carbonyl and amino groups within and across peptide chains mainly through hydrogen bonds [37]. According to Table 1, there is a notable decrease in hydrogen bonds in blended gels as the content of crayfish muscle increases ($p < 0.05$). This decrease may be linked to the lipids in crayfish muscle, which can obstruct the formation of hydrogen bonds, thereby adversely affecting water retention, as corroborated by earlier WHC measurements. This underscores the pivotal role of hydrogen bonds in maintaining water stability within the gel matrix. The addition of TGase appeared to significantly enhance the formation of hydrogen bonds between proteins, possibly due to the rearrangement of hydrogen bonds induced by conformational changes during protein cross-linking [38]. This course of action mitigates the negative effects of crayfish muscle addition on hydrogen bond formation to some extent.

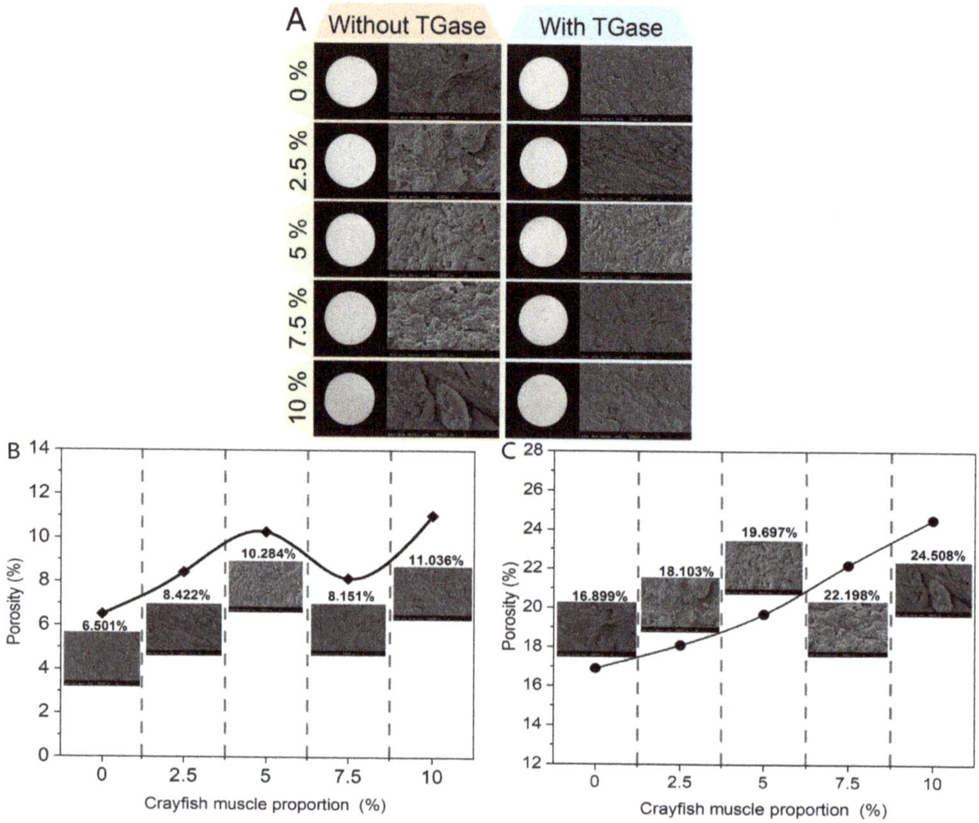

Figure 5. Effect of different crayfish muscle additions (0%, 2.5%, 5%, 7.5%, 10%) and TGase addition on the microstructure (**A**) and porosity ((**B**): with TGase, (**C**): without TGase) of surimi blended gels.

Table 1. The effect of TG on the chemical interactions in blended gels.

Crayfish Muscle Content (%)	Hydrogen Bonds		Ionic Bonds		Hydrophobic Interactions	
	Without TGase	With TGase	Without TGase	With TGase	Without TGase	With TGase
0	1.66 ± 0.04 Ba	2.31 ± 0.12 Aa	0.88 ± 0.03 Ae	0.35 ± 0.04 Be	21.99 ± 0.32 Aa	20.1 ± 0.33 Bd
2.5	1.27 ± 0.10 Bb	2.09 ± 0.11 Ab	1.25 ± 0.08 Ad	0.66 ± 0.02 Bd	20.21 ± 0.23 Bb	21.1 ± 1.04 Ac
5	0.97 ± 0.13 Bc	1.74 ± 0.06 Ac	1.60 ± 0.10 Ac	0.95 ± 0.08 Bc	18.1 ± 0.33 Bc	22.82 ± 1.54 Ab
7.5	0.47 ± 0.16 Bd	1.37 ± 0.03 Ad	2.04 ± 0.15 Ab	1.27 ± 0.05 Bb	16.55 ± 0.28 Bd	25.6 ± 0.74 Aa
10	0.21 ± 0.10 Be	0.85 ± 0.05 Ae	2.34 ± 0.12 Aa	1.69 ± 0.03 Ba	14.16 ± 0.9 Be	25.62 ± 0.36 Aa

Notes: Lowercase letters above each standard deviation bar reveal significant variations among gels with different crayfish muscle contents ($p < 0.05$). Uppercase letters highlight significant differences between samples with and without added TGase ($p < 0.05$).

Ionic bonds are crucial in sustaining the tertiary and quaternary structures of myofibrillar proteins. During gel processing, the disruption of these bonds under various conditions facilitates cross-linking reactions, which are essential for forming a robust gel network [26]. Table 1 illustrates that the concentration of ionic bonds increases with the addition of crayfish muscle. This suggests that with the addition of crayfish muscle, the proteins in the blended gel system had more stable structural characteristics before gelation and were less susceptible to change during the thermogenic gelation process [39]. However, the presence of TGase significantly reduced the concentration of ionic bonds. The reason for this change may have been that the catalytic action of TGase consumed positively charged lysine residues while reducing the exposure of neighboring charged residues (e.g., glutamate, aspartate), resulting in a reduced opportunity for ionic bond pairing [20]. Similarly, ionic bonding was reduced in TGase-enhanced pea protein isolate gels [40]. This further corroborates the positive role played by TGase in promoting the intermolecular cross-linking of proteins.

Regarding hydrophobic interactions, these are essential for a gel's structural integrity and are closely linked to gel strength. In samples with TGase, an increase in crayfish muscle content resulted in enhanced hydrophobic interactions, with no significant difference observed between 7.5% and 10% crayfish content. This may be due to the fact that TGase-induced noncovalent cross-linking shortens the spacing of intermolecular hydrophobic groups, making hydrophobic interactions easier to form [41]. At the same time, the formation of hydrophobic interactions further enhances the intermolecular cross-linking. This process of action was similarly found in TGase-containing cowpea protein isolate gels [42]. Thus, at a certain crayfish muscle threshold, the presence of TGase catalyzed the improvement in hydrophobic interactions, which improved the mechanical properties of the gel, consistent with the observed improvement in gel strength.

Clearly, the incorporation of crayfish muscle through chemical bonding does not favor cross-linking between proteins in surimi blended gels. However, the addition of TGase can promote chemical cross-linking between proteins through indirect effects such as the smaller molecular spacing mediated by catalysis.

2.7. Effect of TGase and Crayfish Muscle on the Total Sulfhydryl Content of Blended Gels

Sulfhydryl groups are the basis for the formation of disulfide bonds, and disulfide bonds are an important force for stabilizing the structure of surimi gels. The effects of crayfish muscle addition and TGase on the sulfhydryl content in the surimi blended gels is shown in Figure 6. As the muscle content of crayfish increased, the sulfhydryl content also tended to increase. It was hypothesized that this could be due to the hydrolysis of proteins in the blended system into small peptides containing free sulfhydryl groups due to the endogenous proteases (e.g., tissue protease) in crayfish muscle [43]. Although this also implies an increase in the concentration of substrate available for disulfide bond formation, it appears that the amount of disulfide bonds does not increase as a result, and the gel properties of the blended gel are instead diminished. This may be due to the change in potential induced by the changes in the peptide segments introduced by hydrolysis, which nay increase intermolecular electrostatic repulsion and consequently inhibit contact between the sulfhydryl groups [44]. However, a notable difference, as observed when TGase was added to this blended system, was that the sulfhydryl content decreased with increasing crayfish muscle content. This is an indirect indication of the enhanced ability to form disulfide bonds, although protein cross-linking catalyzed directly by TGase is mainly maintained by ε-(γ-Gln)-Lys bonds (non-covalent bonds) and does not involve the formation of disulfide bonds (covalent bonds) [20]. However, the resulting reduction in the spacing between protein molecules promotes contact between sulfhydryl groups, which provides a good precondition for disulfide bond formation [45]. As mentioned earlier, the endogenous protease in crayfish muscle provides more available sulfhydryl groups for the blended system, and the electrostatic repulsion is counteracted by TGase, facilitating binding of sulfhydryl groups to each other to form disulfide bonds. This phenomenon is consistent with the elevating effect of TGase on disulfide bond content in myofibrillar protein gels [45]. Thus, although the addition of small dragon muscle decreases the cross-linking of proteins in the blended system due to its inhibitory effect on disulfide bonds, the catalytic effect of TGase counteracts this phenomenon and allows disulfide bond formation to be promoted.

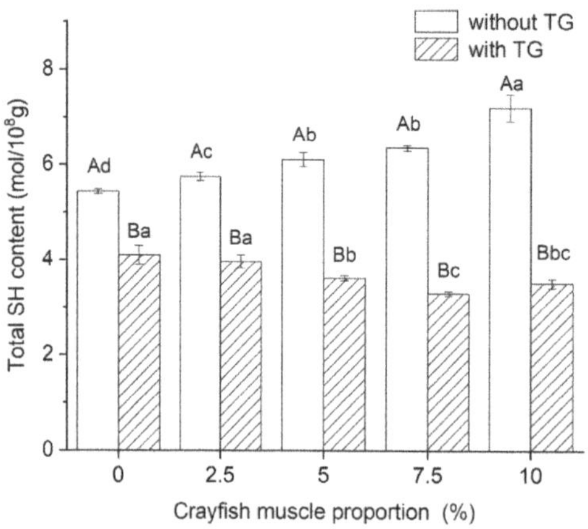

Figure 6. The effect of TG on the total sulfhydryl group content in blended gels. Lowercase letters above each standard deviation bar reveal significant variations among gels with different crayfish muscle contents ($p < 0.05$). Uppercase letters highlight significant differences between samples with and without added TGase ($p < 0.05$).

2.8. Effect of TGase and Crayfish Muscle on the Fourier Transform Infrared Spectroscopy of Blended Gels

Fourier transform infrared (FTIR) spectroscopy is commonly used to analyze the composition of protein secondary structures in surimi gels. The FTIR spectra of the crayfish–surimi blended gel are shown in Figure 7A,B. The blended gels had absorption peaks near 1200 cm^{-1}, 1500 cm^{-1}, 1600–1700 cm^{-1}, 2926–2953 cm^{-1}, and 3100–3300 cm^{-1}, which correspond to the amide III, II, I, B, and A bands of protein molecules, respectively [46,47]. Comparable observations have been reported in cod surimi gels [48]. These wavenumber regions reflect the vibrational modes of various chemical bonds within the protein molecules, providing critical insights into the secondary and tertiary protein structures. As the content of crayfish muscle increases, the peak values of amide A, amide I, and amide II bands tend to decrease. This trend suggests that the addition of crayfish muscle diminishes the hydrogen bonding interactions between protein molecules within the blended gel [49]. Each sample exhibited an absorption peak at 1050 cm^{-1}, where the peak intensity decreased as the crayfish muscle content increased. Generally, gel samples subjected to different processing conditions demonstrated analogous spectral graphs, as depicted in Figure 7A,B. All samples featured a peak at 1050 cm^{-1}, with decreasing amplitude as the proportion of crayfish muscle increased. Studies on the effect of antifreeze on the protein structure of tilapia showed that the peak near 1050 cm^{-1} was related to the -OH group of sucrose in surimi [50].

The amide I band, typically situated in the 1600 to 1700 cm^{-1} range, corresponds to the vibrational modes of carbonyl (C=O) groups in protein molecules, crucial for revealing the protein's secondary structure. For α-helices, β-sheets, β-turns, and random coils, distinct absorption peaks are identified within the respective spectral ranges of 1651–1660 cm^{-1}, 1600–1639 cm^{-1}, 1661–1700 cm^{-1}, and 1640–1650 cm^{-1}, respectively [51]. As illustrated in Figure 7C, the TGase-added group, compared to the group without TGase, displayed a significant reduction in α-helix content ($p < 0.05$) and a rise in the proportion of β-turns ($p < 0.05$), with minimal effects on random coils and β-sheets. The stability of the α-helix relies on continuous hydrogen bonding within the main chain, and TGase-catalyzed cross-linking may interrupt the arrangement of the original hydrogen bonds, leading to deconvolution of the helical structure [52]. The deconvolution of the helical structure can be enhanced after inverse fold accumulation, which provides a good prerequisite for structural transformation and chemical bond formation [53]. The resulting ordered intermolecular cross-linking of proteins imparts good stress and water retention properties to the blended gels. In contrast, higher α-helix content was thought to reduce the mechanical properties and stability of surimi gels [54]. Thus, while the deteriorating effect of crayfish muscle on surimi blended gels may be attributed to the insufficient transformation of the secondary structure, the addition of TGase was able to enhance the cross-linking between proteins in the blended gels through structural changes by promoting the transition from the α-helix to β-turn. This process of action was demonstrated in microbial the transglutaminase-mediated enhancement of the mechanical properties of pork gels [55].

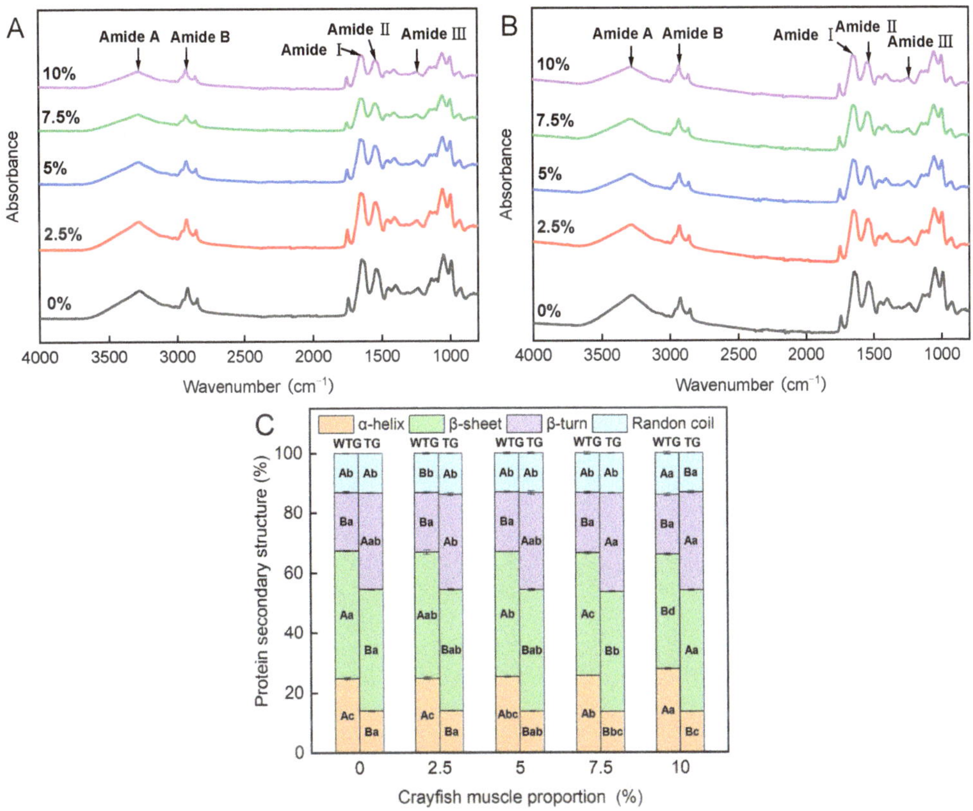

Figure 7. FTIR spectra of surimi gel with varying crayfish muscle contents ((**A**): samples without TGase; (**B**): samples with TGase), and the effect of different crayfish muscle contents on the secondary structure of proteins in the TG and WTG groups of surimi gel (**C**). TG: with TGase; WTG: without TGase. Lowercase letters above each standard deviation bar reveal significant variations among gels with different crayfish muscle contents ($p < 0.05$). Uppercase letters highlight significant differences between samples with and without added TGase ($p < 0.05$).

3. Conclusions

This study systematically elucidates the effects of crayfish muscle incorporation and transglutaminase (TGase) addition on the physicochemical properties, microstructure, and molecular interactions of silver carp–surimi blended gels. Increasing crayfish muscle content (0–10%) in surimi gels without TGase progressively reduced gel strength, WHC, and structural homogeneity. This is attributed to the inhibition of hydrogen bonding forcing insufficient structural transformations to occur between proteins and attenuating conformationally mediated cross-linking between proteins. Meanwhile, the inhibitory effect of crayfish muscle on hydrophobic interactions and disulfide bonds in the surimi blended gel further hindered cross-linking between protein molecules. Ultimately, the protein cross-linking in the mixed system is disordered and is reflected by a loose gel structure. Proteases in crayfish muscle have been suggested as potential factors contributing to the latter's negative effects, and hydrolysis by proteases may alter the peptide composition in the blended system and inhibit protein conformational shifts and the binding of reactive groups through enhanced electrostatic repulsion. The addition of 0.6% TGase counteracted the negative effects of the addition of crayfish muscle. TGase-catalyzed ε-(γ-Gln)-Lys bonds promote protein aggregation while possibly counteracting the electrostatic repulsion induced by crayfish proteases, driving the shift in protein hooking as well as the binding of reactive groups. Under these conditions, protein cross-linking in the blended system becomes more organized. A compact gel structure not only has good water retention capacity, but also has strong stress resistance, which manifests as increased gel strength.

The synergistic use of crayfish muscle ($\leq$7.5%) and TGase represents a viable strategy to develop high-value surimi products with enhanced texture and moisture stability. TGase effectively mitigated the structural defects induced by crayfish muscle; it could thus promote the resource-efficient utilization of underutilized aquatic proteins, and at the same time help to alleviate the problem of product homogenization in crayfish processing.

4. Materials and Methods

4.1. Materials

The frozen silver carp surimi used in the experiment was AAA-grade (Hubei Honghu Jili Aquatic Food Co., Ltd., Honghu, China). Crayfish were purchased from RT-Mart Supermarket (Hefei, China), and crayfish muscle was taken by hand by removing the head and tail. Phosphoric Acid, Urea, Ethanol, Sodium Chloride and other chemicals were all analytically pure and purchased from Sinopharm Chemical Reagent Co., Ltd. (Shanghai, China).

4.2. Preparation of Surimi and Crayfish Muscle Blended Gels

Crayfish muscle was chopped and mixed with surimi, and the amounts of crayfish muscle added were 0%, 2.5%, 5%, 7.5%, and 10%. Subsequently, salt (2.5%, *w*/*w*) and TGase (0.6%, *w*/*w*) were added, and chopping continued for 2 min. The homogeneous surimi mixture was then encased in sausage casings and sealed at both ends using a clipper. The encased surimi–crayfish mixtures underwent a two-stage heating process, being initially submerged in a 40 ± 2 °C water bath for 60 min, followed by undergoing gelation at 90 ± 2 °C for 30 min. The heat-treated sausages were subsequently cooled and stored overnight at 5 ± 3 °C.

4.3. Gel Strength

After reaching equilibrium at room temperature for approximately 60 min, the blended gels were sectioned into cylinders approximately 2 cm in height. Puncture tests were performed with a TA-XT Plus texture analyzer (Stable Micro Systems, Surrey, London, UK) fitted with a P/5S spherical probe, executing five replicates per sample [56]. Test settings were as follows: automatic trigger, trigger force = 5.0 g, both pretest and test speeds at 1.0 mm/s, post-test speed = 1.0 mm/s, pressing displacement = 15 mm, and a return speed of 10.0 mm/s. We calculated the gel strength according to Formula (1):

$$\text{Gel strength (g·cm)} = \text{Breaking force (g)} \times \text{Deformation (cm)} \quad (1)$$

4.4. WHC

Thin slices of approximately 5 g of the blended gel (2 mm) were wrapped in three layers of filter paper and centrifuged at 7040× *g* for 15 min at room temperature (Tianmei Scientific Instrument Co., Ltd., Zurich, Switzerland) [7]. We record the mass of the sample before (m1) and after (m2) centrifugation. Each sample was tested in triplicate, and then we calculated the WHC according to Formula (2):

$$\text{WHC (\%)} = (\text{m2}/\text{m1}) \times 100 \quad (2)$$

4.5. Whiteness

Gel samples were sliced into cylinders about 10 mm thick, and their color attributes were measured using an automatic colorimeter (ZE7700, Nippon Denshoku Industries Co., Ltd., Tokyo, Japan) [7]. The measurements consisted of a* (red/green), b* (yellow/blue) and L* (luminance), and were performed six times for each sample. The whiteness was calculated according to Equation (3). Subsequently, ΔE was calculated to measure the difference in the overall color change in the blended gel before and after TGase addition under the addition of different amounts of crayfish. ΔE was calculated as shown in Equation (4):

$$\text{Whiteness} = 100 - \sqrt{(100 - \text{L}^*)^2 + \text{a}^{*2} + \text{b}^{*2}} \quad (3)$$

$$\Delta \text{E} = \sqrt{(\text{L}^*{}_2 - \text{L}^*{}_1)^2 + (\text{a}^*{}_2 - \text{a}^*{}_1)^2 + (\text{b}^*{}_2 - \text{b}^*{}_1)^2} \quad (4)$$

In Equation (4), $L^*{}_1$, $a^*{}_1$, and $b^*{}_1$ are the color characteristics of the blended gel without TGase with the same amounts of crayfish added, and $L^*{}_2$, $a^*{}_2$, and $b^*{}_2$ are the color characteristics of the blended gel with TGase and with the same amounts of crayfish added.

4.6. Low-Field Nuclear Magnetic Resonance

A low-field nuclear magnetic resonance analyzer (MesoMR23-060H-I, Niumag Electronic Technology Co., Ltd., Suzhou, China) was utilized to evaluate the water relaxation characteristics in surimi gel [2]. After removing the casing, 4 g of the sample was immediately transferred into a 15 mm diameter NMR tube for analysis. A CPMG sequence was utilized to capture transverse relaxation time (T2) signals at a resonance frequency of 18 MHz. The T2 relaxation data were processed through inversion, with each sample undergoing triple testing.

4.7. Magnetic Resonance Imaging

The proton density distribution in samples was analyzed using a multi-slice spin-echo pulse sequence and subsequently transformed into color images using pseudo-color processing [19]. The samples were sectioned into cylinders 20 mm in height and positioned within an NMR tube for imaging. The imaging parameters were set as follows: slice width = 3.0 mm, slice spacing = 1.0 mm, TE = 0.5 ms, and TR = 1800 ms.

4.8. SEM

The surface morphology of blended gel samples was examined. Gel samples were sectioned into 4 mm cubes and immersed in 0.1 M phosphate buffer (pH 7.2) for 20 min. Subsequently, the samples were stabilized in a 2.5% glutaraldehyde solution at 4 °C for a period of 24 h and rinsed five times with the same buffer. The samples then underwent a sequential dehydration process using ethanol solutions at concentrations increasing from 30% to 90% for 15 min each step, followed by dehydration in anhydrous ethanol for 20 min. After the removal of anhydrous ethanol, the samples underwent freeze-drying for a period of 24 h. The samples were surface sprayed with gold (Au-Pd alloy), and the surface features of the blended gels were observed with a scanning electron microscope (EM30+, COXEM Co., Ltd., Daejeon, Republic of Korea) at an operating voltage of 8 kV with a magnification of 2000 times [27]. After converting the raw SEM images into 8-bit grayscale images using ImageJ software (1.54f, National Institutes of Health, Bethesda, MD, USA), the "Threshold" tool (Otsu automatic thresholding) was used to distinguish between pore and matrix areas. By manually fine-tuning the threshold range (gray scale values 0–50 for pores and 51–255 for the matrix), it was ensured that the binarized image accurately reflected the actual pore structure [57]. Three porosity calculations were carried out as shown in Equation (5):

$$\text{Porosity } (\%) = (\text{Pore area}/\text{Total image area}) \times 100\% \quad (5)$$

4.9. Chemical Interactions

Chemical interactions in the blended gels were analyzed by examining the differential solubility of chemical bonds in four different solutions: 0.05 M NaCl (SA), 0.6 M NaCl (SB), 0.6 M NaCl + 1.5 M urea (SC), and 0.6 M NaCl + 8 M urea (SD) [58]. In each experimental run, 10 mL of the designated solution was introduced to 2 g of finely minced gel samples, which were then homogenized for one minute. Subsequently, the mixture was then stirred gently at a steady 4 °C for 2 h before being centrifuged at 8000× *g* for 20 min. Using a detergent-compatible Bradford protein concentration assay kit (Beyotime Biotechnology Co., Ltd., Shanghai, China) to measure the protein content in the solution, absorbance at 595 nm^{-1} was measured with a microplate reader (Bio Tek Instruments Inc., Burlington, VT, USA), while protein concentrations were determined based on the formula provided with the kit. The quantification of ionic interactions (SB–SA), hydrogen bonding (SC–SB), and hydrophobic forces (SD–SC) was facilitated by the observed variations in protein concentration. Each measurement was replicated three times to ensure accuracy and consistency.

4.10. Total Sulfhydryl Content

An amount of 1 g of the blended gel sample was taken and mixed with 10 mL of the extraction solution and centrifuged at 8000× *g* for 10 min. The total sulfhydryl content in the supernatant was quantified using the sulfhydryl detection kit (Beijing Solarbio Science & Technology Co., Ltd., Beijing, China), adhering to the manufacturer's guidelines. Sulfhydryl groups interact with DTNB to produce a yellow derivative that exhibits a prominent absorption peak at 412 nm^{-1}. Absorbance at this wavelength was recorded using a microplate reader (Bio Tek Instruments Inc., Burlington, VT, USA), and the total sulfhydryl content was precisely calculated with the formula provided in the kit.

4.11. FTIR

Samples were sectioned into small fragments, freeze-dried under vacuum conditions, and then finely pulverized to minimize moisture interference during analysis. Spectral analysis was conducted using a Nicolet 6700 Fourier-transform infrared spectrometer (Thermo Fisher Scientific Inc., Waltham, MA, USA) [19]. Spectra were acquired from 500 to 4000 cm^{-1}, with each measurement taken at a 4 cm^{-1} resolution. Air served as the background reference for calibration before experiments. The ATR crystal was cleaned with anhydrous ethanol before and after each measurement. Data accuracy was enhanced by averaging 32 scans. Using PeakFit 4.12 software, the Amide I band spectra (1600–1700 cm^{-1}) were analyzed to ascertain the relative content of the secondary structure by deconvolution, Gaussian curve fitting, and second derivative calculations, represented as a percentage of the peak area.

4.12. Statistical Analysis

Data analysis was performed using SPSS 19.0 software. Data from blended gel samples with different contents of crayfish muscle were analyzed using one-way ANOVA. Data from two groups (without TGase and with TGase) of blended gel samples with the same contents of crayfish muscle were analyzed by independent *t*-tests. The results of data analysis were plotted graphically using Origin 2021 software.

Author Contributions: Conceptualization, H.W. (Hongyi Wang); methodology, H.W. (Hongyi Wang) and Q.L.; software, M.Y.; formal analysis, H.W. (Hong Wang); investigation, H.W. (Hongyi Wang) and Q.L.; writing—original draft preparation, H.W. (Hongyi Wang); writing—review and editing, Q.L. and J.L.; visualization: H.W. (Hongyi Wang) and M.W.; supervision, J.L.; project administration, J.L. and L.L.; funding acquisition, J.L. All authors have read and agreed to the published version of the manuscript.

Funding: This research was funded by the China Agriculture Research System of MOF and MARA (CARS-48), the Anhui Agriculture Research System (AARS-08), and the National Key Research and Development Program of China (2023YFD2401500).

Institutional Review Board Statement: Not applicable.

Informed Consent Statement: Not applicable.

Data Availability Statement: The original contributions presented in the study are included in the article, further inquiries can be directed to the corresponding author.

Conflicts of Interest: The authors declare no conflicts of interest.

References

1. Harlioğlu, M.M.; Harlioğlu, A.G. Threat of Non-Native Crayfish Introductions into Turkey: Global Lessons. *Rev. Fish Biol. Fish.* **2006**, *16*, 171–181. [CrossRef]
2. Bureau of Fisheries Management, Chinese Ministry of Agriculture and Rural Affairs. *China Fishery Statistics Yearbook 2024*; China Agriculture Press: Beijing, China, 2024; pp. 24, 89. ISBN 978-7-109-32126-7.
3. Adegbusi, H.S.; Ismail, A.; Mohd. Esa, N.; Daud, Z.A.M. Effects of Formulated Nigerian Yellow Maize, Soybean, and Crayfish Blends on Some Growth Performance and Physiological Status. *Food Prod. Process. Nutr.* **2023**, *5*, 14. [CrossRef]
4. Zhao, Y.; Wei, K.; Chen, J.; Wei, G.; Li, J.; Zheng, B.; Song, Y.; Gao, P.; Zhou, R. Enhancement of Myofibrillar Protein Gelation by Plant Proteins for Improved Surimi Gel Characteristics: Mechanisms and Performance. *LWT* **2024**, *198*, 116045. [CrossRef]
5. Lee, Y.E.J.; Goh, H.M.; Huang, D.-G. Pea Protein Soften Surimi Gels and Improve In Vitro Digestibility: Potential High Protein Foods for Elderly. *Food Hydrocoll.* **2024**, *159*, 110664. [CrossRef]
6. Liang, F.; Lin, L.; He, T.; Zhou, X.; Jiang, S.; Lu, J. Effect of Transglutaminase on Gel Properties of Surimi and Precocious Chinese Mitten Crab (*Eriocheir sinensis*) Meat. *Food Hydrocoll.* **2020**, *98*, 105261. [CrossRef]
7. Li, Q.; Wang, H.; Lu, J.; Yang, M.; Zhang, M.; Lin, L.; Wang, C. Microwave-Mediated Fibrous Blended Gels of Crab Meat and Surimi: Fibrous Protein Production Method in the Perspective of Structural Remodeling. *Innov. Food Sci. Emerg. Technol.* **2024**, *100*, 103910. [CrossRef]
8. Jiang, Q.; Wang, L.; Gao, P.; Yu, P.; Yang, F.; Yu, D.; Chen, H.; Xia, W. Study on the Effect and Mechanism of Chicken Breast on the Gel Properties of Silver Carp (*Hypophthalmichtys molitrix*) Surimi. *J. Sci. Food Agric.* **2023**, *104*, 1132–1142. [CrossRef]
9. Niu, F.; Li, X.; Lin, C.; Hu, X.; Zhang, B.; Pan, W. The Mechanism of Egg White Protein to Enhance the Thermal Gel Properties of Giant Squid (*Dosidicus gigas*) Surimi. *Food Chem.* **2024**, *469*, 142601. [CrossRef] [PubMed]
10. Rawdkuen, S.; Benjakul, S.; Visessanguan, W.; Lanier, T.C. Chicken Plasma Protein Affects Gelation of Surimi from Bigeye Snapper (*Priacanthus tayenus*). *Food Hydrocoll.* **2004**, *18*, 259–270. [CrossRef]
11. Zhang, L.; Zhang, F.; Wang, X. Effects of Hydrolyzed Wheat Gluten on the Properties of High-Temperature (≥100 °C) Treated Surimi Gels. *Food Hydrocoll.* **2015**, *45*, 196–202. [CrossRef]
12. Shafiee, S.; Goli, M.; Khoshkhoo, Z.; Hosseini, S.E. Optimization of Hydrolysis Conditions (Temperature, Time, and Concentration of Alkalase) of Rainbow Trout Viscera Using the Response Surface Methodology. *J. Food Process. Preserv.* **2021**, *45*, e15456. [CrossRef]
13. Li, Q.; Ye, T.; Zhu, Y.; Xia, L.; Lin, L.; Lu, J. Sustainable Development of Fishery Resources: Precipitation of Protein from Surimi Rinsing Wastewater by Low-Temperature Plasma. *Food Chem.* **2024**, *463 Pt 2*, 141286. [CrossRef]

14. Luan, D.; Wang, C.; Li, S.; Xue, Q.; Wang, X.; Xue, C.; Wang, Y. Improving Gel Properties of Low-Salt Silver Carp Surimi through Single-Mode Microwave-Assisted Processing. *Innov. Food Sci. Emerg. Technol.* **2024**, *97*, 103841. [CrossRef]
15. Zhang, C.; Lu, M.; Ai, C.; Cao, H.; Xiao, J.; Imran, M.; Chen, L.; Teng, H. Ultrasonic Treatment Combined with Curdlan Improves the Gelation Properties of Low-Salt *Nemipterus Virgatus* Surimi. *Int. J. Biol. Macromol.* **2023**, *248*, 125899. [CrossRef]
16. Wang, L.; Wang, X.; Ma, J.; Yang, K.; Feng, X.; You, X.; Wang, S.; Zhang, Y.; Xiong, G.; Wang, L.; et al. Effects of Radio Frequency Heating on Water Distribution and Structural Properties of Grass Carp Myofibrillar Protein Gel. *Food Chem.* **2020**, *343*, 128557. [CrossRef]
17. Tadpitchayangkoon, P.; Park, J.W.; Yongsawatdigul, J. Gelation Characteristics of Tropical Surimi under Water Bath and Ohmic Heating. *LWT* **2012**, *46*, 97–103. [CrossRef]
18. Piao, X.; Li, J.; Zhao, Y.; Guo, L.; Zheng, B.; Zhou, R.; Ostrikov, K. Oxidized Cellulose Nanofibrils-Based Surimi Gel Enhancing Additives: Interactions, Performance and Mechanisms. *Food Hydrocoll.* **2022**, *133*, 107893. [CrossRef]
19. Zhang, C.; Chen, L.; Teng, H. Phase Behavior of the Gelation Process of Myofibrillar Protein-Curdlan Blended System: Discussion Based on Rheology and Gel Properties. *Food Chem.* **2023**, *437*, 137839. [CrossRef]
20. Zhu, S.; Wang, Y.; Ding, Y.; Xiang, X.; Yang, Q.; Wei, Z.; Song, H.; Liu, S.; Zhou, X. Improved Texture Properties and Toughening Mechanisms of Surimi Gels by Double Network Strategies. *Food Hydrocoll.* **2024**, *152*, 109900. [CrossRef]
21. Shi, T.; Yuan, L.; Kong, Y.; Bao, Y.; Zhang, H.; Lu, C.; Jin, W.; Monto, A.R.; Gao, R. Towards Higher-Quality Low Salt Surimi Gels: Significance of the Combinatorial Effects of Chickpea Protein with Transglutaminase on Their Micro-Structures. *LWT* **2024**, *199*, 116103. [CrossRef]
22. Yu, N.; Yang, F.; Gong, H.; Zhou, J.; Jie, C.; Wang, W.; Chen, X.; Sun, L. Gel & Three-Dimensional Printing Properties of Sheep Plasma Protein-Surimi Induced by Transglutaminase. *J. Food Eng.* **2022**, *323*, 111006. [CrossRef]
23. Li, Q.; Yi, S.; Wang, W.; Xu, Y.; Mi, H.; Li, X.; Li, J. Different Thermal Treatment Methods and TGase Addition Affect Gel Quality and Flavour Characteristics of *Decapterus maruadsi* Surimi Products. *Foods* **2021**, *11*, 66. [CrossRef]
24. Martínez, M.A.; Robledo, V.; Velazquez, G.; Ramírez, J.A.; Vázquez, M.; Uresti, R.M. Effect of Precooking Temperature and Microbial Transglutaminase on the Gelling Properties of Blue Crab (*Callinectes sapidus*) Proteins. *Food Hydrocoll.* **2014**, *35*, 264–269. [CrossRef]
25. Huang, P.-H.; Cheng, Y.-T.; Hsieh, H.-C.; Ko, W.-C.; Lu, W.-C.; Li, P.-H. Effects of Transglutaminase on the Physicochemical Properties of Surimi and Kamaboko Prepared by Thermal and Low Level-Pressure Treatments. *LWT* **2023**, *183*, 114863. [CrossRef]
26. Hu, Y.; Shao, Y.; Wu, C.; Yuan, C.; Ishimura, G.; Liu, W.; Chen, S. γ-PGA and MTGase Improve the Formation of ε-(γ-Glutamyl) Lysine Cross-Links within Hairtail (*Trichiurus haumela*) Surimi Protein. *Food Chem.* **2018**, *242*, 330–337. [CrossRef]
27. Zhang, C.; Chen, L.; Lu, M.; Ai, C.; Cao, H.; Xiao, J.; Zhong, S.; Teng, H. Effect of Cellulose on Gel Properties of Heat-Induced Low-Salt Surimi Gels: Physicochemical Characteristics, Water Distribution and Microstructure. *Food Chem. X* **2023**, *19*, 100820. [CrossRef]
28. Liu, Q.; Zhang, N.; Wei, W.; Hu, X.; Tan, Y.; Yu, Y.; Deng, Y.; Bi, C.; Zhang, L.; Zhang, H. Assessing the Dynamic Extrusion-Based 3D Printing Process for Power-Law Fluid Using Numerical Simulation. *J. Food Eng.* **2020**, *275*, 109861. [CrossRef]
29. Terwilliger, N.B.; Ryan, M.C.; Towle, D. Evolution of Novel Functions: Cryptocyanin Helps Build New Exoskeleton in *Cancer magister*. *J. Exp. Biol.* **2005**, *208*, 2467–2474. [CrossRef]
30. Zhao, S.; Liu, Y.; Yuan, X.; Zhao, Y.; Kang, Z.; Zhu, M.; Ma, H. Effect of Low-Frequency Alternating Magnetic Field on the Rheological Properties, Water Distribution and Microstructure of Low-Salt Pork Batters. *LWT* **2022**, *159*, 113164. [CrossRef]
31. Li, J.; Munir, S.; Yu, X.; Yin, T.; You, J.; Liu, R.; Xiong, S.; Hu, Y. Double-Crosslinked Effect of TGase and EGCG on Myofibrillar Proteins Gel Based on Physicochemical Properties and Molecular Docking. *Food Chem.* **2021**, *345*, 128655. [CrossRef]
32. Mi, H.; Zhao, Y.; Li, Y.; Chen, J.; Liu, H.; Yi, S.; Li, X.; Li, J. Combining Effect of Soybean Protein Isolate and Transglutaminase on the Gel Properties of Zhikong Scallop (*Chlamys farreri*) Adductor Muscle. *LWT* **2021**, *138*, 110727. [CrossRef]
33. Luo, H.; Guo, C.; Lin, L.; Si, Y.; Gao, X.; Xu, D.; Jia, R.; Yang, W. Combined Use of Rheology, LF-NMR, and MRI for Characterizing the Gel Properties of Hairtail Surimi with Potato Starch. *Food Bioprocess Technol.* **2020**, *13*, 637–647. [CrossRef]
34. Zhang, T.; Wang, J.; Feng, J.; Liu, Y.; Suo, R.; Jin, J.; Wang, W. Ultrasonic Pretreatment Improves the Gelation Properties of Low-Salt Penaeus Vannamei (*Litopenaeus vannamei*) Surimi. *Ultrason. Sonochem.* **2022**, *86*, 106031. [CrossRef]
35. Du, J.; Zhou, C.; Xia, Q.; Wang, Y.; Geng, F.; He, J.; Sun, Y.; Pan, D.; Cao, J. The Effect of Fibrin on Rheological Behavior, Gelling Properties and Microstructure of Myofibrillar Proteins. *LWT* **2022**, *153*, 112457. [CrossRef]
36. Jiang, Q.; Chen, N.; Gao, P.; Yu, D.; Yang, F.; Xu, Y.; Xia, W. Influence of L-Arginine Addition on the Gel Properties of Reduced-Salt White Leg Shrimp (*Litopenaeus vannamei*) Surimi Gel Treated with Microbial Transglutaminase. *LWT* **2023**, *173*, 114310. [CrossRef]
37. Jia, R.; Jiang, Q.; Kanda, M.; Tokiwa, J.; Nakazawa, N.; Osako, K.; Okazaki, E. Effects of Heating Processes on Changes in Ice Crystal Formation, Water Holding Capacity, and Physical Properties of Surimi Gels during Frozen Storage. *Food Hydrocoll.* **2019**, *90*, 254–265. [CrossRef]

38. Kang, S.; Shao, Y.; Li, Z.; Chang, W.; Song, J.; Hu, Y.; Li, S.; Luan, G. Mechanism of Transglutaminase on Film-Forming and Structural Properties of Soybean Protein and Its Fractions: A Study in Different pH Environments. *Food Hydrocoll.* **2024**, *157*, 110394. [CrossRef]
39. Zhang, L.; Li, Q.; Shi, J.; Zhu, B.; Luo, Y. Changes in Chemical Interactions and Gel Properties of Heat-induced Surimi Gels from Silver Carp (*Hypophthalmichthys molitrix*) Fillets During Setting and Heating: Effects of Different Washing Solutions. *Food Hydrocoll.* **2018**, *75*, 116–124. [CrossRef]
40. Liu, Z.; Li, X.; Guan, Z.; Jia, Z.; Zhang, Z.; Yang, C.; Wang, J. Transglutaminase-Crosslinked Cold-Set Pea Protein Isolate Gels Modified by pH Shifting: Properties, Structure and Formation Mechanisms. *Food Hydrocoll.* **2024**, *154*, 110158. [CrossRef]
41. Zhao, Y.; Han, X.; Hu, N.; Zhao, C.; Wu, Y.; Liu, J. Study on Properties of TGase-Induced Pea Protein–Zein Complex Gels. *J. Food Eng.* **2023**, *354*, 111578. [CrossRef]
42. Zhao, Q.; Hu, X.; Guo, K.; Li, S.; Li, T. Effects of TGase on the Rheological Behaviors, Structural Properties and Molecular Forces of Cowpea Protein Isolate and Cowpea Albumin Gels. *Int. J. Biol. Macromol.* **2024**, *291*, 139154. [CrossRef]
43. Shuai, X.; Gao, L.; Geng, Q.; Li, T.; He, X.; Chen, J.; Liu, C.; Dai, T. Effects of Moderate Enzymatic Hydrolysis on Structure and Functional Properties of Pea Protein. *Foods* **2022**, *11*, 2368. [CrossRef]
44. Li, Q.; Zhang, X.; Tang, S.; Mi, S.; Lu, L.; Zeng, Q.-Y.; Xia, M.; Cai, Z. Improved Effect of Ultrasound-Assisted Enzymolysis on Egg Yolk Powder: Structural Properties, Hydration Properties and Stability Characteristics. *Food Chem.* **2022**, *382*, 132549. [CrossRef]
45. Li, J.; Munir, S.; Yu, X.; Yin, T.; You, J.; Liu, R.; Xiong, S.; Hu, Y. Interaction of Myofibrillar Proteins and Epigallocatechin Gallate in the Presence of Transglutaminase in Solutions. *Food Funct.* **2020**, *11*, 9560–9572. [CrossRef]
46. Priyadarshini, B.; Xavier, K.A.M.; Nayak, B.B.; Dhanapal, K.; Balange, A.K. Instrumental Quality Attributes of Single Washed Surimi Gels of Tilapia: Effect of Different Washing Media. *LWT* **2017**, *86*, 385–392. [CrossRef]
47. Ren, Z.; Li, Z.; Chen, Z.; Zhang, Y.; Lin, X.; Weng, W.; Yang, H.; Li, B. Characteristics and Application of Fish Oil-in-Water Pickering Emulsions Structured with Tea Water-Insoluble Proteins/κ-Carrageenan Complexes. *Food Hydrocoll.* **2021**, *114*, 106562. [CrossRef]
48. Wei, W.; Hu, W.; Zhang, X.-Y.; Zhang, F.-P.; Sun, S.-Q.; Liu, Y.; Xu, C.-H. Analysis of Protein Structure Changes and Quality Regulation of Surimi during Gelation Based on Infrared Spectroscopy and Microscopic Imaging. *Sci. Rep.* **2018**, *8*, 5566. [CrossRef]
49. Fang, Q.; Shi, L.; Ren, Z.; Hao, G.; Chen, J.; Weng, W. Effects of Emulsified Lard and TGase on Gel Properties of Threadfin Bream (*Nemipterus virgatus*) Surimi. *LWT* **2021**, *146*, 111513. [CrossRef]
50. Oujifard, A.; Benjakul, S.; Prodpran, T.; Seyfabadi, J. Properties of Red Tilapia (*Oreochromis niloticus*) Protein Based Film as Affected by Cryoprotectants. *Food Hydrocoll.* **2013**, *32*, 245–251. [CrossRef]
51. Lu, Y.; Zhu, Y.; Ye, T.; Nie, Y.; Jiang, S.; Lin, L.; Lu, J. Physicochemical Properties and Microstructure of Composite Surimi Gels: The Effects of Ultrasonic Treatment and Olive Oil Concentration. *Ultrason. Sonochem.* **2022**, *88*, 106065. [CrossRef]
52. Dong, M.; Sun, Y.; Xiong, D.-M.; Song, Q.; Jia, J.; Liu, X.; Sheng, L.; Duan, X. Comparison of the Effects of pH-Shifting, Acetic Acid Modification, and TGase Treatment on the Physicochemical and Functional Properties of Wheat Gluten Protein. *Food Bioprocess Technol.* **2023**, *17*, 245–256. [CrossRef]
53. Yi, S.; Li, Q.; Qiao, C.; Zhang, C.; Wang, W.; Xu, Y.; Mi, H.; Li, X.; Li, J. Myofibrillar Protein Conformation Enhance Gel Properties of Mixed Surimi Gels with *Nemipterus virgatus* and *Hypophthalmichthys molitrix*. *Food Hydrocoll.* **2020**, *106*, 105924. [CrossRef]
54. Fan, D.; Huang, L.; Li, B.; Huang, J.; Zhao, J.; Yan, B.; Zhou, W.; Zhang, W.; Zhang, H. Acoustic Intensity in Ultrasound Field and Ultrasound-Assisted Gelling of Surimi. *LWT* **2017**, *75*, 497–504. [CrossRef]
55. Herrero, A.M.; Cambero, M.I.; Ordóñez, J.A.; De La Hoz, L.; Carmona, P. Raman Spectroscopy Study of the Structural Effect of Microbial Transglutaminase on Meat Systems and Its Relationship with Textural Characteristics. *Food Chem.* **2008**, *109*, 25–32. [CrossRef]
56. Gao, X.; Xie, Y.; Yin, T.; Hu, Y.; You, J.; Xiong, S.; Liu, R. Effect of High Intensity Ultrasound on Gelation Properties of Silver Carp Surimi with Different Salt Contents. *Ultrason. Sonochem.* **2021**, *70*, 105326. [CrossRef]
57. Ma, Y.; Shan, A.; Wang, R.; Zhao, Y.; Chi, Y. Characterization of Egg White Powder Gel Structure and Its Relationship with Gel Properties Influenced by Pretreatment with Dry Heat. *Food Hydrocoll.* **2021**, *110*, 106149. [CrossRef]
58. Yan, B.; Jiao, X.; Zhu, H.; Wang, Q.; Huang, J.; Zhao, J.; Cao, H.; Zhou, W.; Zhang, W.; Ye, W.; et al. Chemical Interactions Involved In Microwave Heat-induced Surimi Gel Fortified with Fish Oil and Its Formation Mechanism. *Food Hydrocoll.* **2020**, *105*, 105779. [CrossRef]

gels

MDPI

Article

A Mathematical Model of Myosin Heavy Chain Dynamics in the Disintegration of Golden Threadfin Bream *Nemipterus virgatus* Surimi Gel

Ryoko Nakamizo [1,2,†], Tatsuya Hayashi [3,4,†], Yuri Kominami [1,*] and Hideki Ushio [1]

1 Graduate School of Agricultural and Life Sciences, The University of Tokyo, Bunkyo-ku, Tokyo 113-8657, Japan; hushio@g.ecc.u-tokyo.ac.jp (H.U.)
2 Fish Protein Laboratory, Suzuhiro Kamaboko Honten Co., Ltd., Odawara 250-0862, Japan
3 Faculty of Science and Engineering, Yamato University, Suita-shi 564-0082, Japan; hayashi.tatsuya@yamato-u.ac.jp
4 Research and Development Initiative, Chuo University, Bunkyo-ku, Tokyo 112-8551, Japan
* Correspondence: komi.yuri@mail.u-tokyo.ac.jp; Tel.: +81-3-5841-5302
† These authors contributed equally to this work.

Abstract: Surimi gel, a type of hydrocolloidal food, is formed through the gelation of fish meat proteins. Myosin heavy chain (MHC), a key myofibrillar protein, plays a crucial role in the formation of the gel network via both transglutaminase (TGase)-catalyzed and non-enzymatic polymerization. Gel disintegration in surimi is primarily attributed to the proteolytic degradation of MHC. This study focused on golden threadfin bream *Nemipterus virgatus*, a species characterized by low TGase activity and high protease activity at elevated temperatures. We investigated the competition between non-enzymatic polymerization and proteolytic degradation of MHC and their effects on gel mechanical properties using a mathematical model. A mathematical model based on kinetic reactions accurately reflected the changes in MHC observed through SDS-PAGE analysis during *N. virgatus* gel disintegration. Our results indicate that not only unpolymerized but also polymerized MHC was significantly degraded, which substantially contributed to the reduction in the mechanical properties of the *N. virgatus* surimi. Mathematically understanding the dynamics of MHC in surimi during heating helps promote the utilization of noncommercial fish species for surimi processing by enabling better control over surimi gel properties.

Keywords: surimi gel; *Nemipterus virgatus*; gel disintegration; protein gel; mathematical model

1. Introduction

The global demand for fish and fish products has been increasing in parallel with population growth and economic development. To reduce the fishing pressure on commercially important species, the effective utilization of noncommercial or underutilized species is becoming increasingly important. Surimi production represents a practical strategy for utilizing these resources, as surimi can be further processed into a variety of gelled food products, such as fish balls and crab-flavored kamaboko. These surimi-based products are categorized as hydrocolloid gels. During surimi gel production, fish meat is homogenized and its proteins are solubilized under high ionic strength conditions through the addition of salt, followed by gelation mediated by catalytic cross-linking and thermal aggregation of proteins. In surimi gels, myosin heavy chain (MHC), a myofibrillar protein with a

Gels **2025**, *11*, 348 https://doi.org/10.3390/gels11050348

molecular weight of approximately 220 kDa, plays a pivotal role in the formation of the gel network [1].

Species-specific differences in the thermal behavior of surimi gel properties complicate surimi processing, particularly due to variations in the temperature at which gelation or disintegration occurs [2,3]. Transglutaminase (TGase) is known as the enzyme that catalyzes the polymerization reaction between MHC molecules through the formation of ε-(γ-Glu)-Lys isopeptide bonds [4–7]. TGase are widely expressed in the muscle tissues of various fish species, but their activity or temperature dependency is highly species-specific [8]. Conversely, proteases expressed in fish muscle tissues interfere with surimi gelation, and their enzymatic activity and temperature dependency are also highly species-specific [2,3,5,7]. The disintegration of surimi gel due to MHC degradation by proteases can be explained by the following mechanisms: degradation of MHCs proceeds prior to polymerization, known as *modori*, and degradation of polymerized MHCs, referred to as *himodori* [5,9].

To maximize the cross-linking of MHCs catalyzed by TGase and to minimize their proteolytic degradation, a combination of different heating temperatures—known as two-step heating—is commonly employed in the surimi food industry [10]. However, the prediction of surimi gel properties under varying two-step heating conditions remains challenging, as no kinetic model has yet been proposed to explain the dynamics of MHC and the associated mechanical properties during surimi gelation and disintegration. A kinetic model that explains the dynamics of MHC, including the competition among TGase-catalyzed polymerization, non-enzymatic polymerization, and proteolytic degradation, is necessary for a comprehensive understanding of the mechanisms underlying surimi gelation and disintegration. First, it is necessary to develop individual kinetic models that describe *modori* or *himodori* and TGase-catalyzed MHC polymerization, respectively. These models should then be integrated to construct a comprehensive model capable of describing the competitive dynamics among these processes. However, few studies have elucidated the detailed relationship between MHC polymerization/degradation during *himodori* and the mechanical properties of surimi gels, and sufficient empirical data for model construction are still lacking. Therefore, the present study aimed to construct a fundamental model describing the competition between non-enzymatic polymerization and proteolytic degradation of MHC. We focused on golden threadfin bream *Nemipterus virgatus*, which exhibits low TGase activity and high protease activity at elevated temperatures (>50 °C) [3,8,11], and developed a model of MHC dynamics during surimi disintegration caused by *himodori*.

2. Results

2.1. *Effects of the Heating Temperature on the Properties of N. virgatus Surimi Gel*

Figure 1a illustrates the effect of the initial heating temperature on the fracture strength of *N. virgatus* surimi gel. Among the samples, the gel prepared by two-step heating 60–85 exhibited the lowest fracture strength, whereas the gel prepared by two-step heating 50–85 showed the highest.

The separation patterns of the WSP and the reduced USP from the *N. virgatus* surimi gels prepared via two-step heating 50–85, 55–85, and 60–85, as well as single-step heating 85, are shown in Figure 1b,c. In the surimi gel prepared by two-step heating 60–85, the WSP bands appeared smeared, and the band intensity of MHC in the reduced USP was clearly lowest among all samples.

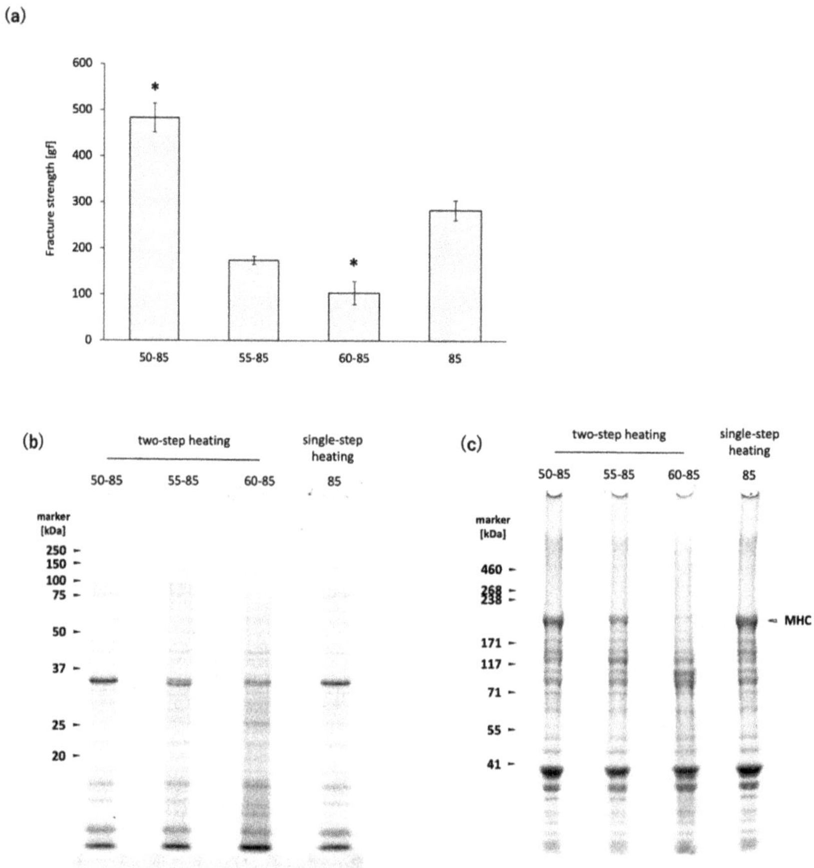

Figure 1. Effects of heating temperature on the properties of N. virgatus surimi gel: (**a**) fracture strength of the surimi gels; (**b**) SDS-PAGE profiles of the WSP; (**c**) SDS-PAGE profiles of the reduced USP. In (**a**), error bars represent the standard error of the mean (SEM) (n = 4). * $p < 0.001$ (Dunnett's test, vs. single-step heating 85).

2.2. Time Course Analysis of *N. virgatus* Surimi Gel Disintegration

Figure 2a shows the temperature curve near the center of the *N. virgatus* surimi during heating at 60 °C. The core temperature of the 5.5 mL surimi-filled jars reached 60 °C at approximately 16 min. Figure 2b shows the changes in maximum compression strength in relation to the duration of heating at 60 °C. The maximum compression strength rapidly increased with heating duration up to 16 min, after which it gradually declined until 60 min.

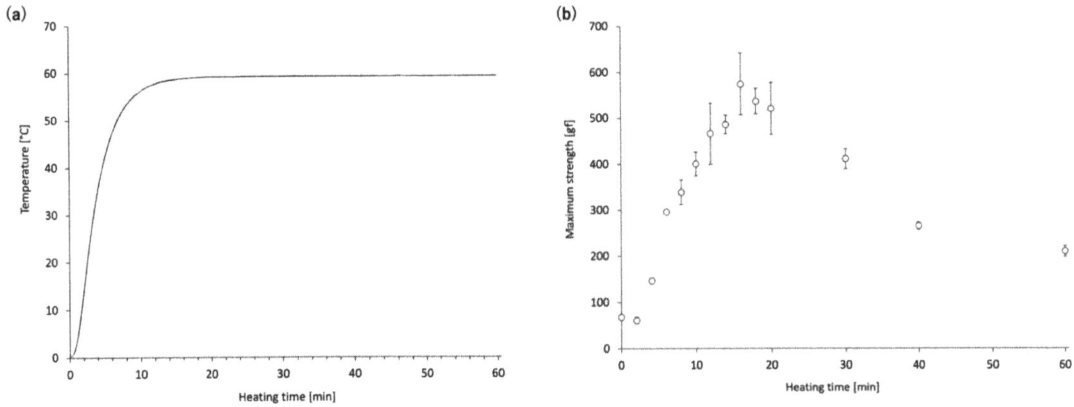

Figure 2. Changes during heating at 60 °C: (**a**) core temperature profile of the surimi sample; (**b**) maximum compression strength as a function of heating duration. Error bars represent the SEM ($n = 3$).

2.3. Time Course Analysis of Protein Changes in *N. virgatus* Surimi During 60 °C Heating

Figure 3a,b,d show the changes in the WSP, reduced USP, and non-reduced USP profiles of the *N. virgatus* surimi during heating at 60 °C. Figure 3c,e illustrate the quantified MHC band intensities in the reduced and non-reduced USP, respectively. In the WSP, both the number and intensity of protein bands decreased with increasing heating time (Figure 3a). Smeared bands were observed in the WSP from the surimi samples heated for 60 min. In the reduced USP, the intensity of MHC (227 kDa) significantly decreased after 20 min of heating, while bands at 38, 51, 83, 126, 144, 157, 173, and 203 kDa increased (Figure 3b,c). However, after 60 min of heating, the 173 and 203 kDa bands showed a marked decrease. In the non-reduced USP, the intensity of MHC decreased after just 2 min of heating (Figure 3e), and the region above 50 kDa became markedly smeared after 30 min of heating (Figure 3d).

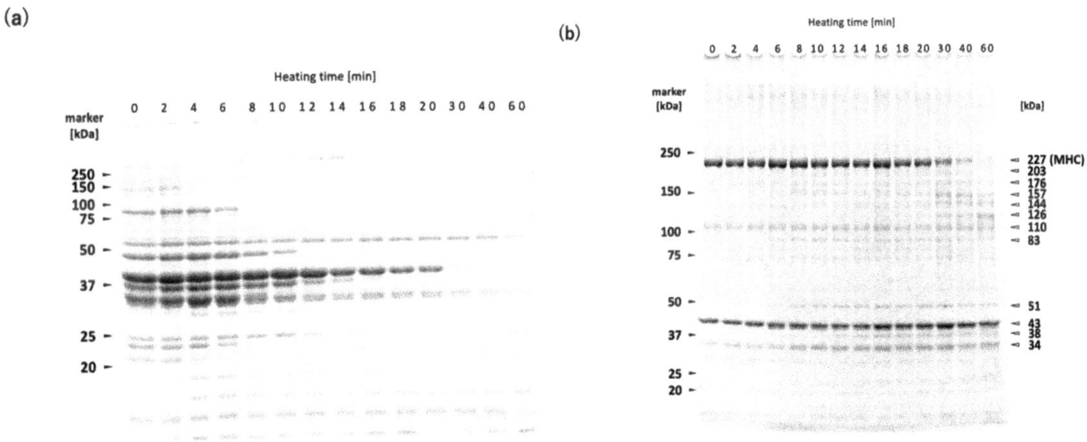

Figure 3. *Cont.*

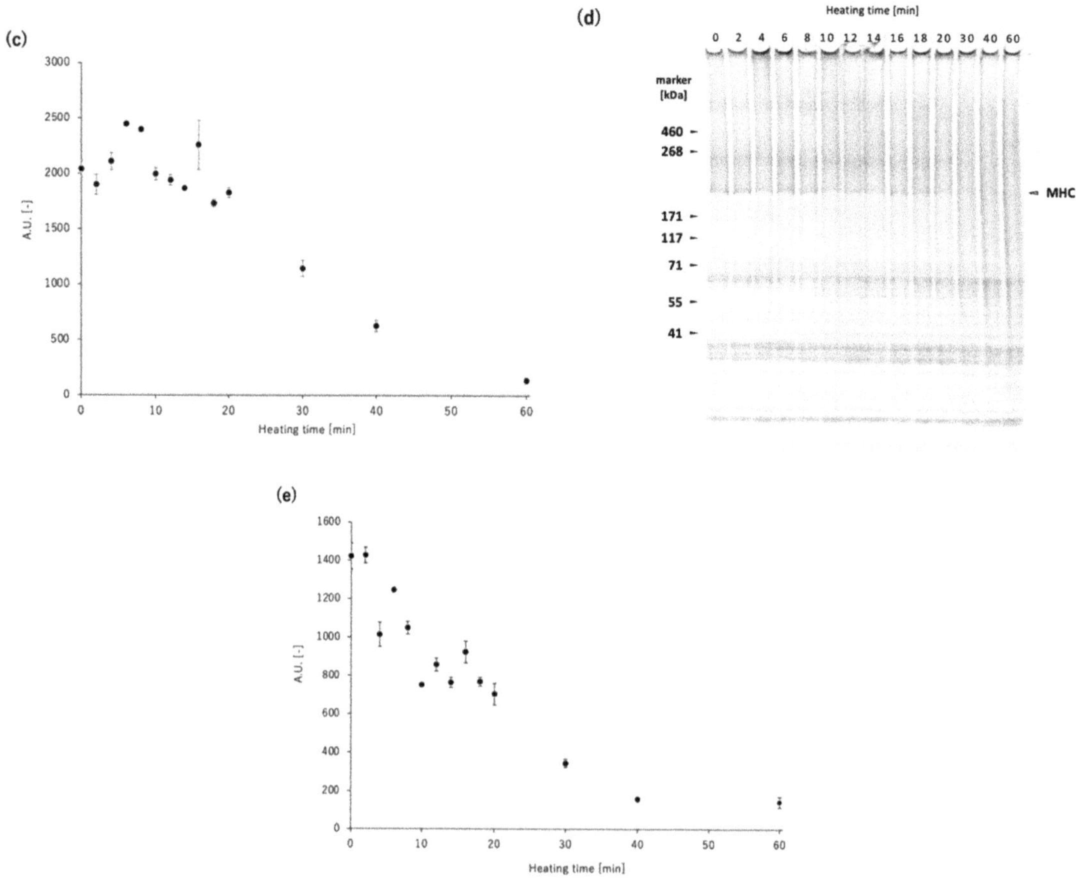

Figure 3. Changes in protein profiles of *N. virgatus* surimi during heating at 60 °C: (**a**) SDS-PAGE profile of the WSP; (**b**) SDS-PAGE profile of the reduced USP; (**c**) MHC band intensities in the reduced USP; (**d**) SDS-PAGE profile of the non-reduced USP; (**e**) MHC band intensities in the non-reduced USP. In (**c**,**e**), error bars represent the standard error of the mean (SEM, $n = 3$), and A.U. denotes arbitrary units (dimensionless).

2.4. Mathematical Modeling of MHC Non-Enzymatic Polymerization and Proteolytic Degradation

Figure 4 compares the calculated amounts of MHC with the measured MHC intensities. The MHC band intensities were normalized by setting the initial value to 1 for both the reducing and the non-reducing USP conditions. The calculated changes in MHC levels matched the measured values in both the reduced and non-reduced USP. The decreases in MHC during heating at 60 °C exhibited a delay of approximately 16–18 min in the reduced USP, whereas in the non-reduced USP, the decrease began as early as 2 min.

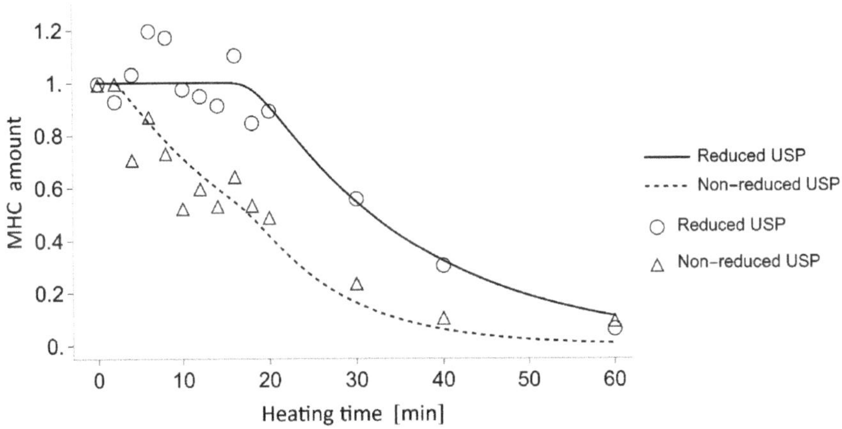

Figure 4. Comparison between the estimated amounts of MHC derived from the mathematical model and the measured MHC band intensities obtained from SDS-PAGE analysis in *N. virgatus* surimi during heating at 60 °C. Solid and dashed lines represent the modeled MHC amounts in the reduced and non-reduced USP, respectively. Circles and triangles indicate the normalized MHC band intensities measured in the reduced and non-reduced USP, respectively.

3. Discussion

The present study investigated the competition between the non-enzymatic polymerization and proteolytic degradation of MHC, and their effects on the gelation of golden threadfin bream *N. virgatus* surimi. *N. virgatus* is one of the important fish species used as a raw material for surimi in Asia. Consequently, insights into the *himodori* phenomenon of *N. virgatus* surimi have been accumulated over a long period [3,12–16], and research aimed at improving its gel properties continues to be actively conducted [11,17–21]. First, we confirmed that proteases in *N. virgatus* surimi are activated by heating at 60 °C, which leads to gel disintegration (Figure 1a) [3,13]. The protease responsible for gel disintegration at 60 °C was purified and identified as a sarcoplasmic serine proteinase with a molecular weight of 77 kDa [13]. A sarcoplasmic serine proteinase with a molecular weight of 540 kDa, which has an optimum temperature of 50 °C, was also purified from *N. virgatus* meat [14]. Kinoshita et al. (1990) [3] referred to the protease activated at 60 °C as Sp-60-MIP, and the one activated at 50 °C as Sp-50-MIP. They reported that the *N. virgatus* meat used for the purification of Sp-50-MIP exhibited gel disintegration when heated at 50 °C, but not at 60 °C [14]. In the present study, the *N. virgatus* surimi exhibited gel disintegration when heated at 60 °C, but not at 50 °C (Figure 1a). The results of SDS-PAGE indicate that MHC degradation was clearly promoted at 60 °C, whereas heating at 50 °C did not induce noticeable degradation (Figure 1c). Seasonal variations and freeze-thaw effects are possible factors contributing to this reversal [22,23]; however, further investigation is needed in future studies.

TGase in white croaker *Pennahia argentata*, a species commonly used as surimi ingredients in Japan due to its strong gelation properties, has been reported to show higher peak activity around 30 °C and little to no activity at 50 °C [5,8]. In contrast, TGase in *N. virgatus* meat exhibits relatively lower activity but operates over a broader temperature range, with an optimum around 40 °C [8]. During two-step heating 50, the slower rate of temperature increase compared to single-step heating 85 results in a longer residence time within the temperature range conductive to TGase activity in the *N. virgatus* surimi. Although the *N. virgatus* surimi gel prepared by two-step heating 50 exhibited higher fracture strength than that prepared by single-step heating 85 (Figure 1a), the MHC band intensities remained at comparable levels (Figure 1c). Since TGase-catalyzed MHC polymerization typically results in a decrease in MHC band intensity, as observed by SDS-PAGE, the observed difference in fracture strength between two-step heating 50 and single-step heating 85 cannot be attributed to TGase activity. These results confirmed the negligible activity of TGase during heating at 50–85 °C in the *N. virgatus* surimi used in the present study. The thermal denaturation point of myosin is approximately 40–50 °C in many fish species [24]. During two-step heating 50, the progression from thermal unfolding to aggregation of MHC occurs more gradually compared to single-step heating 85. It has been reported that purified MHC derived from *Cyprinus carpio* forms multimers at 30 °C through association via subfragment-1 [25]. Based on these findings, it is considered that during heating at 50 °C for 60 min, MHC multimerization progresses in the surimi of *N. virgatus*, leading to the formation of MHC clusters, which subsequently bind to each other at 85 °C. This slower polymerization process may contribute to the formation of a finer gel network, thereby enhancing the fracture strength of the surimi gel.

Next, time-course changes in the mechanical properties and MHC monomer levels of the *N. virgatus* surimi during heating at 60 °C were investigated. The maximum strength of the *N. virgatus* surimi reached its highest value after 16 min of heating (Figure 2b), which corresponded to the time when the core temperature of the surimi sample reached 60 °C (Figure 2a). These results suggest that gelation of the surimi progressed until the core temperature reached approximately 60 °C; however, gel disintegration began once the core temperature reached this point.

The WSP decreased with increasing heating time (Figure 3a). The WSP consists primarily of sarcoplasmic proteins, including glycolytic enzymes. It has been shown that WSP of both rainbow trout *Oncorhynchus mykiss* meat and deep-sea bonefish *Pterothrissus gissu* surimi begin to undergo thermal aggregation above approximately 40 °C [26,27]. Although the thermal denaturation point of sarcoplasmic proteins varies depending on fish species and their living environment, it can be estimated to fall within the range of 40–50 °C in *N. virgatus* surimi. Therefore, the decrease in WSP with increasing heating time at 60 °C can be attributed to thermal aggregation rather than proteolytic degradation (Figure 3a). The bands at 34, 38, 51, and 83 kDa observed in the reduced USP are likely sarcoplasmic proteins that became insoluble through thermal aggregation, as indicated by the increase in their band intensities with prolonged heating (Figure 3b).

A decrease in MHC band intensity in *N. virgatus* surimi during heating at 60 °C was observed in both the reduced and non-reduced USP fractions (Figure 3b,d). The decrease observed in the reduced USP indicated proteolytic degradation of MHC, as it was accompanied by an increase in the staining intensities at 126, 144, and 157 kDa (Figure 3b,c). On the other hand, the decrease observed in the non-reduced USP indicates both non-enzymatic polymerization and proteolytic degradation of MHC (Figure 3d,e). The decrease in MHC intensity in the reduced USP began after approximately 20 min of heating (Figure 3b,c), whereas in the non-reduced USP, the decrease was observed as early as 2 min into heating (Figure 3d,e). Taken together, these results suggest that thermal polymerization of MHCs via disulfide bonding occurred prior to their proteolytic degradation in the *N. virgatus* surimi during heating at 60 °C.

To elucidate the competition between non-enzymatic polymerization and proteolytic degradation of MHC in the *N. virgatus* surimi during heating at 60 °C, the experimentally measured values were compared with those calculated using a mathematical model (Figure 4). The model incorporates only non-enzymatic polymerization and proteolytic degradation of MHC, as TGase activity was confirmed to be negligible during heating at 50–60 °C (Figure 1c), as described above. The decrease in MHC levels in the reduced USP during heating at 60 °C, which exhibited a delay of approximately 16 min, was accurately traced by the estimated values (Figure 4). Thus, this delay can be attributed to the temperature-dependent activation of proteases. In both the measured and calculated data, the decrease in MHC levels in the non-reduced USP began as early as 2 min into heating (Figure 4). This result suggests that the 2 min delay before the decrease in MHC level in the non-reduced USP, due to disulfide bonding, can be explained by the time required to reach the myosin denaturation point. A comparison between the MHC dynamics estimated by mathematical modeling and the measured time-course changes in maximum strength indicates that gelation progresses through the non-enzymatic polymerization of MHC prior to the gel disintegration caused by the proteolytic degradation of MHC in *N. virgatus* surimi. Considering the difference in onset time between non-enzymatic polymerization (~2 min) and proteolytic degradation (~16 min), not only unpolymerized but also polymerized MHC, as described in Equation (1), was significantly degraded, which contributed markedly to the reduction in the mechanical properties of the *N. virgatus* surimi.

It has been suggested that proteolytic degradation of the pre-formed gel network (*himodori*) underlies the mechanism of gel disintegration in *N. virgatus* surimi [3,5,13]. The present study quantitatively elucidated these details using a mathematical model. Elucidating the molecular mechanisms underlying surimi gelation and disintegration enables the prediction of gel properties under varying heating temperatures and with the addition of TGase or protease inhibitors. Therefore, our approach may facilitate the utilization of noncommercial fish species for surimi processing by enabling better control over surimi gel properties. However, it is also necessary to construct a mathematical model that can quantitatively describe the relationship between MHC polymerization and degradation, and the mechanical properties of surimi gels.

4. Conclusions

The present study proposes a model representing the interplay between the non-enzymatic polymerization and the proteolytic degradation of MHC during gel disintegration in golden threadfin bream *N. virgatus* surimi. The reaction rate of MHC was estimated from a densitometric analysis of SDS-PAGE performed under reducing and non-reducing conditions, and its dynamic behavior was represented by a kinetic model. The changes in mechanical strength during the disintegration of *N. virgatus* surimi can be largely explained by the dynamics of MHC, indicating that the approach presented in this study is effective for quantitatively understanding the gelation process of surimi.

5. Materials and Methods

5.1. Material

Fresh specimens of *N. virgatus* were purchased at a local market in Tokyo, Japan. The fish were transported to the laboratory on ice. Filleting was performed in a cold room, and the deboned, skinned flesh was stored at −35 °C until further use.

5.2. Surimi Preparation

N. virgatus surimi was prepared according to the method described by Kominami et al. (2025) [28] with slight modifications. Briefly, frozen flesh was ground with 3% (w/w) NaCl after thawing overnight at 4 °C. The salt-ground surimi was then packed and sealed into plastic jars of either 20 mL (φ42 × 26 mm) or 5.5 mL (φ29 × 16 mm) capacity. All operations were conducted at 4°C.

The surimi-filled 20 mL jars were subjected to a temperature-controlled water bath heating under the following conditions: 50, 55, and 60 °C for 60 min, followed by heating at 85 °C for 20 min (referred to as two-step heating 50–85, 55–85, and 60–85), or a single-step heating at 85 °C for 20 min (referred to as single-step heating 85). Immediately after heating, the jars were rapidly cooled in ice water. Following overnight storage in ice water, the *N. virgatus* surimi gels were subjected to fracture strength measurements.

The 5.5 mL surimi-filled jars were heated in a 60 °C water bath for varying durations (2, 4, 6, 8, 10, 12, 14, 16, 18, 20, 30, 40, and 60 min), then rapidly cooled and stored overnight in ice water. The temperature near the center of the surimi was monitored during 60 °C heating using a coated thermocouple wire (Type T, outer diameter: 0.9 mm, wire diameter: 0.254 mm), and data were recorded using a multichannel recorder (MCR-4TC, T&D Corporation, Nagano, Japan). Some of the 5.5 mL surimi-filled jars were stored in ice water without heating and used as unheated controls. Both unheated and heated *N. virgatus* surimi samples were subjected to constant-speed compression testing.

5.3. Fracture Strength Measurements

The fracture strength of the *N. virgatus* surimi gels was determined using a rheometer, according to the method described by Kominami et al. (2020) [29]. After measurement, the central portion of the sample was excised, flash-frozen in liquid nitrogen, and stored at −80 °C.

5.4. Compression Test

A rheometer (FUDOH RHEOMETER, NRM-2010-CW, Rheotech, Tokyo, Japan) was used to perform compression tests. Since non-gelled surimi lacks shape-retaining properties, it must be evaluated while contained in a plastic jar. Therefore, the evaluation method for gelatin described in the Japanese Industrial Standard K6503 was modified accordingly. In this study, to assess gelation at the center of the jar, a plunger with a diameter (φ5 mm) significantly smaller than that of the jar (φ25 mm) was employed. The 5.5 mL plastic jar with its lid removed was secured onto the sample stage. A cylindrical plunger (φ5 mm) was pressed into the surimi-filled plastic jar at a speed of 1 mm/sec to a depth of 10 mm, and the resulting stress transition of the surimi was recorded using a data logger (NR-HA08, KEYENCE, Osaka, Japan). The maximum stress during compression was defined as the maximum compressive stress (gf) for each sample. After measurement, the central portion of the sample was excised, flash-frozen in liquid nitrogen, and stored at −80 °C.

5.5. Protein Fractionation and SDS-PAGE

Proteins in the *N. virgatus* surimi gel were fractionated into a water-soluble protein (WSP) fraction and a urea-solubilized protein fraction (USP) according to the method described by Kominami et al. (2025) [28] with slight modifications. The frozen surimi sample was ground to a fine powder within the precooled steel cylinders (TH-SPT, Taitec corporation, Saitama, Japan) of an automatic cryogenic crusher (Taitec Freeze Crusher μT-48, Taitec corporation). The frozen powdered sample was suspended in an ice-cold low ionic strength buffer (20 mM KCl, 100 mM Tris-HCl, pH 7.4) at a concentration of 100 mg/mL, followed by centrifugation at 4000 g for 20 min at 4 °C. The supernatant was collected as the WSP fraction. The remaining pellet was washed twice with the low-ionic strength buffer. Subsequently, 40 mg of the resultant pellet was resuspended in 1 mL of either urea-SDS reducing buffer (0.03% bromophenol blue, 3% SDS, 8 M urea, 2 M thiourea, 75 mM dithiothreitol, 50 mM Tris-HCl, pH 6.8) or urea-SDS non-reducing buffer (0.03% bromophenol blue, 3% SDS, 8 M urea, 2 M thiourea, 50 mM Tris-HCl, pH 6.8). The samples were then solubilized overnight at room temperature by gentle inversion using a benchtop tube rotator. Subsequently, they were centrifuged at 12,000 g for 15 min at 20 °C, and the resulting supernatants were collected as the USP fraction.

The WSP fraction was mixed with 4× Laemmli's sample buffer (Bio-Rad, Hercules, CA, USA) at a 3:1 ratio. The reduced USP fraction was diluted 10-fold with the urea−SDS reducing buffer. Both samples were heated at 95 °C for 3 min prior to SDS-PAGE. The

WSP sample was separated using a 12% poly acrylamide gel, while both reduced and non-reduced USP samples were separated using a gradient 2–15% gel (Multi gel II mini 2/15, Cosmo Bio Co., Ltd., Tokyo, Japan). CBB staining and quantitative analysis were performed as described in a previous study [30]. Briefly, images of the SDS-PAGE bands were acquired using an infrared imaging system (Odyssey Fc Imaging System, LI-COR, Lincoln, NE, USA), configured for an excitation wavelength of 680 nm and an emission wavelength of 700 nm. Densitometric analysis of the acquired images was conducted using ImageJ version 1.51t (US National Institutes of Health, MA, USA).

5.6. Data Analysis

Dunnett's test was conducted using the package "DescTools" in R v3.4.37 to compare the fracture strength of the surimi gels versus those prepared by single-step heating [30,31].

5.7. Mathematical Modelling

The polymerization and degradation of MHC were described by a system of ordinary differential equations. Let $X[t]$ denote the amount of MHC monomer, $P[t]$ the amount of MHC polymer, and $T[t]$ the core temperature of the surimi sample at time t. To focus on relative changes, $X[t]$ and $P[t]$ represent normalized amounts of MHC monomer and polymer, respectively. The conceptual framework of the model is depicted in Figure 5. The amount of MHC in the reduced USP corresponds to the sum of $X[t]$ and $P[t]$, whereas in the non-reduced USP, it corresponds to $X[t]$ alone. We assume that the polymerization rate of MHC depends on the surimi sample temperature $T[t]$ through a sigmoid function, which reflects the thermal denaturation behaviour of MHC observed in experiments. Considering that MHC degradation is maximized at an optimal temperature, the degradation rate is assumed to be modeled as a Gaussian function dependent on the surimi sample temperature $T[t]$. The dynamics of MHC polymerization and degradation are described by the following system of differential equations:

$$\begin{aligned} \frac{dX[t]}{dt} &= -\frac{k_1\left(\frac{T[t]}{T_1}\right)^d}{1+\left(\frac{T[t]}{T_1}\right)^d} X[t] - k_2\exp\left(-\frac{(T[t]-T_0)^2}{2\beta^2}\right) X[t] \\ \frac{dP[t]}{dt} &= \frac{k_1\left(\frac{T[t]}{T_1}\right)^d}{1+\left(\frac{T[t]}{T_1}\right)^d} X[t] - k_2\exp\left(-\frac{(T[t]-T_0)^2}{2\beta^2}\right) P[t] \\ \frac{dT[t]}{dt} &= \alpha(T_{\max} - T[t]) \end{aligned} \tag{1}$$

with the initial values $X[0] = 1$, $P[0] = 0$, and $T[0] = 0$. The model contains the following parameters: k_1, k_2, α, β, T_0, T_1, $T_{\max}$, and d. The parameter k_1 represents the maximum polymerization rate of MHC. The parameters T_1 and d are the inflection point and the steepness of the sigmoid function, respectively. The parameter k_2 denotes the maximum rate of temperature-dependent degradation. The parameter T_0 is the optimal temperature for MHC degradation and β is the width of a Gaussian function. $T_{\max}$ is defined as the highest temperature that the surimi sample can reach, i.e., the temperature of the water bath. The parameter α is the rate at which the heating temperature is increased to approach the target temperature $T_{\max}$.

Based on measurements of the thermal denaturation of MHC, the parameters T_1 and d were estimated as 29.61 and 22.34, respectively, using the Simple Fit app (v4.33) in OriginPro 2024 SR1. The parameters α and $T_{\max}$ were set to 0.3 and 60, respectively, from experimental data for the heating temperature. The parameter T_0 was assigned a value of 60, since the optimal temperature for MHC degradation is assumed to be 60 °C, as supported by our findings (Figure 1) and a previous study [13]. The parameters k_1, k_2, and β were determined by grid search to minimize the residual sum of squares between the numerical results and the experimental data. The values of the optimal parameters were found to be $k_1 = 0.044$, $k_2 = 0.053$, and $\beta = 0.214$. Estimation of the parameters (k_1, k_2, β), numerical solutions to the differential equations, and the data visualization for comparison with the experimental data were obtained using Wolfram Mathematica 13.1.

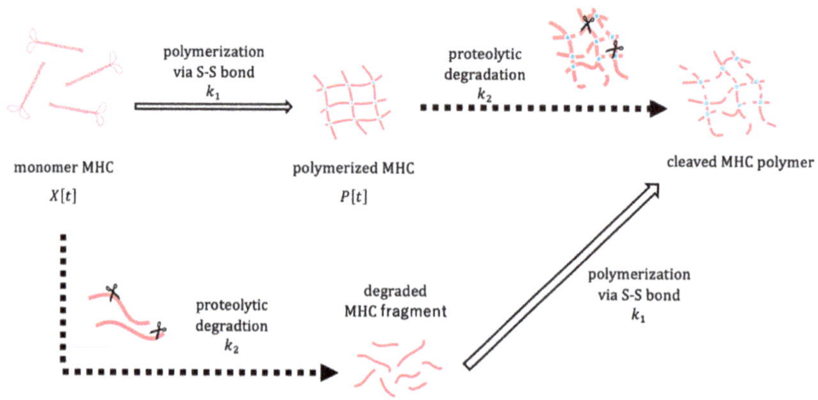

Figure 5. A conceptual illustration of the proposed mathematical model.

Author Contributions: Y.K. and T.H. developed the concept and designed the experimental outline. R.N. and Y.K. performed the experiments and data acquisition. T.H. and Y.K. conducted the in-silico analyses. H.U. supervised the data analysis. All authors wrote the manuscript and approved the final version of the manuscript. All authors have read and agreed to the published version of the manuscript.

Funding: This research was partially supported by The Toyo Suisan Foundation (Academic Research H29) and the Sasakawa Scientific Research Grant from the Japan Science Society (28-741).

Institutional Review Board Statement: Not applicable.

Informed Consent Statement: Not applicable.

Data Availability Statement: The authors confirm that the data supporting the findings of this study are available within the article.

Conflicts of Interest: Author Ryoko Nakamizo is employed by the company Suzuhiro Kamaboko Honten Co., Ltd. The remaining authors declare that the research was conducted in the absence of any commercial or financial relationships that could be construed as a potential conflict of interest.

References

1. Lanier, T.; Yongsawatdigul, J.; Carvajal-Rondanelli, P. Surimi Gelation Chemistry. In *Surimi and Surimi Seafood*, 3rd ed.; Park, J.W., Ed.; CRC Press: Boca Raton, FL, USA, 2013; pp. 101–140. ISBN 1439898588/9781439898581.
2. Shimizu, Y. Species Variations in Gel-Forming Ability of Fish Meat Paste. *Nippon Suisan Gakkaishi* **1974**, *40*, 175–179. [CrossRef]
3. Kinoshita, M.; Toyohara, H.; Shimizu, Y. Diverse Distribution of Four Distinct Types of Modori (Gel Degradation)-Inducing Preoteinases among Fish Species. *Nippon Suisan Gakkaishi* **1990**, *56*, 1485–1492. [CrossRef]
4. Folk, J.E. Transglutaminases. *Annu. Rev. Biochem.* **1980**, *49*, 517–531. [CrossRef] [PubMed]
5. Takeda, H.; Seki, N. Enzyme-Catalyzed Cross-Linking and Degradation of Myosin Heavy Chain in Walleye Pollack Surimi Paste during Setting. *Fish. Sci.* **1996**, *62*, 462–467. [CrossRef]
6. Ni, S.; Nozawa, H.; Seki, N. The Combined Effect of Transglutaminase and Protease Inhibitors on the Thermal Gelation of Actomyosin Sol from Carp and Salmon Muscles. *Fish. Sci.* **1999**, *65*, 606–612. [CrossRef]
7. Shimizu, Y.; Machida, R.; Takenami, S. Species Variations in the Gel-Forming Characteristics of Fish Meat Paste. *Nippon Suisan Gakkaishi* **1981**, *47*, 95–104. [CrossRef]

8. Tsukamasa, Y.; Miyake, Y.; Ando, M.; Makinodan, Y. Total Activity of Transglutaminase at Various Temperatures in Several Fish Meats. *Fish. Sci.* **2002**, *68*, 929–933. [CrossRef]
9. Saeki, H.; Iseya, Z.; Sugiura, S.; Seki, N. Gel Forming Characteristics of Frozen Surimi from Chum Salmon in the Presence of Protease Inhibitors. *J. Food Sci.* **1995**, *60*, 917–921. [CrossRef]
10. Matsuoka, Y.; Wan, J.; Ushio, H.; Watabe, S. Thermal Gelation Properties of White Croaker, Walleye Pollack and Deepsea Bonefish Surimi after Suwari Treatment at Various Temperatures. *Fish. Sci.* **2013**, *79*, 715–724. [CrossRef]
11. Gao, Y.L.; Yoshida, A.; Liu, J.Y.; Yuan, J.; Maeda, S.; Wang, Y.; Jiang, Y.R.; Sun, X.M.; Chen, C.P.; Wang, Y.; et al. Quality Improvement of Threadfin Bream (Nemipterus Virgatus) Surimi-Gel Using Soy Protein as a Natural Food Additive. *Food Chem.* **2024**, *460*, 140423. [CrossRef]
12. Kinoshita, M.; Toyohara, H.; Shimizu, Y. Purification and Properties of a Novel Latent Proteinase Showing Myosin Heavy Chain-Degrading Activity from Threadfin-Bream Muscle. *J. Bio-Chem.* **1990**, *107*, 587–591. [CrossRef]
13. Kinoshita, M.; Toyohara, H.; Shimizu, Y.; Sakaguchi, M. Induction of Modori-Phenomenon (Thermal Gel Degradation)by A Latent Serine Proteinase. *Nippon Suisan Gakkaishi* **1991**, *57*, 1935–1938. [CrossRef]
14. Kinoshita, M.; Toyohara, H.; Shimizu, Y.; Sakaguchi, M. Modori-Inducing Proteinase Active At 50°C In Threadfin Bream Muscle. *Nippon Suisan Gakkaishi* **1992**, *58*, 715–720. [CrossRef]
15. Liu, J.Y.; Yoshida, A.; Gao, Y.L.; Shirota, K.; Shiina, Y.; Osatomi, K. Identification of a Modori-Inducing Proteinase in the Threadfin Bream: Molecular Cloning, Tissue Distribution and Proteinase Leakage from Viscera during Ice Storage. *Food Chem.* **2020**, *330*, 127246. [CrossRef]
16. Liu, J.Y.; Yoshida, A.; Gao, Y.L.; Shirota, K.; Shiina, Y.; Noguchi, E.; Kuwahara, K.; Osatomi, K. Purifi-cation and Characterization of a Sarcoplasmic Serine Proteinase from Threadfin Bream Nemipterus Virgatus Muscle. *Food Chem.* **2019**, *284*, 198–204. [CrossRef]
17. Fang, Q.; Shi, L.; Ren, Z.; Hao, G.; Chen, J.; Weng, W. Effects of Emulsified Lard and TGase on Gel Properties of Threadfin Bream (Nemipterus Virgatus) Surimi. *LWT* **2021**, *146*, 111513. [CrossRef]
18. Ren, Z.; Huang, X.; Shi, L.; Liu, S.; Yang, S.; Hao, G.; Qiu, X.; Liu, Z.; Zhang, Y.; Zhao, Y.; et al. Char-acteristics and Potential Application of Myofibrillar Protein from Golden Threadfin Bream (Nemipterus Virgatus) Complexed with Chitosan. *Int. J. Biol. Macromol.* **2023**, *240*, 124380. [CrossRef] [PubMed]
19. Yu, S.; Pang, G.; Li, S.; Lv, S.; Wei, Z.; Wang, J.; Xiao, H.; Zhu, L. Impacts of Zein-Fucoidan Nanoparticles with and without Curcumin on Gel Properties of Golden Threadfin Bream (Nemipterus Virgatus) Surimi. *Food Chem.* **2025**, *468*, 142415. [CrossRef] [PubMed]
20. Chen, H.; Zhou, A.; Benjakul, S.; Zou, Y.; Liu, X.; Xiao, S. The Mechanism of Low-Level Pressure Coupled with Heat Treatment on Water Migration and Gel Properties of Nemipterus Virgatus Surimi. *LWT* **2021**, *150*, 112086. [CrossRef]
21. Benjakul, S.; Visessanguan, W.; Thummaratwasik, P. Inhibition of Gel Weakening of Threadfin Bream Surimi Using Thai Legume Seed Proteinase Inhibitors. *J. Food Biochem.* **2000**, *24*, 363–380. [CrossRef]
22. Liu, J.-Y.; Yoshida, A.; Shirota, K.; Shiina, Y. Effect of Spawning on Endogenous Proteinases in Abdominal Muscle of Threadfin Bream Nemipterus Virgatus. In Proceedings of the JSFS 85th Anniver-Sary-Commemorative International Symposium "Fisheries Science for Future Generations"; The Japanese Society of Fisheries Science: Tokyo, Japan, 2017; pp. 1–2.
23. Yamashita, M.; Konagaya, S. High Activities of Cathepsins B, D, H and L in the White Muscle of Chum Salmon in Spawning Migration. *Comp. Biochem. Physiol. Part B Comp. Bio-Chem.* **1990**, *95*, 149–152. [CrossRef] [PubMed]
24. Howell, B.K.; Mattews, A.D.; Donnelly, A.P. Thermal Stability of Fish Myofibrils: A Differential Scanning Calorimetric Study. *Int. J. Food Sci. Technol.* **1991**, *26*, 283–295. [CrossRef]
25. Tazawa, T.; Kato, S.; Katoh, T.; Konno, K. Role of Neck Region in the Thermal Aggregation of Myosin. *J. Agric. Food Chem.* **2002**, *50*, 196–202. [CrossRef] [PubMed]
26. Okamoto, Y.; Kominami, Y.; Izawa, S.; Nakamizo, R.; Matsuoka, Y.; Ueki, N.; Wan, J.; Watabe, S.; Ushio, H. The Effects of Length Distribution of the MHC and Its Fragments on Fracture Strength of Fish Meat Gel. In Proceedings of the 19th Food Colloids Conference, Thessaloniki, Greece, 17 April 2024; p. 269.
27. Sakuyama, M.; Kominami, Y.; Hayashi, T.; Ushio, H. Effects of Sous-Vide Cooking Temperature on Protein Changes and Texture of Rainbow Trout (*Oncorhynchus Mykiss*) Meat. *Int. J. Gastron. Food Sci.* 2025; IJGFS-D-25-00323. *under review*.
28. Kominami, Y.; Nakamizo, R.; Matsuoka, Y.; Ueki, N.; Wan, J.; Watabe, S.; Ushio, H. Exploration of Myosin Heavy Chain Fragments within the Protein Network in Disintegrated Deep-Sea Bonefish Pterothrissus Gissu Surimi Gel. *ACS Food Sci. Technol.* **2025**, *5*, 1072–1081. [CrossRef]
29. Kominami, Y.; Nakakubo, H.; Nakamizo, R.; Matsuoka, Y.; Ueki, N.; Wan, J.; Watabe, S.; Ushio, H. Peptidomic Analysis of a Disintegrated Surimi Gel from Deep-Sea Bonefish *Pterothrissus gissu*. *J. Agric. Food Chem.* **2020**, *68*, 12683–12691. [CrossRef]

30. Dunnett, C.W. A Multiple Comparison Procedure for Comparing Several Treatments with a Control. *J. Am. Stat. Assoc.* **1955**, *50*, 1096–1121. [CrossRef]
31. Andri, S. DescTools: Tools for Descriptive Statistics. Available online: https://cran.r-project.org/web/packages/DescTools/index.html (accessed on 2 July 2024).

gels

Article

Development and Application of Anthocyanin-Based Complex Polysaccharide Gels Based on Blueberry Pomace for Monitoring Beef Freshness

Jingxi Zhi [1,2,†], Fuqian Xu [1,†], Shuhuan Yu [1], Jiahui Hao [1], Jie Wang [1] and Ziluan Fan [1,3,*]

1 College of Life Science, Northeast Forestry University, 26 Hexing Road, Xiangfang District, Harbin 150040, China; aulin-zjx@nefu.edu.cn (J.Z.); xufuqian@nefu.edu.cn (F.X.); yushuhuan@nefu.edu.cn (S.Y.); haojiahui5832@163.com (J.H.); wj18365936610@163.com (J.W.)
2 Hangzhou Institute for Advanced Study, School of Life and Health Sciences, University of Chinese Academy of Sciences, 1 Xiangshan Branch Lane, Xihu District, Hangzhou 310024, China
3 Key Laboratory of Forest Food Resources Utilization, Harbin 150040, China
* Correspondence: fzl_1122@nefu.edu.cn; Tel./Fax: +86-0451-82190514
† These authors are considered the first author.

Abstract: This study aimed to develop a green and sustainable composite polysaccharide gel with antioxidant activity and freshness-monitoring properties. Blueberry pomace was repurposed to extract anthocyanins (BA), which were incorporated into chitosan (CS)/polyvinyl alcohol (PVA) and starch (S)/PVA matrices to prepare pH-indicating composite polysaccharide gels. The anthocyanin solution exhibited significant colorimetric responses to pH 2–14 buffer solutions. Comparative analyses revealed distinct performance characteristics: the CS/PVA-BA gel showed optimal elongation at break, low hydration ($8.33 \pm 0.57\%$ water content), and potent antioxidant activity (DPPH: $73.59 \pm 0.1\%$; ABTS: $77.47 \pm 0.1\%$), whereas the S/PVA-BA gel demonstrated superior tensile strength and pH-responsive sensitivity. Structural characterization via FT-IR and SEM confirmed molecular compatibility between BA and polymeric matrices, with anthocyanins enhancing intermolecular hydrogen bonding. Applied to chilled beef (4 °C) freshness monitoring, the CS/PVA-BA gel exhibited color transformations from magenta-red (initial spoilage at 48 h: TVB-N > 15 mg/100 g, TVC > 4.0 lg CFU/g) to bluish-gray (advanced spoilage by day 6), correlating with proteolytic degradation metrics. These findings established a multifunctional platform for real-time food quality assessment through anthocyanin-mediated color changes in the composite gels, coupled with preservation activity, highlighting their significant potential as intelligent active packaging in the food industry.

Keywords: blueberry pomace anthocyanins; complex polysaccharide gels; freshness monitoring; chilled beef

1. Introduction

With the improvement of living standards, the market demand for chilled beef has significantly increased. Beef, rich in protein, vitamins, and minerals, is an essential component of the human diet [1]. However, due to microbial contamination, lipid oxidation, and protein degradation, chilled beef is highly susceptible to spoilage, compromising its quality and commercial value [2,3]. Currently, consumers primarily rely on expiration dates and appearance to assess freshness, which lacks scientific rigor [4]. Therefore, food packaging technology requires further development, and there is an urgent need to create intuitive and rapid methods for monitoring food freshness.

In recent years, intelligent packaging has emerged as a novel solution, particularly in systems that enable the visual judgment of food quality through color changes. pH-sensitive substances blended with polymers can form packaging gels that display distinct colors under varying pH conditions, achieving the real-time monitoring of food freshness [5,6]. However, indicators are categorized into synthetic and natural pigments. Synthetic indicators, such as methyl red, bromocresol green, and bromocresol purple [7], pose safety risks due to potential toxicity [8]. In contrast, natural pigments like anthocyanins have gained prominence. Anthocyanins, as polyphenolic flavonoids, constitute a class of water-soluble natural pigments widely present in plants [9]. Anthocyanins, characterized by their ease of extraction and pH-responsive color changes, were extracted and purified from blueberry pomace in this study to serve as indicators in multifunctional gels. These compounds also exhibit potent antioxidant activity, enabling their utilization in active packaging systems to extend the shelf lives of food products [10]. Blueberry pomace, a processing byproduct typically discarded or used as low-value fertilizer, poses environmental risks when landfilled or incinerated. Direct disposal generates greenhouse gases and leachate, which contaminate soil and water systems [11]. However, blueberry pomace retains significant amounts of underutilized anthocyanins. Utilizing this low-cost byproduct drastically reduces the production costs of anthocyanins. Extracting anthocyanins from blueberry pomace not only achieves a "waste-to-resource" transformation but also aligns with sustainable development goals. Furthermore, compared to synthetic pigments, anthocyanins exhibit higher biological activity and lower costs.

To ensure food safety, freshness-indicating gels are typically fabricated from biodegradable, biocompatible, and non-toxic materials such as starch, chitosan, carboxymethyl cellulose (CMC), and their composites [12]. Natural renewable resources, including polysaccharide-based (e.g., cellulose, starch, and chitosan) and protein-based materials (e.g., corn protein, soy protein isolate, and gelatin), are increasingly utilized due to their low costs and abundant availability [13]. Chitosan, a natural polysaccharide extracted from crustacean shells, is characterized by biodegradability, antibacterial properties, and film-forming ability [14]. Under acidic conditions, the amino groups ($-NH_2$) in chitosan molecules undergo electrostatic interactions with negatively charged molecules to form gels [15]. In food packaging applications, chitosan's antibacterial properties extend the shelf lives of products while its film-forming ability enables the formation of uniform film structures. Starch, a widely available and low-cost natural polymer, exhibits excellent film-forming properties and biodegradability. When starch particles are gelatinized by heating, amylose and amylopectin recombine through hydrogen bonds to form a three-dimensional network structure [16]. This network binds water molecules via hydrogen bonds, conferring viscoelasticity and moisture retention to the gel. For example, Zong et al. [17] used purple potato anthocyanins as indicators to prepare complex polysaccharide gels with starch and gelatin and applied them to the freshness monitoring of the Flammulina velutipes mushroom. Feng [18] fabricated corn starch (CS)/carboxymethyl cellulose (CMC) composite gels with blueberry pomace anthocyanins (BA) as indicators, demonstrating an effective monitoring of beef freshness. These studies validated the feasibility of anthocyanins as functional indicators. Polyvinyl alcohol (PVA), a synthetic water-soluble polymer, offers excellent film-forming ability, mechanical strength, and chemical stability. It exhibits good compatibility with natural components (e.g., anthocyanins and essential oils) and can stably incorporate functional additives [19]. However, pure starch films may suffer from limited mechanical strength or poor water resistance while the biodegradability of PVA requires further enhancement. Blending chitosan with PVA improves the mechanical strength of gels and imparts antimicrobial properties [20]. Barbara Merz et al. [21] investigated chitosan/polyvinyl alcohol (PVA) gels incorporating anthocyanins, analyzing their thickness, microstructure, moisture content, and hydrophobicity, and applied them to monitor shrimp freshness. When starch is mixed with polyvinyl alcohol, its brittleness can be improved. Chen et al. [22] immobilized con go red (CR) and anthocyanins (ATH) in starch/PVA/glycerol gels, highlighting the excellent gel-forming properties of polysaccharide-PVA blends.

This study extracted anthocyanins from blueberry pomace and prepared smart indicator films by combining starch and chitosan with polyvinyl alcohol (PVA) as film-forming matrices, using glycerol as a plasticizer. The films were applied to monitor the freshness of chilled beef, providing a theoretical foundation for the use of smart packaging in meat freshness detection. The resulting gel demonstrated significant potential for application in active food packaging and real-time freshness indication, advancing sustainability goals in the food industry.

2. Results and Discussion

2.1. *Quantification and Characterization of Anthocyanins from Blueberry Pomace*

The total anthocyanin content extracted from blueberry pomace was 29.31 ± 1.18 mg/g. The anthocyanin solution exhibited distinct color changes under varying pH conditions. As shown in Figure 1a, the solution appeared red at pH 2–3, transitioned to pink at pH 4–6 with gradual fading, turned purple at pH 7–8, shifted to blue and green in alkaline conditions (pH 9–11), and finally turned yellowish-brown at pH 12–14 [23]. The color of blueberry pomace anthocyanins varied significantly in different pH ranges. Figure 1b displays the UV-Vis spectra of blueberry pomace anthocyanins. In acidic solutions (pH 2–5), the maximum absorption peak appeared near 523 nm. As the pH increased, the red peak shifted to 540 nm with reduced absorbance. At pH 8, the absorption maximum shifted to 576 nm, while under alkaline conditions (pH 10–14), it stabilized around 600 nm. These spectral shifts corroborated the structural transitions observed in color changes.

This color variation and shift in maximum absorption wavelength arose from the highly conjugated system within anthocyanin molecules, which underwent structural transformations in response to pH changes. Under strongly acidic conditions (pH < 3), anthocyanins primarily exist as red-colored flavylium cations in solution, exhibiting lower maximum absorption wavelengths [24]. As the pH increases, the flavylium cations undergo hydration to form colorless carbinol pseudobases, resulting in diminished red coloration and reduced absorbance [25]. At neutral to slightly alkaline conditions (pH 7–9), anthocyanins adopt a blue quinonoidal base structure. When the pH exceeds 9, further deprotonation leads to the formation of yellowish-brown chalcones [26,27].

The colorimetric analysis data of anthocyanin solutions under different pH conditions are shown in Figure 1c,d. The L value (lightness) of the solution increased under acidic conditions, decreased under neutral conditions (resulting in darker solutions), and gradually rose again as the pH increased to alkaline levels, restoring the solution's brightness. With increasing pH, the a value (redness) gradually decreased, reaching its lowest level at pH 10. Subsequently, as the alkalinity further increased, the a value initially rose and then declined again. This trend aligned with the visual observations shown in Figure 1a, where the redness diminished as the pH increased from 2 to approximately 7. The b value (yellowness) increased under alkaline conditions, accompanied by a color transition of the solution from green to yellowish-brown. ΔE (Delta E) is a metric used to quantify color differences, representing the numerical value of perceptual differences between two colors in a uniform color space. When the ΔE value exceeds 5, the color variation becomes visually perceptible [28]. Since all ΔE values of the indicator were greater than 5, the color changes in the indicator could be distinctly differentiated by the naked eye.

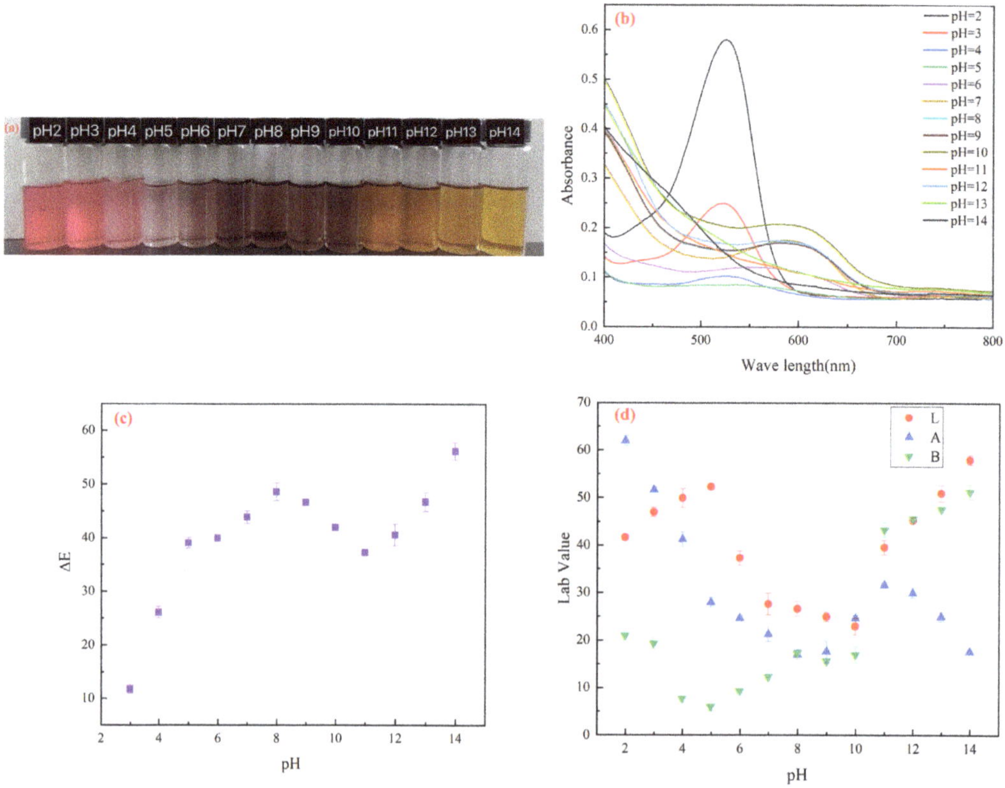

Figure 1. (**a**) Color changes in blueberry pomace anthocyanin solution at pH 2–14; (**b**) UV-VIS spectrum of blueberry pomace anthocyanin solution at pH 2–14; (c) ΔE values and (**d**) lab values of blueberry pomace anthocyanin solution at pH 2–14. Vertical bars represent the standard deviations of three replicates.

2.2. Characterization and Analysis of Complex Polysaccharide Gels

2.2.1. SEM Analysis of Gels

The dispersion state and compatibility of components within a composite gel can be determined through SEM (scanning electron microscopy) images. The surface scanning electron microscopy images of the four gels are shown in Figure 2. Typically, a smooth and flat surface with a uniform structure, free of protrusions or wrinkles, indicates excellent film-forming properties and high compatibility among the gel matrix components [29]. On the contrary, there were obvious protrusions and pores, indicating that the gel surface was rough. Figure 2 shows the surface microstructure of the four gels magnified 1000 times. Compared with the gels without blueberry pomace anthocyanins, the CS/PVA-BA gels and S/PVA-BA gels with blueberry pomace anthocyanins added were tighter and smoother with no obvious bumps and folds. This suggests that blueberry anthocyanins (BA) not only exhibit good compatibility with chitosan (CS), polyvinyl alcohol (PVA), and starch (S) but also enhance intermolecular interactions between CS-PVA and S-PVA matrices [30]. The anthocyanins effectively bind to polysaccharides and disperse uniformly within the film, which aligns with the findings reported by Merz et al. [21].

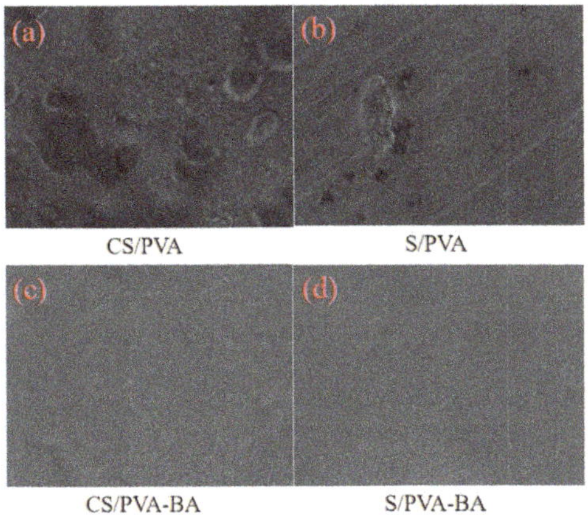

Figure 2. Scanning electron microscopy images (1000×) of surface of (**a**) CS/PVA, (**b**) S/PVA, (**c**) CS/PVA-BA, and (**d**) S/PVA-BA gel.

2.2.2. Infrared Spectrum Analysis of Gel

FT-IR can be used for the qualitative analysis of substances and effectively characterizes the molecular structure and cross-linking of organic compounds. It can be seen in Figure 3 that the absorption peak at 3289–3300 cm^{-1}, indicating that the gel was in the range of 3200–3600 cm^{-1}, belonged to the stretching vibration of free amino ($-NH_2$) in the chitosan (CS) and hydroxyl ($-OH$) of polymers (such as PVA and starch) [31]. Aliphatic C-H vibrations at 2937–2986 cm^{-1} are related to hydrophobic groups in the plasticizer or polymer side chain, which may regulate the gas permeability of a gel [32]. The characteristic front at 1710 cm^{-1} is caused by the stretching vibration of C=C on the aromatic ring skeleton in anthocyanins. In addition, the C-O vibration peak at 1034–1094 cm^{-1} indicates the hydrogen bond between the polysaccharide matrix (CS and starch) and the plasticizer [33], and the low frequency shift of CS/P at 1029 cm^{-1} reflects the strengthening effect of chitosan molecular chain rigidity on the hydrogen bond network [34]. Therefore, it can be seen from the FT-IR spectrum that the structure of the material was mainly affected by the intermolecular force, and the chemical composition of the gel-forming substrate did not change.

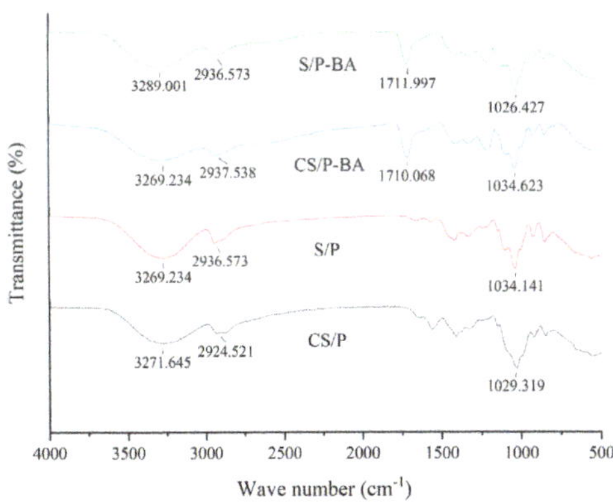

Figure 3. Fourier-transform infrared (FT-IR) spectrum analysis of CS/PVA, S/PVA, CS/PVA-BA, and S/PVA-BA gel.

2.2.3. Analysis of Physical Properties of Gel

It can be seen from Table 1 that there were differences in the thickness of the four gels. In the process of gel forming preparation, the molecular cohesion between gel-forming substrates will lead to differences in gel thickness [35]. The addition of anthocyanins also increased the thickness of the gel.

Table 1. Gel thickness, mechanical properties, moisture content, and water vapor transmission coefficient data.

Complex Polysaccharide Gels Name	Thickness/μm	Tensile Strength/MPa	Elongation at Break/%	Moisture Content/%	Water Vapor Transmission Coefficient × 10^{-4}/(g·mm)/(m^2·h·Pa)
CS/PVA	55.00 ± 2.00^{c}	23.94 ± 1.08^{c}	62.27 ± 1.26^{b}	10.89 ± 0.23^{b}	2.56 ± 0.12^{b}
S/PVA	71.33 ± 1.51^{b}	27.95 ± 1.69^{a}	43.45 ± 1.09^{c}	11.97 ± 0.57^{a}	3.25 ± 0.31^{ab}
CS/PVA-BA	84.48 ± 1.83^{b}	25.42 ± 2.01^{b}	87.84 ± 2.34^{a}	8.33 ± 0.57^{c}	2.68 ± 0.11^{b}
S/PVA-BA	99.00 ± 1.55^{a}	28.31 ± 1.28^{a}	64.52 ± 2.33^{b}	10.12 ± 0.15^{b}	3.81 ± 0.21^{a}

Values are expressed as means ± standard deviations. Different letters in the same column indicate significant differences ($p < 0.05$).

Tensile strength and elongation at break are fundamental indicators for evaluating the performance of packaging materials. Superior mechanical properties enhance resistance to compression during food transportation and improve product protection. Significant differences ($p < 0.05$) were observed in tensile strength and elongation at break among the four indicator films. Chitosan (CS), with its flexible molecular chains, exhibits high elongation at break. However, the incorporation of anthocyanins disrupts the crystalline structure of the film matrix, reducing rigidity. The hydrogen bonding between CS amino groups ($-NH_2$) and polyvinyl alcohol (PVA) hydroxyl groups ($-OH$) may be insufficient in cross-linking density, leading to molecular chain slippage and decreased tensile strength [36]. Starch–PVA films exhibit poor compatibility, resulting in high brittleness. While these films demonstrate higher tensile strength, their elongation at break is significantly lower than that of chitosan-based films [37,38]. The addition of anthocyanins enhances intermolecular interactions, improves structural stability, and partially enhances mechanical properties [39]. Chitosan and PVA films form strong intermolecular interactions, limiting water molecule penetration and maintaining a low water content [40]. The incorporation of anthocyanins further reduces gel hydration due to interactions between chitosan/starch and anthocyanin molecules, which restrict the availability of polysaccharide hydroxyl groups and weaken the water binding capacity.

Water vapor permeability (WVP) is a critical parameter for assessing the moisture barrier performance of food packaging materials. A lower WVP indicates superior resistance to external moisture ingress [41]. As shown in the table, chitosan–PVA gels exhibit lower WVP compared to starch-based gels. However, the addition of anthocyanins increases WVP, likely attributed to the hydrophilic phenolic hydroxyl groups in blueberry anthocyanins (BA), which enhance the film's water adsorption capability [42].

2.3. Antioxidant Properties of the Complex Polysaccharide Gels

The antioxidant properties of the gels were evaluated through DPPH and ABTS radical scavenging assays, as shown in Figure 4. The DPPH and ABTS radical scavenging rates of the CS/PVA gel were 11.69% and 5.98%, respectively, while those of the S/PVA gel were 8.28% and 8.97%. The antioxidant activity of the gels without anthocyanins was relatively low. However, with increasing anthocyanin concentrations, the scavenging rates of CS/PVA (4 mg/mL) reached 73.59% and 77.47%, and those of S/PVA (4 mg/mL) reached 71.69% and 62.46%, demonstrating significant antioxidant capacity. This activity is attributed to the phenolic hydroxyl groups (−OH) in anthocyanins, which donate hydrogen atoms or electrons to neutralize reactive oxygen species (ROS) such as superoxide anions (O_2^-), hydroxyl radicals (·OH), and hydrogen peroxide (H_2O_2), thereby interrupting free radical chain reactions and delaying lipid oxidation and protein denaturation [43]. Notably, the antioxidant efficacy of gels depends on the interaction between active components and the polymer matrix [44]. Factors such as the cross-linking status, moisture content, temperature, and microstructure of the gels may also influence the accessibility and stability of the antioxidant compounds [45].

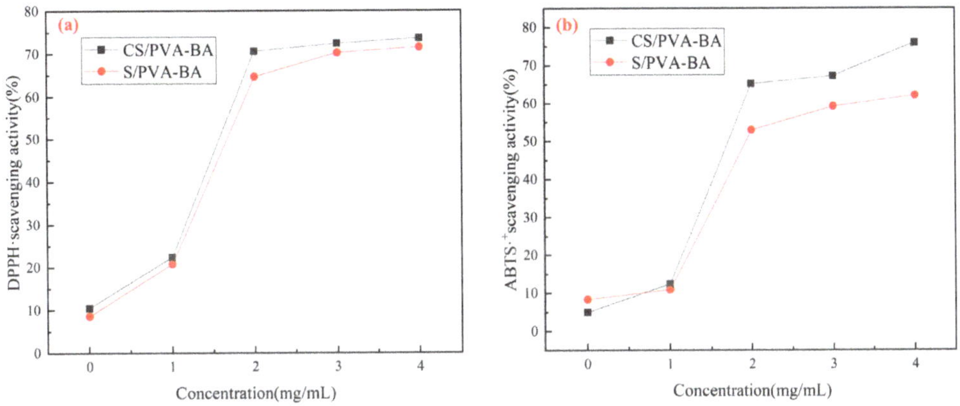

Figure 4. Antioxidant properties of the gels. (**a**) DPPH radical scavenging activity of the gels; (**b**) ABTS radical scavenging activity of the gels.

2.4. Analysis of Sensitivity of Complex Polysaccharide Gels to Ammonia

Volatile nitrogen-containing compounds such as ammonia, trimethylamine, and dimethylamine will be produced in the spoilage process of high-protein foods, resulting in an increase in the pH value in food packaging [46]. Simulating this process with ammonia gas is used to assess the sensitivity of the complex polysaccharide gels to volatile alkaline gases. The release of volatile amines causes the pH in the food packaging environment to rise, and anthocyanins are sensitive to pH changes, resulting in the gel in use showing a significant color change [28]. Figure 5a shows the color changes in the two gels in an ammonia environment in 8 min, and Figure 5b demonstrates the response kinetics and sensitivity of two indicator gels to ammonia. The rate of color change in the gels varied over time. During the initial 4 min, both CS/PVA-BA and S/PVA-BA gels exhibited rapid sensitivity enhancement, indicating their high reactivity and fast response to ammonia. Notably, S/PVA-BA achieved a maximum sensitivity of 69.931% at 8 min while CS/PVA-BA also showed significant responsiveness with a sensitivity of 53.016% at the same time point. The sensitivity enhancement during the reaction primarily resulted from NH_3 molecules interacting with water in the gel matrix, generating an alkaline environment that increased the gel's swelling capacity and thereby amplified its chromatic response.

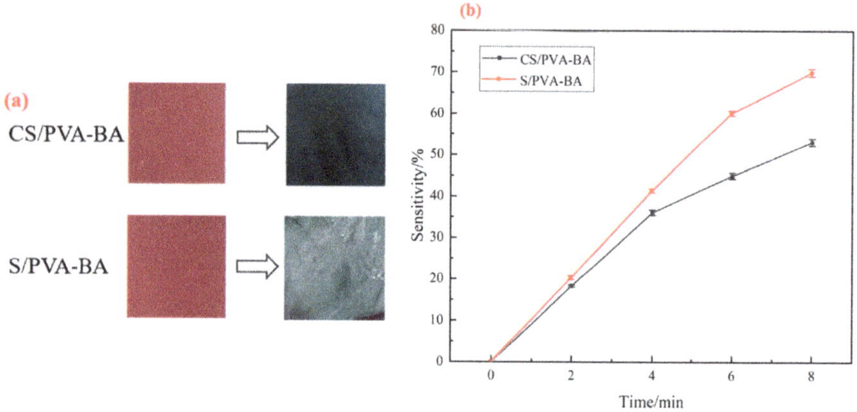

Figure 5. Response of the gels to ammonia. (**a**) Color change in the gels after 8 min of ammonia exposure. (**b**) Sensitivity of gel response to ammonia. Vertical bars represent the standard deviations of five replicates.

2.5. *Application of Complex Polysaccharide Gels in Beef Freshness Monitoring*

2.5.1. Analysis of Changes in Beef Physicochemical Properties

The spoilage of beef during storage at 4 °C was reflected by changes in total volatile basic nitrogen (TVB-N), which indicates protein degradation and microbial metabolic activity. According to national standards, fresh meat should have a TVB-N value $\leq$ 15 mg/100 g [47]. Beef is classified as sub-fresh when TVB-N ranges between 15 and 25 mg/100 g and spoiled when exceeding 25 mg/100 g [48]. As shown in the Figure 6, the initial TVB-N value of beef was 7.65 mg/100 g. After 2 days, it increased to 13.25 mg/100 g, approaching the upper limit for fresh meat. By day 8, the TVB-N value reached 31.25 mg/100 g, indicating complete spoilage with extensive protein breakdown and the production of hydrogen sulfide (H_2S) and amines [49].

The pH evolution of chilled beef is shown in Figure 6a. Freshness criteria classify meat as follows: pH 5.8–6.2 (fresh), pH 6.3–6.6 (sub-fresh), and pH > 6.7 (spoiled) [50]. The initial pH of fresh beef was 5.72. The post-slaughter glycolysis of residual glycogen under anaerobic conditions produced lactic acid, temporarily lowering the pH to 5.52 by day 2 [51]. During this phase, microbial activity (e.g., lactic acid bacteria) further decomposed carbohydrates, generating organic acids (e.g., lactic acid and acetic acid) and slightly reducing the pH [52]. Subsequently, dominant spoilage bacteria (e.g., Pseudomonas) metabolized proteins and amino acids, releasing alkaline metabolites (e.g., ammonia and amines), leading to a significant pH increase to 6.75 by day 8, confirming spoilage.

Microbial growth and metabolism are the primary drivers of meat spoilage [53]. Freshness standards define total viable counts (TVCs) as follows: <4.0 lg (CFU/g) (fresh), 4.0–6.0 lg (CFU/g) (sub-fresh), and >6.0 lg (CFU/g) (spoiled) [54]. A TVC exceeding 7.0 lg (CFU/g) indicates complete deterioration. As shown in Figure 6c, the initial TVC of beef was 2.67 lg (CFU/g). By day 4, it increased to 4.87 lg (CFU/g), categorizing the beef as sub-fresh. After 6 days, the TVC surpassed 6.0 lg (CFU/g), marking the onset of spoilage.

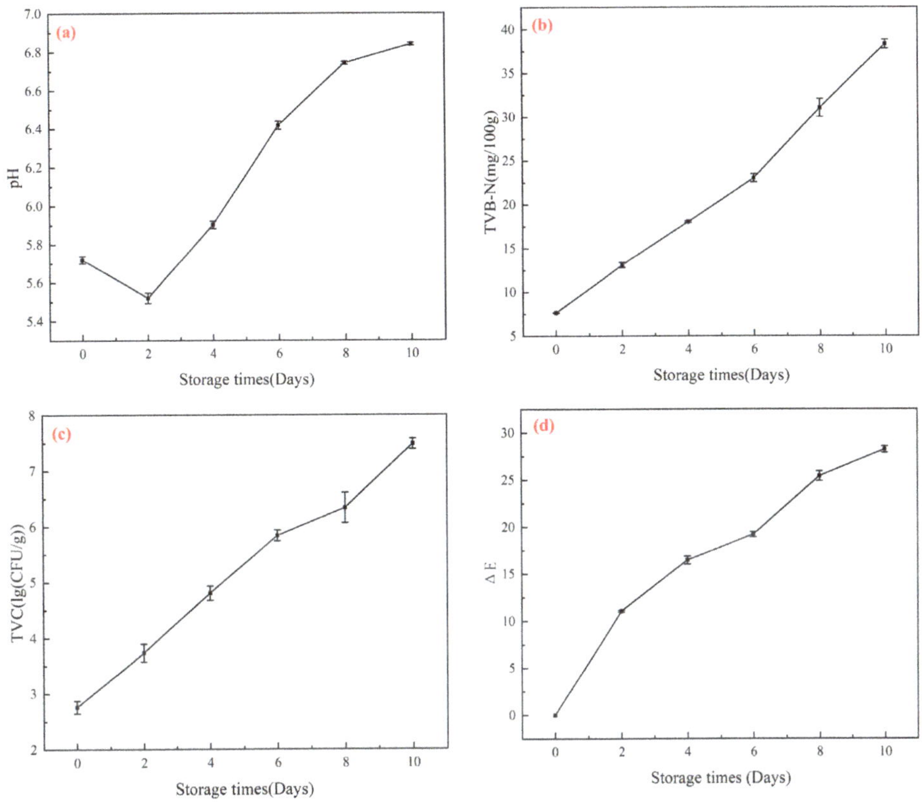

Figure 6. Application of the gels in beef freshness monitoring. (**a**) Changes in pH value during beef storage. (**b**) Changes in TVB-N value during beef storage. (**c**) Changes in the total number of colonies during beef storage. (**d**) Changes in the values during beef storage. Vertical bars represent the standard deviations of five replicates.

2.5.2. Freshness Monitoring Experiment

The CS/PVA-BA gel, selected for its high sensitivity and superior performance, was applied to monitor the freshness of chilled beef. Figure 7a illustrates the chromatic transitions of the gel during the 10-day storage of beef at 4 °C. Initially, the gel exhibited a magenta-red hue, which transitioned to pale magenta-red as storage progressed. Concurrent with the continuous accumulation of volatile nitrogenous compounds, the gel darkened and shifted toward bluish tones, turning bluish-gray by day 8 and fading to light bluish-gray by day 10. These color variations were visually discernible throughout the storage period.

Figure 6d, corresponding to Figure 7a, presents the color difference (ΔE) values of the gel during refrigeration. Higher ΔE values indicate more pronounced chromatic shifts. The CS/PVA-BA gel demonstrated a rapid increase in ΔE, reaching 16.51 by day 4 (visually noticeable color change) and 27.93 by day 8 (bluish-gray appearance). The gel's chromatic evolution closely correlated with beef quality deterioration, validating its applicability in detecting freshness.

A colorimetric indicator label (Figure 7b) was developed based on these transitions, enabling consumers to assess real-time beef freshness through visual comparison. The label features referenced colors corresponding to storage days 0, 4, 6, and 8. Day 0 (magenta-red): Fresh and safe for consumption. Day 4 (pale pink): Sub-fresh; immediate use recommended. Day 6 (pale magenta-red): Early spoilage. Day 8 (bluish-gray): Fully spoiled. This system will allow consumers to visually determine beef freshness, achieving the real-time monitoring of product quality.

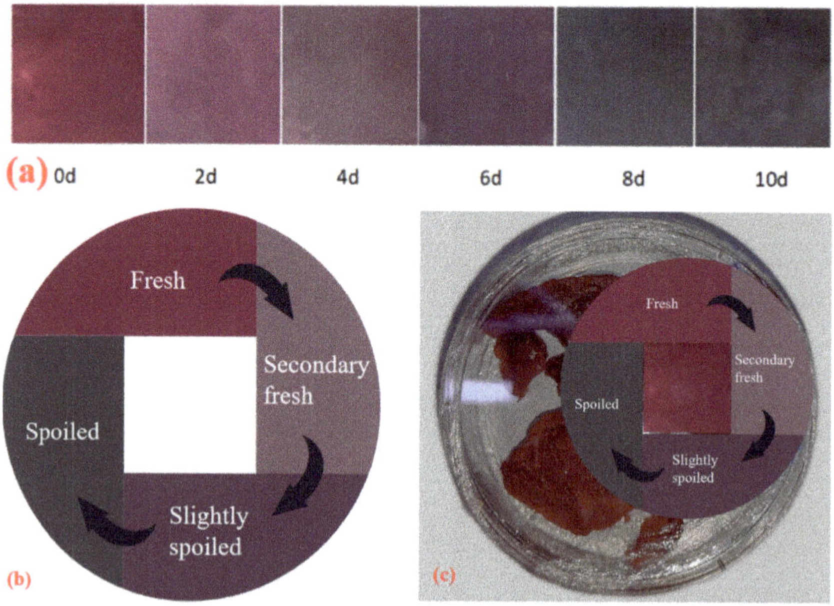

Figure 7. The color changes in the CS/PVA-BA gel for monitoring the freshness of beef stored at 4 °C. (**a**) Color changes in the gels at different storage times. (**b**) Label card image. (**c**) Usage of the label card.

3. Conclusions

Anthocyanins extracted from blueberry pomace were utilized as pH indicators to develop freshness-indicating gels by combining them with film-forming matrices of starch/polyvinyl alcohol (S/PVA) and chitosan/polyvinyl alcohol (CS/PVA). Four types of gels were prepared based on the presence or absence of anthocyanins. Microstructural analysis revealed good compatibility between the matrices and anthocyanins, with anthocyanin incorporation improving the physicochemical properties of the gels. Mechanical and sensitivity evaluations demonstrated that the CS/PVA-BA gel exhibited the highest elongation at break, a low water content (8.33 $\pm$ 0.57%), and reduced water vapor permeability (WVP) while the S/PVA-BA gel showed superior tensile strength and higher pH-responsive sensitivity. DPPH and ABTS radical scavenging assays confirmed the antioxidant capacity of the gels, effectively extending the shelf life of chilled beef. The CS/PVA-BA gel, selected for its optimal performance, was applied to monitor the freshness of chilled beef. Physicochemical analysis indicated initial spoilage at 48 h (TVB-N > 15 mg/100 g, TVC > 4.0 log CFU/g), with the gel transitioning from magenta-red to pale magenta-red. By day 6 of refrigeration, advanced spoilage was marked by a bluish-gray color shift. A smart colorimetric label card was developed to visually indicate freshness stages (0–8 days), enabling consumers to assess beef quality in real time. This composite polysaccharide gel integrates freshness monitoring and preservation functionalities, demonstrating significant potential for intelligent meat packaging. The study has provided guidance for the high-value utilization of blueberry pomace and aligned with current trends in food safety, smart packaging, and sustainable development.

4. Materials and Methods

4.1. Materials and Reagents

Blueberry pomace was provided by the Horticulture Branch of Heilongjiang Academy of Agricultural Sciences (2022); chilled fresh beef was purchased from a local supermarket (Harbin, China); polyvinyl alcohol (PVA) (degree of polymerization = 1700–1800; 88% alcoholysis), chitosan (CS) (deacetylated grade 85%), and starch (S) were obtained from Shanghai Macklin Biochemical Technology Co., Ltd., Shanghai, China; glycerol and citric acid were sourced from Shanghai Pinchen Biotechnology Co., Ltd., Shanghai, China. All other chemicals were analytical-grade and were purchased from Shanghai Macklin Biochemical Technology Co., Ltd., Shanghai, China.

4.2. Extraction and Quantification of Anthocyanins from Blueberry Pomace

Anthocyanins were extracted following the method of Fan [55] with slight modifications. Briefly, 200 g of frozen blueberry pomace was weighed using an electronic balance, mixed with 400 mL of 60% (*v*/*v*) acidified ethanol (pH around 4.5), and soaked at room temperature for 2.5 h. The mixture was homogenized for 5 min in a tissue homogenizer (WBL80Y21, Shanghai Yuanmai Trading Co., Ltd., Shanghai, China) and the homogenate was subjected to vacuum filtration. The filtrate was concentrated to near dryness under reduced pressure at 42 °C using a rotary evaporator (Rotavapor R-300, Buchi Labortechnik AG, Flaville, Switzerland). The concentrate was then adsorbed onto AB-8 macroporous resin, eluted with ethanol, and dried at 40 °C for 24 h to obtain anthocyanins.

The total anthocyanin content (TAC) following the method of Merz [21] in the extract was determined by spectrophotometer (UV-1800, Shimadzu Corporation, Kansai, Japan). Equal aliquots of 0.5 mL sample (1 mg/mL) were dissolved in 4.5 mL of 0.025 M potassium chloride buffer (pH = 1) and 4.5 mL of 0.4 m sodium acetate buffer (pH = 4.5) and placed in different test tubes. The absorbance of the extract was determined at 525 nm (A525 nm) and 700 nm (A700 nm). TAC was calculated using Equation (1):

$$\mathrm{TAC(mg/g)} = \frac{\left[(\mathrm{A}_{525\mathrm{nm}} - \mathrm{A}_{700\mathrm{nm}})_{\mathrm{pH}=1.0} - (\mathrm{A}_{525\mathrm{nm}} - \mathrm{A}_{700\mathrm{nm}})_{\mathrm{pH}=4.5}\right]\mathrm{M}\omega\mathrm{DF}}{\varepsilon\mathrm{L}} \quad (1)$$

Here, M, DF, ε, and L were the molecular weight of anthocyanin-3-glycoside (449.2 g/mol), dilution factor (10), molar extinction coefficient (29,600 L/(cm, mol)), optical path length (1 cm), and blank, respectively.

4.3. pH Response of Anthocyanin Solution

4.3.1. Color Response of Anthocyanins to pH

Buffer solutions (pH 2–14) were prepared and anthocyanin solution was added to each at a final concentration of 0.01% (*w*/*v*). Color changes were observed and photographed. The L*, a*, and b* values of the solution were determined by a colorimeter (CR-400/410, Konica Minolta, Inc., Tokyo, Japan). Total color difference (ΔE) was calculated using Equation (2):

$$\Delta\mathrm{E} = \sqrt{(\mathrm{L}^* - \mathrm{L}_0)^2 + (\mathrm{a}^* - \mathrm{a}_0)^2 + (\mathrm{b}^* - \mathrm{b}_0)^2} \quad (2)$$

Here, L*= the measured brightness value of the sample, a* = the measured red–green value of the sample, b* = the measured yellow–blue value of the sample; L_0 = the initial brightness value of the sample, a_0 = the initial red–green value of the sample, and b_0 = the initial yellow–blue value of the sample.

4.3.2. UV-VIS Spectra of Anthocyanins at Different pH Values

UV-VIS spectra of anthocyanin solutions (pH 2–14) were recorded using a spectrophotometer (UV-1800, Shimadzu Corporation, Japan) at wavelengths of 400–800 nm.

4.4. Preparation of Complex Polysaccharide Gels

Gels were prepared based on Xue's method with modifications [56]. CS, S, and PVA solutions were mixed to form gel substrates. CS solution was prepared by dissolving 1 g of chitosan in 100 mL of citric acid solution under magnetic stirring at 20 °C for 30 min. S solution (1 g starch/100 g solution) and PVA solution (1 g PVA/100 g solution) were prepared by heating at 95 ± 1 °C and 80 ± 1 °C, respectively.

Solutions were mixed at CS/PVA (60/40, %v/v) and S/PVA (40/60, %v/v) ratios. CS/PVA gel was prepared from a suspension of chitosan (0.6, %m/m), polyvinyl alcohol (0.4, %m/m), citric acid solution (0.25, %m/m), and glycerol (1, %m/m); S/PVA gel was prepared from a suspension of starch (0.4, %m/m), polyvinyl alcohol (0.6, %m/m), citric acid solution (0.25, %m/m), and glycerol (1, %m/m), followed by stirring at 20 °C for 30 min [57]. Gels were cast by pouring 18 mL of degassed solution into 9 cm Petri dishes and dried at 36 °C. The gel was colorless and transparent. The resulting gels (CS/PVA, S/PVA) were stored in the dark. The solution was mixed in a CS/PVA ratio of 60/40 (%v/v) and a S/PVA ratio of 40/60 (%v/v). CS/PVA-BA gel was prepared from a suspension of chitosan (0.6, %m/m) and polyvinyl alcohol (0.4, %m/m); S/PVA-BA gel was prepared from a suspension of starch (0.4, %m/m) and polyvinyl alcohol (0.6, %m/m). Citric acid solution (0.25, %m/m) and glycerol (1, %m/m) were added to the mixed solution, and the addition amount of blueberry pomace anthocyanin (pH = 4.5) was 2%. The mixture was stirred at 20 °C with a magnetic stirrer for 30 min. They were named CS/PVA-BA membrane and S/PVA-BA membrane, respectively. The gel was translucent and purplish-red.

4.5. Characterization of Gels

4.5.1. Microstructural Analysis

Gel surfaces were observed using scanning electron microscopy (SEM, 5 kV acceleration voltage, 1000× magnification) (SU1510, Hitachi, Ltd., Tokyo, Japan). Samples were cryo-fractured in liquid nitrogen and sputter-coated with gold.

4.5.2. FT-IR Analysis

Fourier-transform infrared (FT-IR) spectra (Nicolet iS10, Thermo Fisher Scientific Inc., Waltham, MA, USA) were recorded in the range of 4000–500 cm^{-1} with 16 scans per sample [48].

4.5.3. Thickness and Mechanical Properties

Gel thickness was measured using a digital micrometer (average of 8 random points, μm) (CH-1-ST, Shanghai Precision Instrument Co., Ltd., Shanghai, China). Tensile strength (TS) and elongation at break (EB) were tested using a microcomputer-controlled electronic universal testing machine (UTM5205X, Shenzhen Suns Technology Stock Co., Ltd., Shenzhen, China) in accordance with GB/T 1040.3-2006 [58]. Samples (60 mm × 20 mm) were stretched at 36 mm/min with a 40 mm initial grip distance. TS and EB were calculated using Equations (3) and (4):

$$\text{TS(Mpa)} = \frac{F}{\omega d} \tag{3}$$

Here, F = maximum load (N), ω = width (mm), and d = thickness (mm).

$$\text{EB}(\%) = \frac{L_0 - L_1}{L_0} \times 100 \tag{4}$$

Here, L_1= final length at break (mm) and L_0 = initial grip distance (mm).

4.5.4. Moisture Content

Taking Valencia's method and tweaking it slightly [59], samples (m_0) were dried at 105 °C to constant weight (m_1). Moisture content (MC) was calculated using Equation (5):

$$MC(\%) = \frac{m_0 - m_1}{m_0} \times 100 \quad (5)$$

Here, m_0 = indicates the initial mass(g) of the gel and m_1 = indicates the mass (g) of the gel after drying to a constant mass.

4.5.5. Determination of Water Vapor Transmission Coefficient

Taking Zou's method and tweaking it slightly [41], the composite gels were submerged in beakers containing 20 mL of distilled water and placed in a desiccator at 22 °C. Measurements were taken every 2 h, averaging the six weights from three parallel samples. The water vapor transmission coefficient (WVP) was calculated using Equation (6):

$$WVP\left(10^{-4}/(g \cdot mm)/\left(m^2 \cdot h \cdot Pa\right)\right) = \frac{\Delta m \times x}{A \times \Delta p \times t} \quad (6)$$

Here, Δm = the mass of water transmitted (g), x = the thickness of the gel (mm), A = the effective permeable area of the gel (m^2), Δp = the pressure difference between the two sides of the membrane (3179 Pa (22 °C)), and t = the interval time (s).

4.6. Antioxidant Properties

4.6.1. DPPH Radical Scavenging Activity

Gels (0.1 g) were soaked in 10 mL distilled water for 24 h. The extract (1 mL) was mixed with 3 mL of 0.1 mmol/L DPPH-methanol and incubated in the dark for 30 min. Absorbance was measured at 517 nm through spectrophotometry (UV-1800, Shimadzu Corporation, Kansai, Japan). DPPH scavenging rate was calculated using Equation (7) [60]:

$$DPPH(\%) = \left[1 - \frac{A_X - A_{X0}}{A_0}\right] \times 100 \quad (7)$$

Here, A_X, A_{X0}, andA_0 = absorbance of sample, control, and blank, respectively.

4.6.2. ABTS Radical Scavenging Activity

ABTS working solution (absorbance 0.7 ± 0.02 at 734 nm) was prepared by mixing 7.4 mmol/L ABTS and 2.6 mmol/L potassium persulfate (1:1), followed by 12 h incubation. Sample solution (1 mL) was mixed with 4 mL ABTS working solution, incubated for 6 min, and absorbance (A) was measured. Absorbance was measured at 734 nm through spectrophotometry (UV-1800, Shimadzu Corporation, Kansai, Japan). Scavenging activity was calculated using Equation (8) [61]:

$$ABTS(\%) = \frac{A_0 - A}{A_0} \times 100 \quad (8)$$

Here, A = the light absorption value of the sample and A_0 = the light absorption value of the blank.

4.7. Ammonia Sensitivity

According to the method of Kuswandi with appropriate modifications [62], in a 50 mL centrifuge tube containing 45 mL of 8 mmol/L ammonia solution, the gel (2 cm × 2 cm) was fixed within the culture medium and inverted on the bottle mouth. After 8 min, gel images were collected under room temperature conditions. Five sets of RGB values were extracted from the gel every two minutes using Photoshop 2018 color sampler, with mean values calculated to determine gel sensitivity. The calculation formula is shown in Equation (9):

$$S(\%) = \frac{|R - R_0| + |G - G_0| + |B - B_0|}{R + G + B} \times 100 \quad (9)$$

Here, R_0, G_0, and B_0 are the initial values before this gel reaction; R, G, and B are the values after the reaction.

4.8. Application in Beef Freshness Monitoring

4.8.1. Color Change

Gels were fixed on Petri dish lids and stored with beef at 4 °C. Color parameters (L^*, a^*, b^*) were measured every 2 days using a colorimeter (CR-400/410, Konica Minolta, Inc., Tokyo, Japan). Total color difference (ΔE) was calculated using Equation (10):

$$\Delta E = \sqrt{(L^* - L_0)^2 + (a^* - a_0)^2 + (b^* - b_0)^2} \quad (10)$$

Here, L^* = the measured brightness value of the sample, a^* = the measured red–green value of the sample, b^* = the measured yellow–blue value of the sample; L_0 = the initial brightness value of the sample, a_0 = the initial red–green value of the sample, and b_0 = the initial yellow–blue value of the sample.

4.8.2. TVB-N Determination

Total volatile base nitrogen (TVB-N) was measured according to GB 5009.228-2016 [63]. Sample Preparation: A quantity of 10 g of beef was added to 50 mL of water. The beef sample was pre-homogenized into a fine paste, thoroughly mixed, and oscillated for 30 min. The mixture was then filtered and the filtrate was transferred to a conical flask for subsequent analysis. Determination Procedure: A condenser was inserted into a receiving flask containing 10 mL of 2% boric acid absorption solution, which was supplemented with a mixed indicator solution (0.2% methyl red ethanol solution and 0.1% methylene blue aqueous solution, mixed in a 1:1 ratio). Quantities of 5 mL of the filtered sample solution and 5 mL of a 1% magnesium oxide suspension were combined and introduced into the reaction chamber of the distillation apparatus. The chamber was sealed and distilled water was added. Steam distillation was initiated, with timing starting upon collection of the first condensate droplet. The distillation was terminated after 5 min. The absorption solution was titrated with 0.01 mol/L hydrochloric acid (HCl) until the endpoint (transition to a bluish-purple hue) was reached. A blank experiment was performed in parallel. Total volatile base nitrogen (TVB-N) was calculated using Equation (11):

$$X(mg/100g) = \frac{(V_1 - V_2) \times c \times 14}{m} \times 100 \quad (11)$$

Here, V1 = the volume of hydrochloric acid consumed by the sample (mL), V2 = the volume of hydrochloric acid consumed by the blank (mL), c = the concentration of hydrochloric acid (mol/L), and m = the mass of the sample (g).

4.8.3. pH Measurement

Beef (10 g) was homogenized with 50 mL distilled water. The pH of the supernatant was measured after 30 min.

4.8.4. Total Bacterial Count

Total viable count (TVC) was determined using GB 4789.2-2016 [64] (plate count method). Under aseptic conditions, 10 g of ground beef sample was placed into a sterile bag, mixed with 90 mL of sterile saline solution, and homogenized for 120 s. The homogenate was then serially diluted at a 1:10 ratio. Three appropriate dilution gradients were selected, and three plates were prepared for each dilution, along with blank controls. The spread plate method was performed using plate count agar (PCA). The plates were incubated at 37 °C in a constant-temperature incubator for 48 h. The total viable count was calculated, and the results were expressed in lg CFU/g (logarithmic colony-forming units per gram). Three replicates were performed for each sample.

4.9. Statistical Analysis

Data were analyzed using SPSS26 and Origin2021 software. Results are expressed as means ± standard deviations. ANOVA was performed, and ($p < 0.05$) indicated statistical significance.

Author Contributions: J.Z.: writing—original draft; writing—review and editing; methodology; software; data curation; investigation; formal analysis; validation; funding acquisition; resources. F.X.: writing—original draft; methodology, investigation, formal analysis, data curation. S.Y.: investigation, formal analysis, resources. J.H.: visualization, methodology, formal analysis. J.W.: writing—review and editing. Z.F.: writing—review and editing, resources, project administration, funding acquisition, formal analysis, data curation. All authors have read and agreed to the published version of the manuscript.

Funding: This research was funded by the Provincial Innovation Training Program of Northeast Forestry University (S202310225145), the Heilongjiang Science Fund (LH2020C035), and the China Postdoctoral Science Foundation (2016M600239).

Institutional Review Board Statement: Not applicable.

Informed Consent Statement: Not applicable.

Data Availability Statement: All data are available within the manuscript.

Conflicts of Interest: The authors declare no conflicts of interest.

References

1. Xu, Y.; Dai, L. Analysis of texture parameters and determination of freshness index of beef in low temperature storage. *Trans. Chin. Soc. Agric. Eng.* **2016**, *32*, 267–272.
2. Zhang, R.; Xu, G.; Su, Y.; Rao, S. Potential Application of Ovalbumin Gel Nanoparticles Loaded with Carvacrol in the Preservation of Fresh Pork. *Gels* **2023**, *9*, 941. [CrossRef] [PubMed]
3. Tian, T.; Kang, Y.; Liu, L.J.; Wang, X.H. The effect of super-chilled preservation on shelf life and quality of beef during storage. *Food Sci. Technol.* **2022**, *42*, e73222. [CrossRef]
4. Pitirollo, O.; Messinese, E.; Grimaldi, M.; Barbanti, D.; Cavazza, A. Effects of a Biobased Antioxidant Gel on Meat Shelf-Life: Oxidative Stability and Color as Quality Parameters. *Gels* **2025**, *11*, 279. [CrossRef] [PubMed]
5. Guo, C.; Li, Y.; Zhang, H.; Zhang, Q.Y.; Wu, X.D.; Wang, Y.; Sun, F.D.; Shi, S.; Xia, X.F. A review on improving the sensitivity and color stability of naturally sourced pH-sensitive indicator films. *Compr. Rev. Food Sci. Food Saf.* **2024**, *23*, e13390. [CrossRef]
6. Kim, D.; Lee, S.; Lee, K.; Baek, S.; Seo, J. Development of a pH indicator composed of high moisture-absorbing materials for real-time monitoring of chicken breast freshness. *Food Sci. Biotechnol.* **2017**, *26*, 37–42. [CrossRef]
7. Choi, I.; Lee, J.Y.; Lacroix, M.; Han, J. Intelligent pH indicator film composed of agar/potato starch and anthocyanin extracts from purple sweet potato. *Food Chem.* **2017**, *218*, 122–128. [CrossRef]
8. Zheng, L.; Liu, L.; Yu, J.; Shao, P. Novel trends and applications of natural pH-responsive indicator film in food packaging for improved quality monitoring. *Food Control* **2022**, *134*, 108769. [CrossRef]
9. Wang, P.K.; Liu, J.N.; Zhuang, Y.H.; Fei, P. Acylating blueberry anthocyanins with fatty acids: Improvement of their lipid solubility and antioxidant activities. *Food Chem. X* **2022**, *15*, 100420. [CrossRef]

10. Ayvaz, H.; Cabaroglu, T.; Akyildiz, A.; Pala, C.U.; Temizkan, R.; Agcam, E.; Ayvaz, Z.; Durazzo, A.; Lucarini, M.; Direito, R.; et al. Anthocyanins: Metabolic Digestion, Bioavailability, Therapeutic Effects, Current Pharmaceutical/Industrial Use, and Innovation Potential. *Antioxidants* **2023**, *12*, 48. [CrossRef]
11. Lei, L.; Zhou, J.; Huang, Y.; Zhao, H.; Zhao, M.; Chen, J. Optimization of Extraction Conditions of Anthocyanin from Blueberry Pomace and Its Antioxidant Activity. *Sci. Technol. Food Ind.* **2018**, *39*, 178–184.
12. Akhtar, H.M.S.; Ahmed, S.; Olewnik-Kruszkowska, E.; Gierszewska, M.; Brzezinska, M.S.; Dembinska, K.; Kalwasinska, A. Carboxymethyl cellulose based films enriched with polysaccharides from mulberry leaves (*Morus alba* L.) as new biodegradable packaging material. *Int. J. Biol. Macromol.* **2023**, *253*, 127633. [CrossRef] [PubMed]
13. Wang, X.; Huang, L.; Zhang, C.; Deng, Y.; Xie, P.; Liu, L.; Cheng, J. Research advances in chemical modifications of starch for hydrophobicity and its applications: A review. *Carbohydr. Polym.* **2020**, *240*, 116292. [CrossRef]
14. Li, S.Z.; Ren, Y.; Hou, Y.J.; Zhan, Q.P.; Jin, P.; Zheng, Y.H.; Wu, Z.G. Polysaccharide-Based Composite Films: Promising Biodegradable Food Packaging Materials. *Foods* **2024**, *13*, 3674. [CrossRef]
15. Zhou, Y.H.; Fan, M.Z.; Luo, X.L.; Huang, L.L.; Chen, L.H. Acidic ionic liquid catalyzed crosslinking of oxycellulose with chitosan for advanced biocomposites. *Carbohydr. Polym.* **2014**, *113*, 108–114. [CrossRef]
16. Liu, W.M.; Chen, L.; McClements, D.J.; Peng, X.W.; Xu, Z.L.; Jin, Z.Y. Development of starch film to realize the value-added utilization of starch in food and biomedicine. *Food Biosci.* **2024**, *57*, 103521. [CrossRef]
17. Zong, Z.; Liu, M.; Chen, H.; Farag, M.A.; Wu, W.; Fang, X.; Niu, B.; Gao, H. Preparation and characterization of a novel intelligent starch/gelatin binary film containing purple sweet potato anthocyanins for *Flammulina velutipes* mushroom freshness monitoring. *Food Chem.* **2023**, *405*, 134839. [CrossRef]
18. Feng, Q.; Wang, L. Preparation and Application of Blueberry Anthocyanin Intelligent Indicator Films. *J. Chin. Inst. Food Sci. Technol.* **2022**, *22*, 281–290.
19. Antunes, J.C.; Tavares, T.D.; Teixeira, M.A.; Teixeira, M.O.; Homem, N.C.; Amorim, M.T.P.; Felgueiras, H.P. Eugenol-Containing Essential Oils Loaded onto Chitosan/Polyvinyl Alcohol Blended Films and Their Ability to Eradicate *Staphylococcus aureus* or *Pseudomonas aeruginosa* from Infected Microenvironments. *Pharmaceutics* **2021**, *13*, 195. [CrossRef]
20. Al-Tayyar, N.A.; Youssef, A.M.; Al-Hindi, R.R. Antimicrobial packaging efficiency of ZnO-SiO_2 nanocomposites infused into PVA/CS film for enhancing the shelf life of food products. *Food Packag. Shelf Life* **2020**, *25*, 100523. [CrossRef]
21. Merz, B.; Capello, C.; Leandro, G.C.; Moritz, D.E.; Monteiro, A.R.; Valencia, G.A. A novel colorimetric indicator film based on chitosan, polyvinyl alcohol and anthocyanins from jambolan (*Syzygium cumini*) fruit for monitoring shrimp freshness. *Int. J. Biol. Macromol.* **2020**, *153*, 625–632. [CrossRef] [PubMed]
22. Chen, H.Z.; Zhang, M.; Bhandari, B.; Yang, C.H. Novel pH-sensitive films containing curcumin and anthocyanins to monitor fish freshness. *Food Hydrocoll.* **2020**, *100*, 105438. [CrossRef]
23. Chen, S.Y.; Qian, Y.F.; Wang, Y.; Lyu, L.; Zhou, X.H. EVOH encapsulating blueberry anthocyanin dispersed in PVA nanofibers for real-time shrimp freshness monitoring based on pH-responsive colorimetric. *J. Text. Inst.* **2024**, *12*, 1–16. [CrossRef]
24. Prietto, L.; Pinto, V.Z.; El Halal, S.L.M.; de Morais, M.G.; Costa, J.A.V.; Lim, L.T.; Dias, A.R.G.; Zavareze, E.D. Ultrafine fibers of zein and anthocyanins as natural pH indicator. *J. Sci. Food Agric.* **2018**, *98*, 2735–2741. [CrossRef]
25. Zhao, Y.; Zhang, M.; Zhang, P.; Zhang, B.; Qi, Y.; Xian, L.; Wang, W.; Chen, L. Advances in smart pH complex membranes based on anthocyanins from different extracts. *Food Ferment. Ind.* **2025**, *51*, 373–380.
26. Yao, B.; Zhao, H.; Wu, W.; Li, W. Comparative Stability of Anthocyanins from Five Blueberry Cultivars. *Food Sci.* **2017**, *38*, 142–147.
27. Zou, X.B.; Zhang, J.J.; Shi, J.Y.; Jiang, C.P.; Zhai, X.D.; Wang, S.; Zhao, H.; Liang, N.N. Intelligent Indicator Film Based on Roselle Anthocyanins for Monitoring Pork Freshness. *Food Sci.* **2017**, *38*, 243–248.
28. Mohammadalinejhad, S.; Almasi, H.; Moradi, M. Immobilization of *Echium amoenum* anthocyanins into bacterial cellulose film: A novel colorimetric pH indicator for freshness/spoilage monitoring of shrimp. *Food Control* **2020**, *113*, 107169. [CrossRef]
29. Kochkina, N.E.; Lukin, N.D. Structure and properties of biodegradable maize starch/chitosan composite films as affected by PVA additions. *Int. J. Biol. Macromol.* **2020**, *157*, 377–384. [CrossRef]
30. Dong, Y.; Li, W.; Sun, W.; Jia, L.; Li, L.; Jin, Z.; Sun, W. Porous poly-L-lactic acid nanofiber indicator membrane loaded with anthocyanins for non-destructive testing of mutton freshness. *Food Ferment. Ind.* **2024**, *50*, 278–283.
31. Wang, J.; Zhang, Q.; Zhang, Z.; Li, Z. International journal of biological macromolecules. *Int. J. Biol. Macromol.* **2024**, *280*, 135553.
32. Pereira, V.A., Jr.; de Arruda, I.N.Q.; Stefani, R. Active chitosan/PVA films with anthocyanins from *Brassica oleraceae* (Red Cabbage) as Time–Temperature Indicators for application in intelligent food packaging. *Food Hydrocoll.* **2015**, *43*, 180–188. [CrossRef]
33. Gutiérrez, T.J.; Toro-Márquez, L.A.; Merino, D.; Mendieta, J.R. Hydrogen-bonding interactions and compostability of bionanocomposite films prepared from corn starch and nano-fillers with and without added Jamaica flower extract. *Food Hydrocoll.* **2019**, *89*, 283–293. [CrossRef]
34. Silva-Pereira, M.C.; Teixeira, J.A.; Pereira-Júnior, V.A.; Stefani, R. Chitosan/corn starch blend films with extract from *Brassica oleraceae* (red cabbage) as a visual indicator of fish deterioration. *LWT-Food Sci. Technol.* **2015**, *61*, 258–262. [CrossRef]

35. Yong, H.; Liu, J.; Kan, J.; Liu, J. Active/intelligent packaging films developed by immobilizing anthocyanins from purple sweetpotato and purple cabbage in locust bean gum, chitosan and κ-carrageenan-based matrices. *Int. J. Biol. Macromol.* **2022**, *211*, 238–248. [CrossRef]
36. Mohamed, S.A.A.; El-Sakhawy, M.; El-Sakhawy, M.A.M. Polysaccharides, Protein and Lipid -Based Natural Edible Films in Food Packaging: A Review. *Carbohydr. Polym.* **2020**, *238*, 116178. [CrossRef]
37. Wu, J.Y.; Ooi, C.W.; Song, C.P.; Wang, C.Y.; Liu, B.L.; Lin, G.Y.; Chiu, C.Y.; Chang, Y.K. Antibacterial efficacy of quaternized chitosan/poly (vinyl alcohol) nanofiber membrane crosslinked with blocked diisocyanate. *Carbohydr. Polym.* **2021**, *262*, 117910. [CrossRef]
38. Liang, J.; Huang, F.; Yu, Y.; Ouyang, Y.; Hu, Y.; Xiang, D.; Shu, Y. Preparation and Properties of Starch/Polyvinyl Alcohol Composite. *Plastics* **2020**, *49*, 5.
39. Ma, Q.; Wang, L. Preparation of a visual pH-sensing film based on tara gum incorporating cellulose and extracts from grape skins. *Sens. Actuators B Chem.* **2016**, *235*, 401–407. [CrossRef]
40. Ghanbarzadeh, B.; Almasi, H.; Entezami, A.A. Improving the barrier and mechanical properties of corn starch-based edible films: Effect of citric acid and carboxymethyl cellulose. *Ind. Crops Prod.* **2011**, *33*, 229–235. [CrossRef]
41. Zou, J.H.; Zhong, H.M.; Jiang, C.; Zhu, G.C.; Lin, X.E.; Huang, Y.Y. *Ginkgo biloba* leaf polysaccharide-stabled selenium nanozyme as an efficient glutathione peroxidase mimic for the preservation of bananas and cherry tomatoes. *Food Chem.* **2024**, *459*, 140443. [CrossRef] [PubMed]
42. Yong, H.M.; Liu, J. Recent advances in the preparation, physical and functional properties, and applications of anthocyanins-based active and intelligent packaging films. *Food Packag. Shelf Life* **2020**, *26*, 100550. [CrossRef]
43. Geoffroy, T.R.; Meda, N.R.; Stevanovic, T. Suitability of DPPH spiking for antioxidant screening in natural products: The example of galloyl derivatives from red maple bark extract. *Anal. Bioanal. Chem.* **2017**, *409*, 5225–5237. [CrossRef]
44. Xue, S.; Li, C.; Xiong, Z. Preparation of Complex Polysaccharide Gels with *Zanthoxylum bungeanum* Essential Oil and Their Application in Fish Preservation. *Gels* **2024**, *10*, 533. [CrossRef]
45. Gemili, S.; Yemenicioglu, A.; Altinkaya, S.A. Development of antioxidant food packaging materials with controlled release properties. *J. Food Eng.* **2010**, *96*, 325–332. [CrossRef]
46. Alizadeh-Sani, M.; Tavassoli, M.; Mohammadian, E.; Ehsani, A.; Khaniki, G.J.; Priyadarshi, R.; Rhim, J.-W. pH-responsive color indicator films based on methylcellulose/chitosan nanofiber and barberry anthocyanins for real-time monitoring of meat freshness. *Int. J. Biol. Macromol.* **2021**, *166*, 741–750. [CrossRef]
47. Chen, R.; Li, H.; Wang, J.; He, Z. Preparation of antioxidant activity packaging film and its application in meat products. *Food Ferment. Ind.* **2021**, *47*, 287–294.
48. Moradi, M.; Tajik, H.; Almasi, H.; Forough, M.; Ezati, P. A novel pH-sensing indicator based on bacterial cellulose nanofibers and black carrot anthocyanins for monitoring fish freshness. *Carbohydr. Polym.* **2019**, *222*, 115030. [CrossRef]
49. Vareltzis, K.; Koufidis, D.; Gavriilidou, E.; Papavergou, E.; Vasiliadou, S. Effectiveness of a natural rosemary (*Rosmarinus officinalis*) extract on the stability of filleted and minced fish during frozen storage. *Z. Für Leb. Und-Forsch. A* **1997**, *205*, 93–96. [CrossRef]
50. Delbarre-Ladrat, C.; Chéret, R.; Taylor, R.; Verrez-Bagnis, V. Trends in postmortem aging in fish: Understanding of proteolysis and disorganization of the myofibrillar structure. *Crit. Rev. Food Sci. Nutr.* **2006**, *46*, 409–421. [CrossRef]
51. Lin, Y.D.; Ma, J.; Cheng, J.H.; Sun, D.W. Visible detection of chilled beef freshness using a paper-based colourimetric sensor array combining with deep learning algorithms. *Food Chem.* **2024**, *441*, 138344. [CrossRef]
52. Liu, X.X.; Song, X.S.; Gou, D.J.; Li, H.L.; Jiang, L.; Yuan, M.L.; Yuan, M.W. A polylactide based multifunctional hydrophobic film for tracking evaluation and maintaining beef freshness by an electrospinning technique. *Food Chem.* **2023**, *428*, 136784. [CrossRef]
53. Liu, H.; Zhu, L.; Ji, Z.T.; Zhang, M.; Yang, X.T. Porphyrin fluorescence imaging for real-time monitoring and visualization of the freshness of beef stored at different temperatures. *Food Chem.* **2024**, *442*, 138420. [CrossRef]
54. Bhadury, D.; Nadeem, H.; Lin, M.; Dyson Jennifer, M.; Tuck Kellie, L.; Tanner, J. Application of on-pack pH indicators to monitor freshness of modified atmospheric packaged raw beef. *Food Qual. Saf.* **2024**, *8*, 710–719. [CrossRef]
55. Hao, J.; Wang, J.; Wang, J.; Wang, S.; Zhao, C.; Fan, Z. A novel multifunctional bioactive film based on *Gelidium amausli* polysaccharide incorporated with lingonberry anthocyanin for fish freshness monitoring and preservation. *LWT* **2025**, *215*, 117173. [CrossRef]
56. Xue, J. Anthocyanin-Based Intelligent Packaging Material for Monitoring Salmon Freshness. Master's Thesis, Jiangsu University, Zhenjiang, China, 2019.
57. Golasz, L.B.; Silva, J.d.; Silva, S.B.d. Filme com antocianinas como indicador da deterioração de carne suína refrigerada. *Food Sci. Technol.* **2013**, *33*, 155–162. [CrossRef]
58. *GB/T 1040.3-2006*; Plastics—Determination of Tensile Properties—Part 3: Test Conditions for Films and Sheets. Standardization Administration of the People's Republic of China: Beijing, China, 2006.
59. Valencia, G.A.; Luciano, C.G.; Lourenço, R.V.; Bittante, A.; Sobral, P.J.D. Morphological and physical properties of nano-biocomposite films based on collagen loaded with laponite®. *Food Packag. Shelf Life* **2019**, *19*, 24–30. [CrossRef]

60. Liu, D.; Chen, X.; Huang, M.; Zhou, G. Antioxidant activity of peptides in postmortem aged duck meat as affected by cooking and in vitro digestion. *Int. J. Food Prop.* **2019**, *22*, 727–736. [CrossRef]
61. Dodange, S.; Shekarchizadeh, H.; Kadivar, M. Real-time tracking of fish quality using a cellulose filter paper colorimetric indicator incorporated with prickly pear fruit betacyanins. *LWT* **2024**, *205*, 116523. [CrossRef]
62. Kuswandi, B.; Restyana, A.; Abdullah, A.; Heng, L.Y.; Ahmad, M. A novel colorimetric food package label for fish spoilage based on polyaniline film. *Food Control* **2012**, *25*, 184–189. [CrossRef]
63. *GB 5009.228-2016*; National Food Safety Standard—Determination of Volatile Basic Nitrogen in Foods. National Health and Family Planning Commission of the People's Republic of China: Beijing, China, 2016.
64. *GB 4789.2-2016*; National Food Safety Standard—Microbiological Examination of Food—Enumeration of Total Aerobic Bacterial Count. National Health and Family Planning Commission of the People's Republic of China: Beijing, China, 2016.

gels

MDPI

Article

Development, Characterization, and Stability of Margarine Containing Oleogels Based on Olive Oil, Coconut Oil, Starch, and Beeswax

Bárbara Viana Barbosa Naves [1], Thais Lomonaco Teodoro da Silva [2], Cleiton Antônio Nunes [2], Felipe Furtini Haddad [2] and Sabrina Carvalho Bastos [1,*]

[1] Nutrition Department, Federal University of Lavras, Lavars 37203-202, Minas Gerais, Brazil; vianaba9@gmail.com
[2] Food Science Department, Federal University of Lavras, Lavars 37203-202, Minas Gerais, Brazil; thaissilva@ufla.br (T.L.T.d.S.); cleiton.nunes@ufla.br (C.A.N.); felipe.haddad@ufla.br (F.F.H.)
* Correspondence: sabrinabastos@ufla.br

Abstract: The removal of partially hydrogenated fats, as well as the substitution of saturated fats with healthier alternatives, has become increasingly common due to their well-established association with adverse health effects. As a result, the demand for alternative formulations in the food industry has driven the development of a promising emerging technology: oleogels. Oleogels are a semi-solid material made by trapping liquid oil within a three-dimensional network formed by structuring agents. Within this context, this study aimed to develop and characterize margarines prepared with oleogels formulated from extra virgin olive oil, coconut oil, starch, and beeswax at varying concentrations. The proposed oleogel-based formulations exhibited a high melting temperature range and lower enthalpy. Although lipid oxidation levels differed between the commercial and oleogel-based margarines, they remained within acceptable limits. A significant difference in color was observed, with the oleogel formulations imparting a slight greenish hue compared to the commercial margarine. In terms of microstructure, the commercial margarine presented smaller and more uniformly distributed water droplets. Oleogel-based margarines demonstrated technological feasibility. Considering consumers' growing interest in food innovation and health-conscious products, olive oil-based oleogel margarines represent a promising alternative, particularly due to the nutritional benefits associated with olive oil.

Keywords: trans-fatty acid; beeswax; saturated fatty acid; lipid oxidation; organogel

1. Introduction

The human diet plays a key role in the prevalence of chronic noncommunicable diseases, such as cardiovascular diseases. Both the quantity and the quality of the food consumed, especially fats, are extremely important in health [1]. Therefore, the elimination of partially hydrogenated fats is a key goal in the food industry due to their well-supported association with increased cardiovascular risk, and several guidelines recommend their exclusion from the diet [2].

Hydrogenated and interesterified fats are widely used in food products because they confer desirable sensory and technological properties, including consistency, specific melting characteristics, flavor, cost reduction, and extended shelf life [1]. However, the removal of these components—often rich in saturated and trans-fatty acids—presents a significant challenge for the processed food sector. As such, there is a need for fat substitutes that are technologically viable, offer sensory acceptance, maintain stability during and after processing, and meet texture and spreadability requirements [3].

In recent years, significant advances have been made in the study of oleogels, which can be used to replace trans and saturated fats in processed foods such as pastes, bars, margarines, cakes, and ice creams [4]. Oleogels are semi-solid systems formed by entrapping liquid oils within a three-dimensional network created by structuring agents [5,6].

The preparation of oleogels requires careful selection of both oils and structuring agents, with their proportions tailored to the specific food application. A successful formulation must combine components in a way that enhances functional performance. Nevertheless, challenges remain, particularly in achieving and maintaining the desired physical properties of the final product [7].

Olive oil, known for its well-documented health benefits, is widely used as a base oil in oleogel formation [8–10]. It is rich in monounsaturated fatty acids and is associated with numerous health benefits, including reduced risk of cardiovascular disease and cancer, lower body mass index with long-term moderate consumption, and anti-inflammatory properties [11].

Coconut oil, although predominantly composed of saturated medium-chain fatty acids, has an intermediary melting point and is extensively used in the food industry due to its structural and stabilizing properties at room temperature [12,13]. Its medium-chain triglycerides (MCTs) yield different metabolic effects compared to long-chain saturated fats, with some studies reporting increased energy expenditure and fat oxidation. Despite its high saturated fat content, moderate coconut oil consumption has been linked to favorable lipid profiles, such as increased HDL cholesterol without a significant rise in LDL cholesterol [14,15]. However, few studies have explored the combination of olive oil and coconut oil in oleogel systems, particularly in health-oriented food formulations.

Margarine is a fat-based product used both for direct consumption and as an ingredient in baking, confectionery, and homemade preparations [16,17]. While previous studies on Brazilian margarines report low trans-fat levels, they often contain high amounts of saturated fats [18]. Reducing saturated fat content improves the nutritional profile by increasing unsaturated fatty acid levels and lowering atherogenic and thrombogenic indices, thereby reducing cardiovascular risk [18]. According to the 2015–2020 Dietary Guidelines for Americans, replacing trans and saturated fats with polyunsaturated fats significantly lowers blood cholesterol and cardiovascular risk [19].

Similarly, the 2017 Brazilian Guidelines for Dyslipidemia and Atherosclerosis Prevention [20] recommend eliminating trans fats and partially replacing saturated fats with mono- and polyunsaturated fats. These recommendations are echoed in the 2019 European guidelines for dyslipidemia management [21], and the World Health Organization advises limiting trans-fat intake to less than 1% of total energy intake, while encouraging replacement of saturated fats with polyunsaturated fats [22].

Given consumers' growing interest in healthier foods and the innovative potential of oleogel systems, margarine based on olive oil oleogels represents a promising alternative. However, a consistent challenge with wax-based oleogels is their tendency to produce an undesirable waxy mouthfeel [16,23,24]. Moreover, achieving both low saturation and zero trans fats remains difficult when reformulating water-in-oil emulsions such as margarine [18]. In this context, the present study aimed to develop a novel margarine using a multi-component oleogel with extra virgin olive oil, coconut oil, starch, and beeswax.

2. Results and Discussion

2.1. Effect of Oleogelators on Margarine Formation and Physical Properties

To optimize the texture of the margarines, the values of firmness, adhesiveness, and consistency of oleogel-based margarines were targeted to match those of a commercial margarine [25]. These individual responses were transformed into desirability values, which consolidate multiple textural attributes into a single score ranging from 0 to 1. A desirability value closer to one indicates a texture more closely resembling that of the commercial reference. Table 1 presents the experimental design matrix used to evaluate the effects of the oleogel components on texture desirability.

Table 1. Experimental design matrix used to evaluate the effects of the oleogel components on texture desirability of the formulated margarines. Experimental values are shown in parentheses.

Formulation	CS (g/100 g)	BW (g/100 g)	OO/CO (g/100 g)	Average Desirability
1	−1.00 (1.80)	−1.00 (2.80)	−1.00 (36.10/39.30)	0.00
2	−1.00 (1.80)	−1.00 (2.80)	1.00 (53.90/21.50)	0.85
3	−1.00 (1.80)	1.00 (5.20)	−1.00 (36.10/36.90)	0.40
4	−1.00 (1.80)	1.00 (5.20)	1.00 (53.90/19.10)	0.82
5	1.00 (4.20)	−1.00 (2.80)	−1.00 (36.10/36.90)	0.46
6	1.00 (4.20)	−1.00 (2.80)	1.00 (53.90/19.10)	0.36
7	1.00 (4.20)	1.00 (5.20)	−1.00 (36.10/45.50)	0.53
8	1.00 (4.20)	1.00 (5.20)	1.00 (53.90/16.70)	0.46
9	−1.68 (0.98)	0.00 (4.00)	0.00 (45.00/30.02)	0.87
10	1.68 (5.02)	0.00 (4.00)	0.00 (45.00/25.98)	0.76
11	0.00 (3.00)	−1.68 (1.98)	0.00 (45.00/30.02)	0.65
12	0.00 (3.00)	1.68 (6.02)	0.00 (45.00/25.98)	0.81
13	0.00 (3.00)	0.00 (4.00)	−1.68 (30.03/42.97)	0.52
14	0.00 (3.00)	0.00 (4.00)	1.68 (59.97/13.03)	0.47
15	0.00 (3.00)	0.00 (4.00)	0.00 (45.00/28.00)	0.20
16	0.00 (3.00)	0.00 (4.00)	0.00 (45.00/28.00)	0.10
17	0.00 (3.00)	0.00 (4.00)	0.00 (45.00/28.00)	0.00

CS: corn starch, BW: beeswax, OO: olive oil, CO: coconut oil.

The concentrations of corn starch (CS), beeswax (BW), extra virgin olive oil (OO), and coconut oil (CO) were modeled using a quadratic regression to evaluate their effects on the desirability associated with the texture of a commercial margarine. The quality of the model fit was assessed using pure error analysis. The regression coefficients and results of the analysis of variance (ANOVA) are presented in Tables 2 and 3, respectively. Among the model terms, only the interaction between CS and BW was not statistically significant ($p > 0.05$), indicating that this interaction did not have a meaningful effect on texture. All other main effects and interactions were statistically significant ($p < 0.05$), demonstrating that variations in these components had a direct influence on the textural attributes of the margarine, either enhancing or diminishing overall desirability.

According to Table 3, although the model exhibited a significant lack of fit, it still showed a satisfactory coefficient of determination ($R^2 = 0.76$), and the regression was statistically significant ($p < 0.05$). This indicates that the model reasonably described the relationship between the independent and dependent variables. Furthermore, the small difference between R^2 and the adjusted R^2 (0.08) reinforces the adequacy of the model fit.

In general, an increase in corn starch concentration was associated with a reduction in the overall desirability, indicating a negative effect on the target texture parameters. Beeswax (BW) acted as the primary oleogelator, contributing significantly to the structure and texture of the oleogel-based margarine. This trend is consistent with previous findings, such as those reported for oleogels structured with potato starch and candelilla wax, where higher starch levels led to reduced stability, and in the absence of wax, self-sustaining gels could not be formed [26,27].

Table 2. Regression coefficients of the quadratic model used to evaluate the effects of the oleogel components on texture desirability of the formulated margarines.

Coefficient	Value	Error	t	*p*
b_0	0.12	0.03	4.22	0.000464
CS	−0.03	0.01	−2.54	0.019896
BW	0.06	0.01	4.55	2.19×10^{-4}
OO/CO	0.07	0.01	5.70	1.70×10^{-5}
CS × BW	−0.03	0.02	−1.49	0.152451
CS × OO/CO	−0.18	0.02	−10.68	1.81×10^{-9}
BW × OO/CO	−0.05	0.02	−3.02	7.04×10^{-3}
CS^2	0.20	0.01	14.16	1.51×10^{-11}
BW^2	0.17	0.01	12.07	2.35×10^{-10}
OO/CO^2	0.09	0.01	6.30	4.79×10^{-6}

CS: corn starch, BW: beeswax, OO: olive oil, CO: coconut oil.

Table 3. ANOVA for the quadratic model used to evaluate the effects of the oleogel components on texture desirability of the formulated margarines.

	GL	SQ	QM	F	*p*-Value
Regression	2.04	8	0.26	9.90	4.23×10^{-6}
Residue	0.64	25	0.03		
Lack of fit	0.56	6	0.09	20.58	2.27×10^{-7}
Pure error	0.09	19	0.00		
Total	2.69	33			
R^2	0.76				
R^2 adjusted	0.68				

Among the formulations tested, those that achieved high desirability values (>0.8)—notably formulations 2, 4, and 9—had lower concentrations of corn starch, while formulation 12 attained a similar desirability through a markedly high BW content. These results align with other studies using oleogel systems based on wax and monoglycerides, where the mechanical properties of the gels were predominantly influenced by wax concentration rather than by other components [28,29]. However, for oleogel emulsions, the partial substitution of wax with starch appears to promote the formation of a softer and more uniform oleogel system capable of absorbing up to 45 times more water. This results in a gel with a solid-like appearance, reduced candelilla wax content, and improved spreadability [27]. These findings support the complementary structuring roles of corn starch and wax—where corn starch more effectively structures the aqueous phase and wax structures the lipid phase—with beeswax being particularly efficient due to the higher oil content in the formulation.

Based on the results described above, a numerical optimization was conducted to determine the proportions of corn starch, beeswax, extra virgin olive oil, and coconut oil that would yield a desirability value of one in the response surface model. This target desirability reflects textural attributes (firmness, adhesiveness, and consistency) closest to those of commercial margarine. As a result, three optimized margarine formulations were generated for further analysis:

Margarine formulation 1 (M1): 1.8% CS, 4.0% BW, and 73.9% combined OO and CO, with a ratio of 60% OO to 40% CO.

Margarine formulation 2 (M2): 1.8% CS, 5.2% BW, and 72.7% combined OO and CO, with a ratio of 57.5% OO to 42.5% CO.

Margarine formulation 3 (M3): 1.0% CS, 5.2% BW, and 73.5% combined OO and CO, with a ratio of 45% OO to 55% CO.

In addition to the optimized formulations prepared for further analyses, a commercial margarine sample was also evaluated to serve as a reference standard, enabling comparison of the parameters of the newly developed products with those of an existing market option.

2.2. Melting Behavior

The melting profiles of the margarine formulations (M1, M2, M3) and the commercial margarine (MC) are presented in Figure 1. The commercial margarine exhibited a lower melting temperature range, with a peak melting temperature (T_p) of 34.46 °C, compared to all oleogel-based margarines. Initially, both commercial and oleogel margarines had similar T_p values, as also reported in previous studies [16,23]. However, the oleogel margarines presented a broader and higher melting range, approximately between 42 °C and 55 °C (Tp ranging from 46 °C to 49 °C). This elevated melting temperature is likely attributable to the melting point of beeswax, which acts as the primary structuring agent in the oleogel formulations [16,30,31].

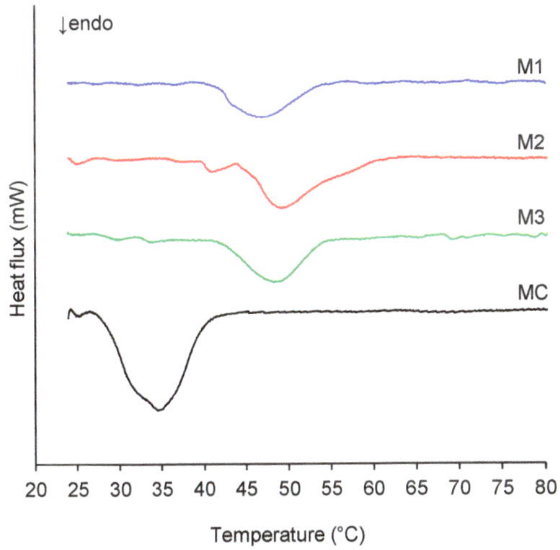

Figure 1. Melting profile (DSC) of the margarine formulations expressed in milliwatts (mW) as a function of temperature.

The formulations containing oleogels exhibited onset melting temperatures (T_{onset}) between 42 and 45 °C, with maximum melting peak temperatures of 46.69 °C, 48.26 °C, and 49.21 °C for M1, M3, and M2, respectively. The higher T_p values for M3 and M2 correspond to their greater beeswax content. M1, which had the highest proportion of olive oil and the lowest wax concentration among the oleogel-based formulations, showed a comparatively lower melting range. A notable difference was also observed in the enthalpy values: the commercial margarine had an enthalpy of 5.20 mW, whereas the oleogel margarines ranged from 1.9 to 2.8 mW. This reduced enthalpy in the oleogel samples is commonly attributed to their higher unsaturated oil content [23], with M1, containing the most olive oil, presenting the lowest enthalpy (1.9 mW). From a practical perspective, oleogel-based margarines demonstrate greater stability at room temperature, enabling them to remain unrefrigerated for longer periods without melting or losing creaminess—qualities that are highly desirable for consumers in everyday use [16,23].

2.3. Color

Visual attributes are among the primary and most noticeable characteristics perceived in a product [32]. Consequently, color is a crucial factor influencing consumer acceptance of new products. Figure 2 illustrates the visual appearance of the margarine formulations evaluated in this study.

Figure 2. Visual appearance of the evaluated margarine formulations.

Initially, formulations M1, M2, and M3 exhibited similar color characteristics, which appeared slightly lighter than those of the commercial margarine—an observation confirmed by instrumental analysis. Figure 3 presents the results for lightness (L*), the green-red axis (a*), and the blue-yellow axis (b*), highlighting the differences among all formulations as well as their changes over time.

The lightness parameter (L*) ranged from 77 to 86 across all formulations. Significant differences in L* values were observed among the samples at each time point. The ascending order of lightness was M2 < M3 < M1 < MC, suggesting that beeswax had a substantial influence on luminosity. All formulations exhibited a significant increase in lightness after 30 and 60 days of storage, followed by a decrease at 90 days, indicating a gradual loss of lightness during the shelf life of both the commercial and oleogel-based margarines.

Similarly, the a* values varied significantly among formulations at all time points. The oleogel-based margarines had negative a* values, indicating a shift toward green tones, whereas the commercial margarine exhibited positive a* values, reflecting a redder hue. This difference is likely attributable to the presence of extra virgin olive oil in the oleogel formulations [9,16].

Regarding the b* coordinate, although significant differences were observed among samples, all values were positive, ranging from 27 to 37, indicating a consistent yellow hue—characteristic of margarine products. Among the oleogel formulations, M2 showed the least variation in b* values over time, while M3's b* values closely matched those of the commercial sample throughout the study.

Comparable results were reported in previous studies for the b* coordinate; however, the a* values differed, as all their formulations exhibited positive a* values [23]. The variations in a* among the oleogel formulations in the present study may be due to experimental variability or interactions with other formulation components whose concentrations differed. Despite minor shifts in color parameters, the product remained within acceptable visual quality standards [23].

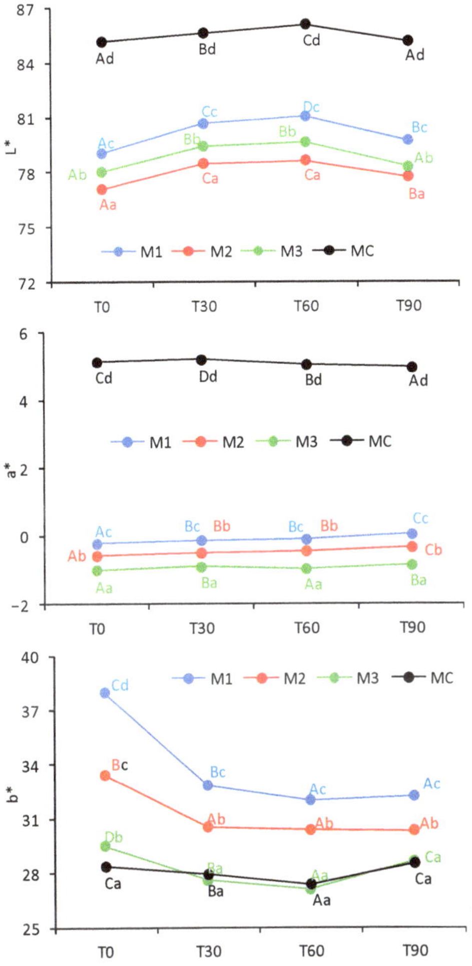

Figure 3. Color parameters of the margarines over time: L* (lightness), a* (red-green), and b* (yellow-blue), shown from top to bottom. Uppercase letters indicate statistical differences within each formulation over time, while lowercase letters denote comparisons between formulations at each time point.

2.4. Thermal Cyclization

The formulations were subjected to thermal cycling to evaluate their stability under elevated temperatures and fluctuating storage conditions. Figure 4 presents the applied storage conditions and depicts the appearance of the formulations at the conclusion of the experiment.

Based on the presented images, the standard commercial sample exhibited the lowest stability, as evidenced by substantial oil exudation in the central and upper regions of the product. This exudation progressively increased, reaching approximately 4 milliliters (mL) by the conclusion of the experiment (Figure 4). Furthermore, the commercial margarine was unable to regain its structure upon returning to 5 °C, regardless of the duration of exposure. In contrast, the oleogel-based margarines did not exhibit any oil or water exudation. Overall, a very slight onset of exudation was observed after 24 h at 35 °C; however, following the final stage of thermal cycling (72 h at 5 °C), the samples reverted to their initial condition. These findings are consistent with the melting profiles of the margarines and reinforce the observation that the commercial sample was more susceptible to temperature fluctuations, particularly during extended storage periods.

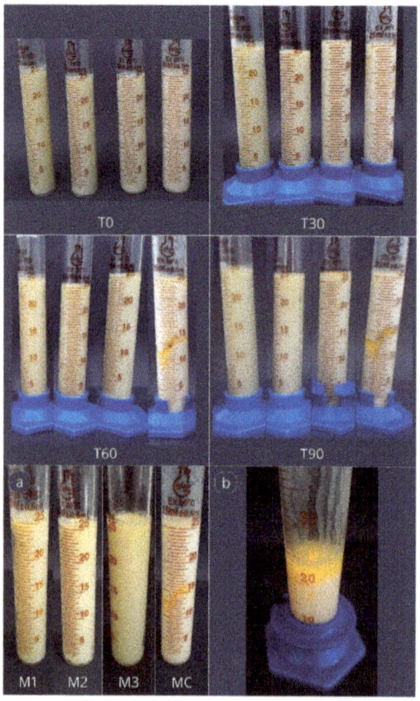

Figure 4. Margarine formulations (from left to right: M1, M2, M3, MC) stored in test tubes at times T0, T30, T60, and T90, and after the final cyclization stage: (**a**) Formulations M1, M2, M3, and MC at the end of the storage period (T90). (**b**) Oil exudation observed in sample MC at T90.

The commercial margarine sample also demonstrated lower resistance to temperature fluctuations, which led to oil exudation in several instances [23,33]. In some cases, emulsion breakdown was observed. This instability may be attributed to the absence of a robust crystalline network, as found in the oleogel-based formulations. The oleogels likely formed a strong crystal network capable of effectively entrapping oil through the action of the structuring agent and other formulation components, thereby preventing phase separation even under thermal stress. These findings, in combination with the melting profile analyses, suggest that margarines formulated with oleogels exhibited greater stability than the commercial margarine, making them less susceptible to temperature variations such as those associated with alternating storage between refrigeration and ambient conditions.

2.5. Microstructure

The microstructure of the margarine samples was assessed both qualitatively and quantitatively by measuring the mean diameter of water droplets. Staining confirmed the formation of a water-in-oil (W/O) emulsion, with water droplets dispersed throughout the lipid phase. The microstructural characteristics of the samples are presented in Figure 5.

At the initial time point, a clear distinction was observed between the microstructure of the oleogel-based margarines and that of the commercial margarine. In all the formulations, a general trend was noted wherein the number of water droplets decreased while their size increased over time. However, the commercial margarine consistently displayed smaller droplets in higher numbers compared to the oleogel-based samples. These findings are supported by the data shown in Figure 6, which present the average droplet diameter (DD). Although variations in DM were observed throughout the storage period, these changes were not statistically significant ($p > 0.05$).

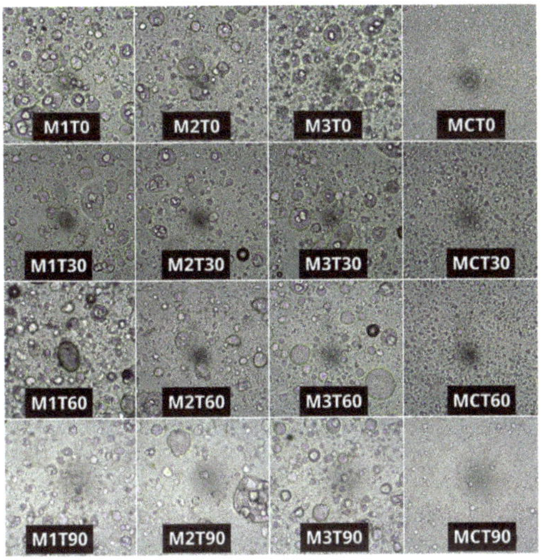

Figure 5. Microstructure of the margarine formulations observed over 90 days of storage.

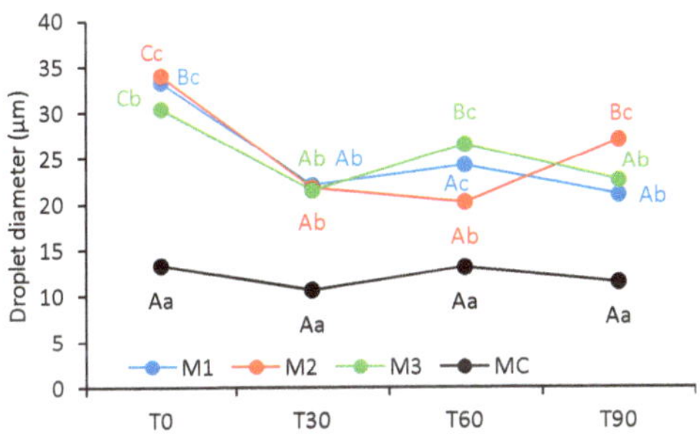

Figure 6. Average droplet diameter of the margarine formulations over time. Uppercase letters indicate statistical differences within each formulation over time, while lowercase letters compare different formulations at the same time point.

In general, the formulations M1, M2, and M3 had mean diameters between 20 and 34 µm, while the commercial margarine had an average of 12.16 µm, possibly due to the difference in the margarine production process. Nevertheless, the margarines formulated with oleogels remained stable, with similar textures and without oil exudation. Additionally, in M1, M2, and M3, the water droplets were more irregular, larger, and more dispersed in the visualized space. A disorganized aqueous phase characterized by large droplets has been previously reported in nonpolarized microstructure analyses [23]. This confirms, once again, a greater uniformity, including structurally, of the commercial sample, which probably did not negatively influence its stability over time, as seen in the thermal and oxidative analyses. These irregularities became more pronounced over time, with visually larger and more agglomerated droplets observed. Nevertheless, the mean droplet diameter did not change significantly over the storage period (Figure 6).

Although the oleogel-based formulations appeared less regularly structured and less homogeneous under microscopic observation, they were not less stable, contrary to what might be expected given the typical correlation between microstructure and macroscopic properties. In fact, as demonstrated in the thermal stability analyses, these formulations exhibited greater resistance to thermal fluctuations. This enhanced stability may be attributed to the synergistic effect of the CS-BW-CO combination. When used alone at similar concentrations, beeswax has been reported to produce margarines with structural instability [28]. Previous studies have shown that combining BW with a second wax can improve structural integrity; however, such combinations may also intensify a waxy mouthfeel. In contrast, the use of starch in combination with waxes may result in improved sensory properties and overall product acceptability [34].

2.6. Oxidation Evaluation

Foods with high lipid content are particularly susceptible to spoilage, primarily due to oxidative processes [35]. The peroxide value (PV) is a key indicator used to assess fat rancidity, reflecting the extent of primary lipid oxidation. In the food industry, oxidation and the resulting rancidity negatively impact product quality and compromise various functional properties [23]. The PV measurements for the margarine samples are presented in Figure 7a.

With the exception of the measurement at T60, the commercial margarine exhibited a significantly lower PV compared to the other samples, with an average PV of 1.74 mEq O_2/kg. In contrast, the PV of the oleogel-based margarines ranged from 1.74 to approximately 9.73 mEq O_2/kg. The higher PV observed in the oleogel margarines is likely attributable to the inclusion of extra virgin olive oil, for which a PV $\leq$ 20 mEq O_2/kg is generally considered acceptable [36]. The initial PV (T0) was higher in samples with greater amounts of olive oil, a trend explained by the higher unsaturated fat content of olive oil relative to coconut oil or the commercial fat blend used in the commercial margarine.

A substantial increase in PV was observed during the first 30 days of storage, particularly in the oleogel-based formulations, followed by a subsequent decline. This decrease may be attributed either to the activity of antioxidants halting the propagation of peroxides or to the progression of oxidation into secondary stages. In the commercial margarine, multiple antioxidants and additional preservatives are included, whereas the formulations developed in this study contained only a single antioxidant and no industrial preservatives. In a previous study [37], significantly lower PVs and no notable changes in oleogel samples during storage were reported—findings attributed to the natural antioxidant compounds present in extra virgin olive oil.

Margarine contains water, which can influence oxidative stability during storage [37]. When peroxide value analysis was conducted on samples subjected to the thermal cycling process, the resulting values were slightly elevated but did not exceed the maximum recorded in Figure 7a, which was approximately 5.77 mEq O_2/kg. Nevertheless, as previously discussed, although the oleogel-based margarines exhibited higher PVs than the commercial sample, the values remained within acceptable limits, particularly considering the inclusion of extra virgin olive oil in the formulations.

Figure 7b presents the acidity values measured throughout the storage period. The commercial margarine consistently showed significantly lower acidity compared to all other formulations, with an average value of 0.53 mg KOH/g. In contrast, the oleogel-based margarines exhibited some variation in acidity over the 90-day storage period; however, these fluctuations were gradual and not statistically significant at specific time points. Notably, the acidity values of the oleogel margarines remained stable, ranging from 1.20 to 1.70 mg KOH/g, indicating good chemical stability and the absence of spoilage. Furthermore, the thermal cycling treatment did not significantly alter the acidity index, suggesting that this process preserved the chemical integrity of the samples. These results indicate that while the incorporation of oleogels may influence initial acidity, all formulations maintained acceptable acidity levels throughout storage.

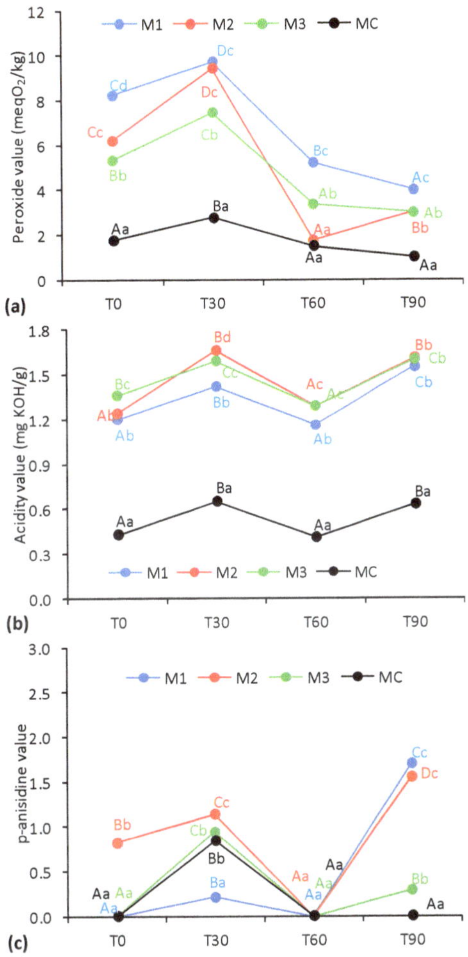

Figure 7. Lipid stability based on (**a**) peroxide value, (**b**) acidity value, and (**c**) p-anisidine value of the margarine formulations over time. Uppercase letters indicate statistical differences within each formulation across time points, and lowercase letters indicate differences between formulations at the same time point.

Secondary oxidation products, resulting from lipid degradation, were assessed using the p-anisidine value, as shown in Figure 7c. The average p-anisidine values were low, ranging from 0 to 1.70, indicating minimal formation of secondary oxidation products during storage, especially considering that values below 10 are commonly regarded as acceptable. Similar to the PV trend, the p-anisidine values declined after day 30 (T30), with all formulations approaching a value of 0. This reduction may be attributed to antioxidant activity, preventing the formation of secondary oxidation compounds. However, these slight variations remain within a narrow range. Among the oleogel formulations, M3—containing the lowest percentage of olive oil—exhibited the smallest increase in p-anisidine value after T60. This suggests that the observed decrease in PV for formulations M1 and M2 during storage was likely due to the progression of lipid oxidation into secondary stages.

In samples subjected to thermal cycling, a further decrease in the p-anisidine values of M1 and M2 was observed, while an increase occurred in M3 and MC.

3. Conclusions

The production of margarines using oleogels based on extra virgin olive oil, coconut oil, starch, and wax is a feasible approach, particularly due to the favorable appearance, thermal stability, and oxidative stability of the resulting products. The presence of bioactive compounds in extra virgin olive oil likely contributed to enhanced oxidative stability, as the oleogel-based formulations did not exhibit elevated oxidation values throughout the study. Additionally, the structuring agents may have played a role in protecting the lipid components of olive oil from oxidation during storage. As expected, attributes such as color were only minimally influenced by the ingredients used, which is unlikely to negatively affect consumer perception. The oleogel-based margarines met established identity and quality standards, making them a promising and appealing alternative for the food industry—especially in response to growing consumer demand for healthier fat sources.

4. Materials and Methods

4.1. Materials

Extra virgin olive oil was supplied by Irarema Farm (Poços de Caldas, Minas Gerais, Brazil). The monoglyceride VEROL N-90 was provided by Lasenor (Potim, São Paulo, Brazil), and the butter flavoring was obtained from Doremus (Guarulhos, São Paulo, Brazil). The antioxidant butylhydroxytoluene (BHT) was acquired from Mix das Essências (Belo Horizonte, Minas Gerais, Brazil), while beeswax was purchased from Fenix Ceras (São Paulo, São Paulo, Brazil). Coconut oil, powdered milk, salt, corn starch, commercial margarine, and turmeric (used as a coloring agent) were procured from a local market in Lavras, Minas Gerais, Brazil. Methylene blue and Sudan III were used as dyes for microstructural analysis. Reagents employed in the oxidation analyses included sodium thiosulfate, starch, potassium hydroxide (KOH), and phenolphthalein, all obtained from Êxodo Científica. All ingredients were stored according to the manufacturers' recommendations until use in margarine preparation.

4.2. Formulation of the Oleogel-Based Margarines

The final margarine formulations were prepared according to the compositions outlined in Table 4. Each formulation consisted of 80% lipid phase and 20% aqueous phase. A simplified production method, adapted for the incorporation of oleogels, was employed based on the procedure described by da Silva [23].

Table 4. Formulation of the oleogel-based margarine used in the study.

Ingredients	g/100 g
Aqueous phase 20%:	
Water *	16.2
Salt *	2.0
Powdered milk *	1.8
Lipid phase 80%:	
Extra virgin olive oil	45–60
Coconut oil	40–55
Beeswax	4–5.2
Corn starch	1–1.8
Monoacylglycerol *	0.2
Turmeric *	0.03
Butter scent *	0.04
Antioxidant *	0.03

* fixed amount.

4.2.1. Experimental Design

The concentrations of the structural components were initially defined through preliminary testing, followed by a Central Composite Rotational Design (CCRD) to establish the minimum and maximum concentration limits required to obtain spreadable margarines. The CCRD was employed as an optimization tool. Seventeen margarine formulations (Table 1) were prepared following the procedure detailed in Section 4.2.2. These formulations, along with a commercial reference sample, were stored under refrigeration and characterized using a texture analyzer (TA-XT.plus, Stable Microsystems, England). The optimized variables included the contents of corn starch (CS), beeswax (BW), extra virgin olive oil (OO), and coconut oil (CO), while the remaining formulation components were kept constant.

The optimization objective was to achieve a texture most similar to that of commercial margarine. This target was quantified using a desirability function with the criterion "Nominal-The-Best" [25]. Firmness, adhesiveness, and consistency values were transformed into desirability scores ranging from 0 to 1, with a value of 1 indicating the closest match to the desired texture. All texture measurements were performed in duplicate. The CCRD data were analyzed using Chemoface software version 1.71 [38], applying a quadratic model based on pure error estimation.

4.2.2. Margarine Preparation

All ingredients were weighed according to the formulations presented in Table 4. The aqueous phase ingredients were manually mixed and heated to 60 °C in a water bath (HH-S3 Warmnest, Brazil). Simultaneously, the lipid-phase components were premixed and heated to 80 °C in a separate water bath for approximately 10 min until fully solubilized. The two phases were then combined and homogenized at 1400 revolutions per minute (rpm) using an overhead agitator (713D, Fisatom, Brazil) for 20 min to ensure complete emulsification.

Following homogenization, the emulsions underwent a rapid dynamic cooling process (~10 °C/min) to room temperature (25 ± 1 °C) using a water bath maintained at 10 ± 1 °C, while being stirred with the overhead agitator at 200 rpm for 5 min. The resulting margarines were transferred into plastic containers, sealed with aluminum foil, appropriately labeled, and stored under refrigeration at 5 ± 1 °C until further analysis.

Samples were evaluated over a 90-day storage period at intervals of 0 (T0), 30 (T30), 60 (T60), and 90 (T90) days. All subsequent analyses were performed at each storage interval, with the exception of differential scanning calorimetry (DSC), which was conducted only at the start of the experiment.

4.3. Thermal Behavior

Differential Scanning Calorimetry (DSC) was performed to evaluate the melting behavior of the margarine samples. The analysis was conducted using a DSC-60A calorimeter equipped with a TA-60WS data acquisition system and an FC-60A furnace flow and atmosphere controller (Shimadzu, Japan). Approximately 5 mg of each margarine sample was weighed into a non-sealed aluminum pan. The samples were introduced into the DSC at room temperature (~25 °C), which was considered the onset temperature (T_{onset}). The heating protocol involved a linear temperature increase from 25 °C to 90 °C at a rate of 2 °C/min. The melting peak temperature (T_p) was recorded and used as the primary parameter for comparing the samples. All measurements were conducted under a nitrogen (N_2) atmosphere to prevent oxidative degradation during the analysis.

4.4. Color Analysis

Color analysis of the margarine samples was performed using a spectrophotometer (CM-5, Konica Minolta, Japan). Samples were placed in 50 mL graduated glass beakers, and color measurements were conducted based on the CIELAB color space, recording the L*, a*, and b* values. The L* value represents lightness, ranging from 0 (black) to 100 (white). The a* coordinate indicates chromaticity along the green (negative) to red (positive) axis, while the b* coordinate reflects the blue (negative) to yellow (positive) axis. Measurements were carried out in triplicate using a D65 standard illuminant and a 10° observer angle.

4.5. Thermal Stability by Cyclization

The margarines were evaluated for emulsion stability—specifically, oil or water exudation—at various storage times and temperatures using a modified thermal cycling method adapted from [33]. Triplicate samples of approximately 25 g were placed in test tubes. To simulate temperature fluctuations, the samples were initially stored at 5 °C (T-34, Thelga, Brazil) for 48 h to ensure complete crystallization. They were then subjected to a temperature of 35 °C for 24 h. Following this period, the samples were returned to 5 °C for an additional 72 h. Visual inspections and, when possible, quantitative measurements of oil or water exudation were performed after each stage of the thermal cycle.

4.6. Microstructure Analysis

The microstructure of the margarine samples was analyzed using an optical microscope (BA210E, Motic, Spain) equipped with a 40× objective lens. To confirm the formation of an emulsion, a dual-staining technique was employed: methylene blue was used to stain the aqueous phase, and Sudan III was used to stain the lipid phase. These dyes were applied to the samples and examined microscopically to verify the presence of both phases, thereby confirming the formation of a water-in-oil (W/O) emulsion.

Following confirmation, the microstructure was further evaluated in the unstained, natural state. Small droplets were collected from various regions of each sample and placed onto glass slides, which were then sealed with coverslips. These slides were analyzed both qualitatively and quantitatively to assess the diameter of dispersed water droplets over time. For each sample, three slides were prepared, and quantitative measurements were conducted using ImageJ software (version 1.53, National Institutes of Health, Bethesda, MD, USA).

4.7. Lipid Stability and Oxidation

To evaluate lipid stability, margarine samples were preheated in an oven (MD1.1, Medicate, Brazil) at approximately 40 °C. Prior to analysis, the samples were filtered through filter paper to separate the lipid and aqueous phases; only the lipid phase was subjected to further analysis.

4.7.1. Peroxide Index

The peroxide index (PI) was determined in triplicate using iodometric titration. Sodium thiosulfate was used as the titrant and starch as the indicator, following the AOCS official method Cd 8-53 [39].

4.7.2. Acidity Index

The acidity index (AI) was measured in triplicate through acid–base titration, using potassium hydroxide (KOH) as the titrant and phenolphthalein as the indicator, in accordance with AOCS method Cd 3d-63 [39].

4.7.3. Anisidine Value

The p-anisidine value was determined in triplicate based on a modified version of the AOCS method Cd 18-90 [40]. Approximately 1 g of each sample was diluted to a final volume of 10 mL in a volumetric flask prior to analysis.

4.8. Statistical Analysis

For statistical analysis, analysis of variance (ANOVA) was performed, followed by the Scott–Knott test. Both analyses were conducted using Sisvar version 5.6 [41], with a level of significance of 5%. In the figures, uppercase letters indicate statistical differences within each formulation over time, while lowercase letters represent differences among formulations at each specific time point.

Author Contributions: B.V.B.N.: Conceptualization, Methodology, Investigation, Formal Analysis, Visualization, Validation, Writing—Original draft; T.L.T.d.S.: Investigation, Formal Analysis, Writing & Editing, preparation; C.A.N.: Formal Analysis, Visualization, Writing—Original draft preparation; F.F.H.: Investigation, Formal Analysis; S.C.B.: Conceptualization, Methodology, Validation, Visualization, Supervision, Project administration, Writing—Original draft preparation. All authors have read and agreed to the published version of the manuscript.

Funding: This research was funded by Fundação de Amparo à Pesquisa do Estado de Minas Gerais (FAPEMIG).

Institutional Review Board Statement: Not applicable.

Informed Consent Statement: Not applicable.

Data Availability Statement: The original contributions presented in this study are included in the article. Further inquiries can be directed to the corresponding author.

Acknowledgments: The authors would like to acknowledge the support of the Minas Gerais Research Foundation (FAPEMIG), as well as the assistance of ChatGPT version 4.5 with grammar and English spelling.

Conflicts of Interest: The authors declare no conflicts of interest.

References

1. de Oliveira Izar, M.C.; Lottenberg, A.M.; Giraldez, V.Z.R.; Santos Filho, R.D.D.; Machado, R.M.; Bertolami, A.; Assad, M.H.V.; Saraiva, J.F.K.; Faludi, A.A.; Moreira, A.S.B.; et al. Position statement on fat consumption and cardiovascular health-2021. *Arq. Bras. Cardiol.* **2021**, *116*, 160–212. [CrossRef]
2. Federal Register: Final Determination Regarding Partially Hydrogenated Oils. Available online: https://www.federalregister.gov/documents/2015/06/17/2015-14883/final-determination-regarding-partially-hydrogenated-oils (accessed on 7 May 2025).
3. Chaves, K.F.; Barrera-Arellano, D.; Ribeiro, A.P.B. Potential application of lipid organogels for food industry. *Food Res. Int.* **2018**, *105*, 863–872. [CrossRef] [PubMed]
4. Singh, A.; Auzanneau, F.I.; Rogers, M.A. Advances in edible oleogel technologies—A decade in review. *Food Res. Int.* **2017**, *97*, 307–317. [CrossRef] [PubMed]
5. Dassanayake, L.S.K.; Kodali, D.R.; Ueno, S.; Sato, K. Physical properties of rice bran wax in bulk and organogels. *JAOCS J. Am. Oil Chem. Soc.* **2009**, *86*, 1163–1173. [CrossRef]
6. Vintiloiu, A.; Leroux, J.C. Organogels and their use in drug delivery—A review. *J. Control. Release* **2008**, *125*, 179–192. [CrossRef]
7. Pușcas, A.; Mureșan, V.; Socaciu, C.; Muste, S. Oleogels in Food: A Review of Current and Potential Applications. *Foods* **2020**, *9*, 70. [CrossRef]
8. Lupi, F.R.; Gabriele, D.; Facciolo, D.; Baldino, N.; Seta, L.; de Cindio, B. Effect of organogelator and fat source on rheological properties of olive oil-based organogels. *Food Res. Int.* **2012**, *46*, 177–184. [CrossRef]
9. Öğütcü, M.; Yilmaz, E. Oleogels of virgin olive oil with carnauba wax and monoglyceride as spreadable products. *Grasas Aceites* **2014**, *65*, e040. [CrossRef]

10. Zampouni, K.; Mouzakitis, C.K.; Lazaridou, A.; Moschakis, T.; Katsanidis, E. Physicochemical properties and microstructure of bigels formed with gelatin and κ-carrageenan hydrogels and monoglycerides in olive oil oleogels. *Food Hydrocoll.* **2023**, *140*, 108636. [CrossRef]
11. Gaforio, J.J.; Visioli, F.; Alarcón-de-la-Lastra, C.; Castañer, O.; Delgado-Rodríguez, M.; Fitó, M.; Hernández, A.F.; Huertas, J.R.; Martínez-González, M.A.; Menendez, J.A.; et al. Virgin Olive Oil and Health: Summary of the III International Conference on Virgin Olive Oil and Health Consensus Report, JAEN (Spain) 2018. *Nutrients* **2019**, *11*, 2039. [CrossRef]
12. Rocha, K.D.C.; Ferreira, M.S.; Garcia, C.E.R. Óleo de Coco: Características e aplicações fisiológicas. In *Compostos Bioativos e Suas Aplicações*; Mérida Publishers: Canoas, Brasil, 2021. [CrossRef]
13. Cheng, J.; Kan, Q.; Cao, J.; Dudu, O.E.; Yan, T. Interfacial compositions of fat globules modulate coconut oil crystallization behavior and stability of whipped-frozen emulsions. *Food Hydrocoll.* **2021**, *114*, 106580. [CrossRef]
14. Jadhav, H.B.; Annapure, U.S. Triglycerides of medium-chain fatty acids: A concise review. *J. Food Sci. Technol.* **2023**, *60*, 2143–2152. [CrossRef] [PubMed]
15. Nimbkar, S.; Leena, M.M.; Moses, J.A.; Anandharamakrishnan, C. Medium chain triglycerides (MCT): State-of-the-art on chemistry, synthesis, health benefits and applications in food industry. *Compr. Rev. Food Sci. Food Saf.* **2022**, *21*, 843–867. [CrossRef] [PubMed]
16. Yilmaz, E.; Öğütcü, M. Oleogels as spreadable fat and butter alternatives: Sensory description and consumer perception. *RSC Adv.* **2015**, *5*, 50259–50267. [CrossRef]
17. Zulfiqar, A.; Shabbir, M.A.; Tahir, F.; Khan, M.R.; Ahmed, W.; Yıkmış, S.; Manzoor, M.F.; Abdi, G.; Aadil, R.M. Development of oleogel by structuring the blend of corn oil and sunflower oil with beeswax to replace margarine in cookies. *Food Chem. X* **2024**, *23*, 101676. [CrossRef]
18. Silva, T.J.; Barrera-Arellano, D.; Badan Ribeiro, A.P. The impact of fatty acid profile on the physicochemical properties of commercial margarines in Brazil. *JAOCS J. Am. Oil Chem. Soc.* **2022**, *99*, 469–483. [CrossRef]
19. US Department of Health and Human Services; U.S. Department of Agriculture. *2015–2020 Dietary Guidelines for Americans*, 8th ed.; U.S. Department of Health and Human Services and the U.S. Department of Agriculture: Washington, DC, USA, 2015. Available online: https://odphp.health.gov/sites/default/files/2019-09/2015-2020_Dietary_Guidelines.pdf (accessed on 10 March 2024).
20. Faludi, A.A.; de Oliveira Izar, M.C.; Saraiva, J.F.K.; Chacra, A.P.M.; Bianco, H.T.; Afiune Neto, A.; Bertolami, A.; Pereira, A.C.; Lottenberg, A.M.; Sposito, A.C.; et al. Update of the Brazilian guidelines on dyslipidemias and prevention of atherosclerosis-2017. *Arq. Bras. Cardiol.* **2017**, *109*, 1–76.
21. Mach, F.; Baigent, C.; Catapano, A.L.; Koskinas, K.C.; Casula, M.; Badimon, L.; Chapman, M.J.; De Backer, G.G.; Delgado, V.; Ference, B.A.; et al. 2019 ESC/EAS Guidelines for the management of dyslipidaemias: Lipid modification to reduce cardiovascular risk. *Eur. Heart J.* **2020**, *41*, 111–188. [CrossRef]
22. World Health Organization. *Saturated Fatty Acid and Trans-Fatty Acid Intake for Adults and Children: WHO Guideline*; World Health Organization: Geneva, Switzerland, 2023; p. 117. Available online: https://www.who.int/publications/i/item/9789240073630 (accessed on 10 March 2024).
23. da Silva, T.L.T.; Chaves, K.F.; Fernandes, G.D.; Rodrigues, J.B.; Bolini, H.M.A.; Arellano, D.B. Sensory and Technological Evaluation of Margarines With Reduced Saturated Fatty Acid Contents Using Oleogel Technology. *JAOCS J. Am. Oil Chem. Soc.* **2018**, *95*, 673–685. [CrossRef]
24. Winkler-Moser, J.K.; Anderson, J.A.; Hwang, H.S. Texture and flavor evaluation of peanut butter stabilized with natural waxes. *J. Food Sci.* **2022**, *87*, 1851–1864. [CrossRef] [PubMed]
25. Costa, N.R.; Lourenço, J.; Pereira, Z.L. Desirability function approach: A review and performance evaluation in adverse conditions. *Chemom. Intell. Lab. Syst.* **2011**, *107*, 234–244. [CrossRef]
26. Gao, Y.; Wu, S. Development and evaluation of a novel oleogel system based on starch–water–wax–oil. *Food Funct.* **2020**, *11*, 7727–7735. [CrossRef] [PubMed]
27. Gao, Y.; Wu, S. Thermal and oxidation stability of functional oleogels formed by edible wax/starch and Schisandra chinensis oil. *Food Funct.* **2019**, *10*, 8056–8068. [CrossRef]
28. Chai, X.; Zhang, Y.; Shi, Y.; Liu, Y. Crystallization and Structural Properties of Oleogel-Based Margarine. *Molecules* **2022**, *27*, 8952. [CrossRef]
29. Lomonaco, T.; da Silva, T.; Deschamps Fernandes, G.; Arellano, D.B. The combination of monoglycerides, wax and hardfat on oleogels structuration. *Braz. J. Food Technol.* **2022**, *25*, e2021137. [CrossRef]
30. Shi, Y.; Zhang, M.; Bhandari, B. Effect of addition of beeswax based oleogel on 3D printing of potato starch-protein system. *Food Struct.* **2021**, *27*, 100176. [CrossRef]
31. Winkler-Moser, J.K.; Anderson, J.; Felker, F.C.; Hwang, H.S. Physical Properties of Beeswax, Sunflower Wax, and Candelilla Wax Mixtures and Oleogels. *JAOCS J. Am. Oil Chem. Soc.* **2019**, *96*, 1125–1142. [CrossRef]

32. Rupini, R.V.; Nandagopal, R. A study on the influence of senses and the effectiveness of sensory branding. *Afr. J. Psychiatry* **2015**, *18*, 2. [CrossRef]
33. Garcia, R.K.A.; Moreira Gandra, K.; Barrera-Arellano, D. Development of a zero-trans margarine from soybean-based interesterified fats formulated using artificial neural networks. *Grasas Aceites* **2013**, *64*, 521–530. [CrossRef]
34. Hu, J.; Gao, Y.; Lu, Q.; Jiang, Y.; Jin, H.; Li, Q.; Yu, X. Development and evaluation of a novel margarine using starch hydrogel combined edible wax oleogel bigels. *J. Food Eng.* **2025**, *388*, 112360. [CrossRef]
35. da Silva, R.F.; Ascheri, J.L.R.; de Souza, J.M.L. Influência do processo de beneficiamento na qualidade de amêndoas de castanha-do-brasil. *Ciência Agrotecnol.* **2010**, *34*, 445–450. [CrossRef]
36. FAO/WHO. *Joint FAO/WHO Codex Alimentarius: Fats and oils*, 2nd ed.; Food and Agriculture Organization (FAO): Geneva, Switzerland, 2001; Volume 8.
37. Yılmaz, E.; Öğütcü, M. Comparative analysis of olive oil organogels containing beeswax and sunflower wax with breakfast margarine. *J. Food Sci.* **2014**, *79*, E1732–E1738. [CrossRef] [PubMed]
38. Nunes, C.A.; Freitas, M.P.; Pinheiro, A.C.M.; Bastos, S.C. Chemoface: A novel free user-friendly interface for chemometrics. *J. Braz. Chem. Soc.* **2012**, *23*, 2003–2010. [CrossRef]
39. AOCS—American Oil Chemists' Society. *Official Methods and Recommended Practices of the American Oil Chemists' Society*, 4th ed.; AOCS: Champaign, IL, USA, 1990.
40. AOCS—American Oil Chemists' Society. Method Cd 18-90. P-anisidine value. In *Official Methods and Recommended Practices of the American Oil Chemists' Society*, 5th ed.; AOCS: Champaign, IL, USA, 2011.
41. Ferreira, D.F. SISVAR: A program for statistical analysis and teaching. *Rev. Cient. Symp.* **2008**, *6*, 36–41. Available online: https://des.ufla.br/~danielff/meusarquivospdf/art63.pdf (accessed on 7 May 2025).

gels

MDPI

Article

Encapsulation of *Lactobacillus reuteri* in Chia–Alginate Hydrogels for Whey-Based Functional Powders

Alma Yadira Cid-Córdoba [1], Georgina Calderón-Domínguez [2], María de Jesús Perea-Flores [3], Alberto Peña-Barrientos [3], Fátima Sarahi Serrano-Villa [2], Rigoberto Barrios-Francisco [1], Marcela González-Vázquez [4] and Minerva Rentería-Ortega [1,*]

[1] Tecnológico Nacional de México/TES de San Felipe del Progreso, San Felipe del Progreso 50640, Mexico; yadiscica19@gmail.com (A.Y.C.-C.); rigoberto.bf@sfelipeprogreso.tecnm.mx (R.B.-F.)
[2] Departamento de Ingeniería Bioquímica, Escuela Nacional de Ciencias Biológicas, Instituto Politécnico Nacional, Ciudad de México 07738, Mexico; gcalderon@ipn.mx (G.C.-D.); fserranov2001@alumno.ipn.mx (F.S.S.-V.)
[3] Centro de Nanociencias y Micro y Nanotecnologías, Instituto Politécnico Nacional, Ciudad de México 07738, Mexico; mpereaf@ipn.mx (M.d.J.P.-F.); apenab@ipn.mx (A.P.-B.)
[4] Instituto de Farmacología, Universidad de la Cañada, Carretera Teotitlan—San Antonio Nanahuatipan Km 1.7 s/n, Paraje Titlacuatitla, Teotitlan de Flores Magón, Oaxaca 68540, Mexico; marcelaglezvaz89@hotmail.com
* Correspondence: minerva.ro@sfelipeprogreso.tecnm.mx

Abstract: This study aimed to develop a functional powder using whey and milk matrices, leveraging the protective capacity of chia–alginate hydrogels and the advantages of electrohydrodynamic spraying (EHDA), a non-thermal technique suitable for encapsulating probiotic cells under stress conditions commonly encountered in food processing. A hydrogel matrix composed of chia seed mucilage and sodium alginate was used to form a biopolymeric network that protected probiotic cells during processing. The encapsulation efficiency reached 99.0 ± 0.01%, and bacterial viability remained above 9.9 $\log_{10}$ CFU/mL after lyophilization, demonstrating the excellent protective capacity of the hydrogel matrix. Microstructural analysis using confocal laser scanning microscopy (CLSM) revealed well-retained cell morphology and homogeneous distribution within the hydrogel matrix while, in contrast, scanning electron microscopy (SEM) showed spherical, porous microcapsules with distinct surface characteristics influenced by the encapsulation method. Encapsulates were incorporated into beverages flavored with red fruits and pear and subsequently freeze-dried. The resulting powders were analyzed for moisture, protein, lipids, carbohydrates, fiber, and color determinations. The results were statistically analyzed using ANOVA and response surface methodology, highlighting the impact of ingredient ratios on nutritional composition. Raman spectroscopy identified molecular features associated with casein, lactose, pectins, anthocyanins, and other functional compounds, confirming the contribution of both matrix and encapsulants maintaining the structural characteristics of the product. The presence of antioxidant bands supported the functional potential of the powder formulations. Chia–alginate hydrogels effectively encapsulated *L. reuteri*, maintaining cell viability and enabling their incorporation into freeze-dried beverage powders. This approach offers a promising strategy for the development of next-generation functional food gels with enhanced probiotic stability, nutritional properties, and potential application in health-promoting dairy systems.

Keywords: encapsulation; *Lactobacillus reuteri DSM 17938*; electrospraying; whey

1. Introduction

In recent years, the primary purpose of foods has evolved from fulfilling their basic nutritional role to providing health and well-being benefits, thereby driving the growth of the functional food and probiotic industry [1]. In this sense, the International Life Sciences Institute (ILSI) defines a functional food as "A food that contains a component, nutrient, or non-nutrient, with an effect on one or more functions of the body, with an added effect beyond its nutritional value" [2]. Within this context of development, the incorporation of probiotic strains into food matrices has gained significant importance as they must withstand processing conditions to ensure a sufficient quantity of viable cells ($\geq 10^6$ CFU/g) at the time of consumption, thus guaranteeing their functional effect. Functional foods often incorporate probiotics, as the inclusion of beneficial live microorganisms is one of the most effective ways to enhance the beneficial properties of food products.

Considering the above, some probiotic microorganisms, such as *Lactobacillus reuteri*, contribute to the balance of the intestinal microbiota, immune modulation, and the prevention of gastrointestinal disorders, thus reinforcing the functional attributes of the food products that deliver them. Accordingly, probiotics have become a central element in the formulation of dairy products, beverages, and functional supplements, representing a rapidly growing sector of the food industry. However, their integration into food matrices requires strategies that guarantee their viability and stability during processing and storage. However, achieving high levels of viability represents a challenge for the food industry due to the sensitivity of probiotics to adverse factors such as temperature and oxygen during processing [3]. Therefore, the use of encapsulation techniques that protect probiotic cells and promote their stability and functionality has become essential.

Encapsulation is a technique that has been established as a strategy to improve the survival of probiotic microorganisms against adverse conditions in food and during their passage through the gastrointestinal tract, enhancing cell viability by retaining the bacteria within polymeric matrices that act as semipermeable physical barriers, regulating the interaction between the probiotic and the environment. Currently, extrusion, ionic gelation, and spray drying are the most commonly used methods for probiotic encapsulation. However, these methods present several limitations due to poor control of particle size, low uniformity in the microcapsules, and exposure to high temperatures during the process.

Despite the growing demand for functional foods enriched with probiotics, the industry continues to face major challenges related to the survival of these microorganisms during food processing, storage, and gastrointestinal transit. Conventional encapsulation methods often expose probiotics to high temperatures or produce irregular microcapsules with poor protective capacity. At the same time, there is an urgent need to incorporate sustainable ingredients such as whey, a dairy by-product, into value-added formulations. Therefore, this study seeks to address these limitations by applying electrohydrodynamic atomization (EHDA) to encapsulate *L. reuteri* in chia–alginate hydrogels, aiming to preserve viability and functionality in a stable powdered beverage system.

Electrohydrodynamic atomization (EHDA), also known as electrospraying, has established itself as an emerging technology that allows the formation of monodisperse microcapsules at room temperature, making it a particularly suitable technique for thermosensitive compounds such as enzymes and probiotics [4–6]. EHDA is a non-thermal process that uses a high-voltage electric field to disperse a polymeric solution into fine droplets. These droplets are then solidified by cross-linking or dehydration, enabling the formation of uniform microparticles with tunable size and morphology. EHDA is especially advantageous in food applications for protecting labile bioactives under mild conditions. This technique has gained increasing attention due to its ability to generate particles with high homogeneity and dimensional control [7]. In this regard, EHDA is considered a promising technique for the encapsulation of various thermolabile compounds, such as enzymes and probiotic strains, since it operates without the use of heat and enables the formation of uniform microcapsules, thereby maintaining the viability of probiotics and improving their stability against gastrointestinal stress [8,9]. Studies have demonstrated its efficacy for strains such as *Lactobacillus plantarum*, *Bifidobacterium animalis*, *L. rhamnosus*, and *Lactobacillus fermentum*, helping to improve survival during gastrointestinal digestion, promoting adhesion to the intestinal mucosa, and allowing for a controlled release in the intestine [8,10,11]. Furthermore, the use of natural coating materials, such as sodium alginate and chia seed mucilage, contributes to improving the protective barrier and modulating the release kinetics. *Salvia hispanica* seed mucilage and sodium alginate in this study were carefully selected based their proven biocompatibility, film forming ability, and synergistic gelling properties. These polymers have demonstrated excellent performance in forming hydrogel networks that entrap probiotics and preserve their viability under gastrointestinal stress and processing conditions. Studies have reported that their combination improves encapsulation efficiency, modulates release kinetics, and enhances the mechanical stability of the capsules, making them suitable carriers for functional food applications [12].

In addition to the technological advantages of EHDA, it is crucial to consider the use of sustainable food matrices. In this context, the dairy industry is positioned as one of the main generators of by-products, with whey being one of the most abundant. According to Silva e Alves et al. (2018) [13], it is estimated that the global dairy industry produces approximately 200 million tons of whey per year, which has a high nutritional value containing lactose, calcium, phosphorus, folic acid, vitamins, and high-quality proteins [14], making it a functional matrix with great potential for the development of new fermented or functional beverages. In addition to the above, in recent years, various whey-enriched beverages have been developed, valued not only for their nutritional profile but also for their low cost and potential sustainability. Within this framework, the present study aimed to evaluate the encapsulation efficacy of *Lactobacillus reuteri DSM 17938* electrohydrodynamic spraying (EHDA) and dripping mode, with sodium alginate and chia seed mucilage as coating materials. The resulting encapsulates were added to a whey-based functional beverage flavored with pear and red berries, which was subsequently dehydrated by lyophilization for preservation and characterization.

Although prior research has demonstrated the potential of probiotic encapsulation through various techniques, the present study introduces a novel integration of electrosprayed *L. reuteri* into a real powdered beverage system based on dairy by-products, a combination not previously explored in this depth. Moreover, structural and functional characterization using advanced spectroscopic tools, such as Raman spectroscopy, remains scarce in this context. Therefore, this study fills a methodological and application gap by evaluating not only the encapsulation process but also the behavior of the encapsulated probiotics within a functional matrix. This dual focus on process optimization and matrix-level analysis distinguishes our approach from prior works and contributes new knowledge to both probiotic delivery and sustainable product development.

Unlike previous studies, this research presents an integrated approach that combines the encapsulation of *Lactobacillus reuteri* in chia–alginate hydrogels using electrohydrodynamic atomization (EHDA) with the development of freeze-dried functional powders based on whey and milk matrices. The innovation lies in the application of encapsulated probiotics to a complete food system, which was structurally characterized by Raman spectroscopy to assess the impact of matrix components and drying conditions. Furthermore, the use of whey as a sustainable ingredient not only adds value to dairy by-products but also enhances the functional properties of the formulation. This approach bridges the gap between encapsulation technology and real food applications, offering a promising strategy for the design of next-generation functional beverages.

2. Results and Discussion

2.1. Encapsulation Efficiency (EE)

As shown in Table 1, the results obtained showed no significant differences ($p > 0.05$) between the two microencapsulation methods (drip mode (DM) and electrohydrodynamic atomization (EHDA), with both achieving a viability of 9.9 log CFU/mL and an average EE of 99 ± 0.16%. The encapsulation efficiency (EE) values are expressed as mean ± standard deviation, based on three independent replicates (n = 3). No significant differences ($p > 0.05$) were found between the two encapsulation methods. These results suggest that both EHDA and DM provided a highly protective matrix for *Lactobacillus reuteri* during the encapsulation process. These values suggest that both encapsulation methods are effective in protecting and preserving *L. reuteri* cells during the microencapsulation process. This is possibly related to the characteristics of the polymers used (MC-AlgNa) and the type of structural network they form, which positively contributed to the microorganism's survival. In addition, the nature of the coating may have facilitated some interactions, such as ionic or hydrogen bonds, which not only increased cell retention within the polymer matrix but also protected LR from adverse conditions during powder formulation. In this regard, Lee et al. (2003) [15] reported that the high effectiveness of these biopolymers is associated with an increased number of active calcium-binding sites in the polymer chains, resulting in a greater degree of crosslinking.

Table 1. Encapsulation efficiency (EE) and viability of *Lactobacillus reuteri DSM 17938* using different encapsulation methods. Results are expressed as mean ± standard deviation (n = 3). Different letters in the same column indicate significant differences ($p < 0.05$).

Encapsulation Method	Viability (log CFU/mL)	EE (%)
Dripping mode (DM)	9.90 ± 0.05 [a]	99.00 ± 0.16 [a]
Electrohydrodynamic spraying (EHDA)	9.91 ± 0.04 [a]	99.00 ± 0.01 [a]

According to the above, several studies have reported that sodium alginate is a highly effective material for encapsulating probiotic bacteria, achieving high encapsulation efficiencies. In this sense, Perea-Flores et al. (2020) [12] reported that the combination of chia seed mucilage and sodium alginate as wall materials exhibited high entrapment efficiency (~94%) and provided a controlled release profile of the encapsulated compound. Although their study focused on rhodamine B, their findings highlight the potential of this biopolymer blend to protect sensitive compounds, suggesting it could be promising for applications involving probiotic encapsulation and protection under adverse conditions. Additionally, Popović et al. (2021) [16] reported a higher number of viable L. reuteri B2 cells encapsulated in sodium alginate combined with another biopolymer (sodium maleate), which improved their stability and viability.

2.2. Micromorphology Analysis

2.2.1. Confocal Laser Scanning Microscopy (CLSM) Analysis

The micrographs obtained (Figure 1a,b) show the interaction between *Lactobacillus reuteri DSM 17938* cells, the coating material (MC-AlgNa) and the fluorochromes (propidium iodide and calcofluor), where the uniform red fluorescence observed indicates that the bacteria remained embedded within the polymeric matrix of the biopolymers (MC-AlgNa), highlighting their typical bacillus morphology, with well-defined edges and a regular periphery. These morphological characteristics demonstrate that the encapsulation methods evaluated (EHDA and DM) did not cause deformations or damage to the bacterial cells. The smooth and uniform cell surface indicates an adequate interaction between the polymeric matrix and the cells, contributing to the protection of the bacteria against processing conditions.

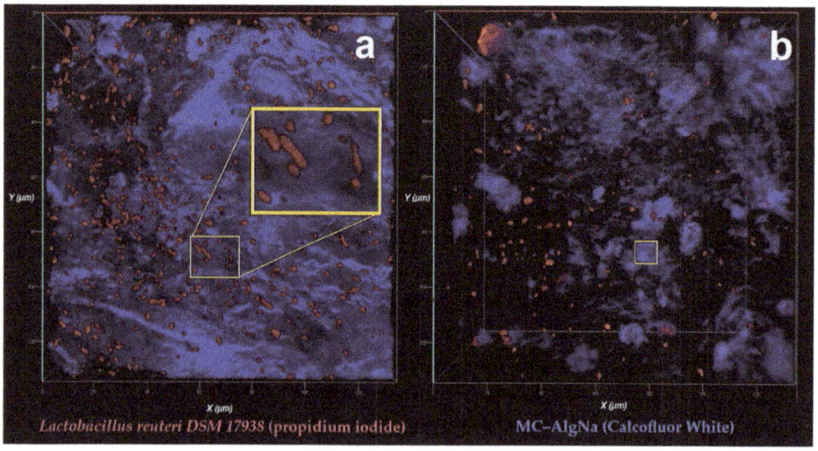

Figure 1. Confocal laser scanning microscopy (CLSM) images showing the distribution of *Lactobacillus reuteri DSM 17938* entrapped within the polysaccharide matrix composed of mucilage–alginate (MC–AlgNa). (**a**) Cells stained in red with propidium iodide; the orange box highlights a magnified region to emphasize bacteria embedded within the matrix. (**b**) Polymer matrix stained in blue with Calcofluor White.

In micrograph (a), a more uniform distribution of the bacteria in the matrix is observed. In contrast, in (b), the bacteria appear to be more dispersed, and some are grouped in specific regions. This difference is probably due to the rheological properties of the wall material during the formation of the capsules or to the specific operating conditions of each encapsulation method, as reported by Dragoni-Rosado, (2014) [17], who encapsulated *Lactobacillus casei ATCC 393* by extrusion and atomization methods, finding that extrusion tended to generate capsules with a more dispersed and clustered distribution of the bacteria, while atomization promoted a more homogeneous distribution, associating it with the forces applied during the process, which directly influenced the final distribution of the cells within the polymeric matrix. In addition, Huang et al. (2015) [18] demonstrated that similar biopolymeric matrices offer high biocompatibility and good cell retention. Furthermore, the absence of cell surface irregularities or deformations suggests that the processing conditions (EHDA, DM) did not negatively impact cell viability, which is crucial for prebiotic applications.

2.2.2. Micromorphology Analysis by SEM

Morphological analysis of chia mucilage and sodium alginate microcapsules containing *Lactobacillus reuteri* (MC-AlgNa-LR) was performed by scanning electron microscopy (SEM) (Figure 2). The encapsulates prepared by electrohydrodynamic spraying (EHDA) had an average diameter of 1 ± 0.02 mm, while those obtained by the drip method showed a diameter of 2.5 ± 0.01 mm. Significant differences ($p < 0.05$) were observed between the two encapsulation methods ($p < 0.05$), which can be attributed to the conditions of microcapsule formation. After the lyophilization process, a reduction in particle size was observed, reaching values of 252 ± 0.02 μm for the microcapsules obtained by EHDA and 515 ± 0.003 μm for those prepared by dripping.

Figure 2. Scanning electron microscopy (SEM) images of microcapsules formulated with *Lactobacillus reuteri DSM 17938*. (**a1–a4**) Microcapsules obtained by the dripping method. (**b1–b4**) Microcapsules produced by electrohydrodynamic spraying (EHDA). Scale bars correspond to 200 μm in images (**a1**,**b1**), 100 μm in images (**a2**,**b2**), 50 μm in images (**a3**,**b3**), and 10 μm in images (**a4**,**b4**).

The encapsulates obtained by both methods (DM and EHDA) presented a spherical shape, with a rough surface and some wrinkles, characteristics typically associated with the partial collapse of the polymer gel network during dehydration by lyophilization [19]. Similarly, irregular structures were observed in both encapsulates, likely due to the formation of an agglomerate with variable textures, an effect attributed to moisture loss during the dehydration process. Figure 2a1–a4 (DM) display corrugated structures with fractures, similar to those observed in electrohydrodynamic spraying (Figure 2b1–b4). However, a significant difference in terms of uniformity and porosity was observed, as this parameter appeared to be affected in the encapsulates made by EHDA. This may be because the fractions of the hydrocolloids used as wall materials first gelled on the surface due to the interaction of calcium ions, slightly increasing the roughness. This result is consistent with data reported by Nasiri et al. (2021) [20], who encapsulated *Lactobacillus casei* in alginate microcapsules containing wild sage seed mucilage using the emulsion technique. Their results demonstrated spherical beads with a rough surface due to the incorporation of wild sage seed mucilage.

On the other hand, greater porosity was observed in the encapsulates prepared by EHDA (Figure 2b4), which was associated with the applied voltage, creating an electrostatic force that stretches the material and generates much finer, electrostatically charged droplets, thus promoting a much faster gelation compared to DM. Moreover, during lyophilization, these structures tend to retain and amplify the porosity channels initially formed. While in DM, the droplets are larger and more compact, with less interaction with the gelling solution, resulting in a denser network with fewer pores, as reported by Hernández San José, C. (2018) [21]. Humidity plays a critical role in the structural integrity and stability of microcapsules. Although the lyophilized powders obtained in this study exhibited low residual moisture, the porous structure observed especially in EHDA-derived microcapsules may increase their susceptibility to water uptake under high humidity conditions. Biopolymers such as alginate and chia mucilage are hygroscopic and tend to swell upon moisture absorption, which can lead to capsule deformation, reduced mechanical strength, and premature release or degradation of encapsulated probiotics. Similar phenomena have been reported in alginate-based hydrogel systems stored under uncontrolled humidity, where microcapsule stability deteriorates over time [22,23]. In agreement, Rajam et al. (2012) [24], demonstrated that probiotic-loaded microcapsules stored under humid conditions showed significant structural collapse, decreased cell retention, and early release of encapsulated *Lactobacillus plantarum* during storage. These findings underscore the importance of controlling environmental moisture to preserve probiotic functionality. Therefore, to maintain probiotic viability and microcapsule performance, it is essential to store these powders in low-humidity environments using moisture barrier packaging.

2.3. Physicochemical Properties of the Functional Beverages

Figure 3 shows response surface graphs for different physicochemical properties of beverages made from whey and milk, flavored with red fruits (columns A, C, E, G) and pear (columns B, D, F, H). Each graph evaluates a specific property: total lipids, carbohydrates, total fiber, and protein, depending on the concentrations of whey and milk. The graphs were generated from a statistically validated 3^2 factorial design using Design Expert ® v10 software, with milk and whey concentrations as independent variables. Although the regression parameters are not displayed in the figure, the models showed strong predictive ability ($R^2 > 0.90$), supporting the interpretation of the plotted surfaces. Regarding the total lipids present in the beverages (Figure 3A,B), it was observed that for red fruits (A), there was a linear increase with the proportion of milk, confirming that milk contributes the greatest amount of fat to the beverage. In contrast, the pear beverage (B) showed a peak in the intermediate combinations of milk and whey, suggesting that a possible synergistic effect between whey and milk components, or better homogenization, might have influenced the total lipid content. In addition, these differences are likely associated with the intrinsic composition of the fruits and their interaction with the liquid bases.

Similar studies have been reported by Chacón-Gurrola (2017) [25] in the physicochemical analysis of whey and milk, obtaining values of 0.3 g of fat content, while Shiby et al. (2013) [26] reported a value of 0.2 g in the development of energy drink mixes based on whey and Blue Grapes (*Vitis vinifera*) juice. On the other hand, Sepúlveda Valencia & Peña Alvarez, (2002) [27] reported a fat content of 2.64% in a beverage made from sweet whey (a fermented drink added with passion fruit), which is higher than that observed in the pear formulation, which showed a percentage of 1.02%. This difference is possibly associated with the type of milk used in both studies. In addition, these differences can be attributed to other factors such as the composition of the fruit, since red fruits (strawberry, raspberry and blueberry) have a very low fat content (<1%) [28], whereas pears, although also low in fat, contain pectins and soluble fibers that could interact with the lipids present in the dairy matrix, altering their distribution. Pectin is also widely used to stabilize acidified milk products, thanks to electrostatic interactions between its negatively charged chains and positively charged milk proteins, which reduce syneresis and help maintain a homogeneous texture [13]. However, this interaction is less significant in berry beverages, where pectins and other polysaccharides are present in lower concentrations. According to previous studies, the interaction between the dairy matrix and soluble fibers can promote the formation of more stable emulsions in formulations containing fruits, such as pears [29,30]. This phenomenon may explain the peak observed in the intermediate concentrations of milk and whey in pear beverages.

Similar behavior was observed regarding total carbohydrates, with distinct trends between the two formulations (red fruits and pear), depending on the proportions of milk and whey used. In Figure 3C, the carbohydrate content increased linearly with the increase in the proportion of milk, suggesting that the lactose present in milk is the primary source of carbohydrates in this formulation. In contrast, in Figure 3D, total carbohydrates are lower compared to drinks with red fruits, although they also increase with the proportion of milk. These results are possibly associated with the composition of the fruits. In this regard, red fruits have a significant carbohydrate content, primarily in the form of simple sugars such as glucose, fructose, and sucrose.

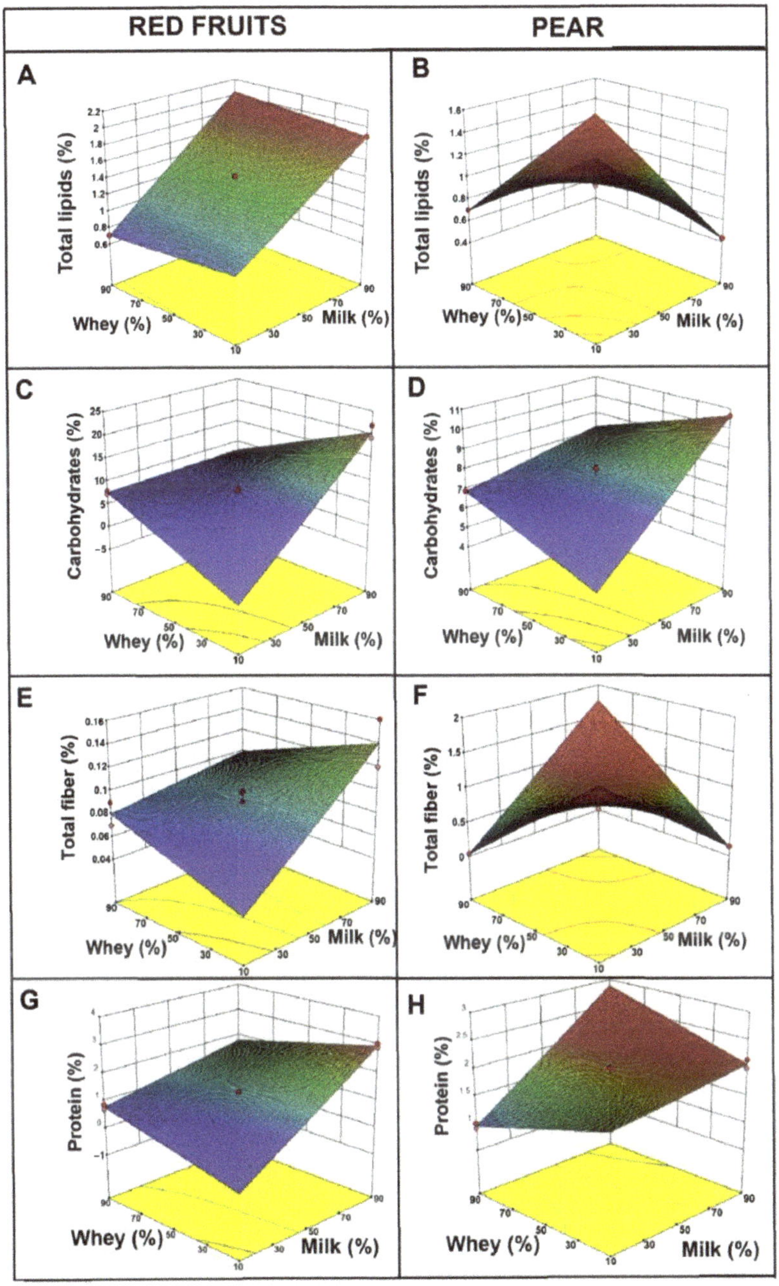

Figure 3. Response surface plots showing the effect of whey and milk concentrations on the physicochemical properties of beverages flavored with red fruits (**A**,**C**,**E**,**G**) and pear (**B**,**D**,**F**,**H**): total lipids (**A**,**B**), carbohydrates (**C**,**D**), total fiber (**E**,**F**), and protein content (**G**,**H**).

Furthermore, lactose, the main carbohydrate present in milk, although constant in concentration, can form complexes or interact with polysaccharides (such as pectins from pear), potentially limiting the availability of free carbohydrates in pear-flavored beverages, which could explain the lower apparent concentrations observed [31]. Moreover, Liang et al. (2020) [32] report that pectin binds to casein micelles in acidified milk systems, reducing syneresis and modifying the release and apparent concentration of carbohydrates in the matrix.

In pear-based beverages, pectin could form networks with proteins and other components of the dairy matrix, retaining part of the sugars and decreasing their measured concentration. In this sense, similar studies have shown that the addition of strawberry or berry pulps modifies the carbohydrate profile in dairy formulations [33]. Moreover, Silva et al. (2020) [13] found that beverages made with pectin-rich fruits, such as pears, may exhibit a reduction in available carbohydrates due to the interaction of the protein and fiber matrix. Additionally, Arcila & Mendoza (2006) [34] reported a content of 7.62% total carbohydrates in their drink based on whey and amaranth, attributed the higher carbohydrate content in the whey they used in its preparation.

Regarding fiber, a linear increase was observed in the red fruit drinks (Figure 3E) as the proportion of milk increased, which is associated with a greater retention of soluble fibers in the dairy matrix. On the other hand, in the pear drinks (Figure 3F), the highest amount of fiber was observed at intermediate concentrations of milk and whey; this effect may be attributed to the higher pectin content in pears and its ability to form networks with milk proteins [35].

In addition to the above, it was observed that the protein percentage in the red fruit drinks showed a linear trend, indicating that proteins were directly derived from the proportion of added milk. However, in pear drinks, a non-linear effect was observed due to the presence of pectins and other soluble polysaccharides that can form three-dimensional networks with proteins, improving their stability. In this sense, Laurent and Boulenguer (2003) [36] demonstrated that the addition of pectin-rich fruits to dairy products enhances protein–polymer interactions, thereby increasing the stability and viscosity of the beverages. On the other hand, Zamani et al. (2020) [30] found that interactions between pectins improve the stability of pear drinks.

2.4. Determination of Moisture and Color

In order to determine the moisture content of the powders used to prepare berry and pear-flavored beverages, a significant change was observed ($p < 0.05$) in both powders. For the berry powder, the higher the whey concentration (10/90 milk/whey) and the lower the milk concentration, the higher the moisture content, with an average value of 3.60 ± 0.20%. The opposite effect was observed at lower whey concentrations (90/10 milk/whey), with a percentage of 2.53 ± 0.31%. A similar trend was observed in the pear flavored beverages, where the highest moisture content was reported at higher whey concentrations (4.13 ± 0.12%), and the lowest moisture content corresponded to the formulation with the highest proportion of milk (3.98 ± 0.13%) (Table 2). This behavior can likely be attributed to the high water content of whey, which is approximately 93–94% [37].

Furthermore, the higher whey concentration likely favored an increase in free water and, consequently, moisture content in the powders. The opposite effect occurred with increasing milk concentration, as the higher fat content tended to limit water retention by displacing available water within the matrix. Likewise, differences in the solids content of the fruits used may have influenced the results; in this sense, red fruits provide a higher amount of total soluble solids, which may contribute to a reduction in moisture, while pears, with a lower solids content, may favor water retention [29].

Table 2. Humidity and color parameters (L, a, b*) of functional beverage powders flavored with pear and red berries containing encapsulated *Lactobacillus reuteri DSM 17938*. Different letters within the same row indicate significant differences ($p < 0.05$).

Parameter	Concentration (Milk/Whey)	Pear Flavor	Red Fruits
Moisture (%)	10/90	4.13 ± 0.12 [a]	3.6 ± 0.2 [b]
Color *(L*)*		76.16 ± 10.47 [a]	44.06 ± 5.44 [b]
(a)*		1.92 ± 0.56 [b]	11.21 ± 7.27 [a]
(b)*		14.87 ± 11.66 [a]	1.43 ± 1.44 [b]
Moisture (%)	50/50	4.6 ± 0.13 [a]	3.8 ± 0.6 [b]
Color *(L*)*		74.91 ± 5.70 [a]	41.58 ± 5.34 [b]
(a)*		7.20 ± 9.00 [a]	7.2 ± 8.87 [a]
(b)*		0.53 ± 0.79 [a]	0.53 ± 0.79 [a]
Moisture (%)	90/10	3.98 ± 0.13 [a]	2.53 ± 0.31 [b]
Color *(L*)*		71.43 ± 7.20 [a]	44.11 ± 4.07 [b]
(a)*		5.90 ± 4.10 [b]	18.54 ± 4.57 [a]
(b)*		1.00 ± 0.60 [a]	0.63 ± 1.56 [b]

With respect to the color of the powders, the formulations with red fruits presented statistically different lightness (L), red–green (a) and yellow–blue (b) values than those of the pear drink ($p < 0.0500$) (Table 2). In the 10/90 formulation (milk/whey), the powders with red fruits reached L = 44.06 ± 5.44, while the pear drink reached L = 76.16 ± 10.47, evidencing a considerably darker color in the former. Similarly, the a parameter was higher in red fruits (11.21 ± 7.27) compared to pear (1.92 ± 0.56), indicating a more intense red blue. For its part, the b value was lower in red fruits (1.43 ± 1.44) compared to the pear drink (14.87 ± 11.66), reflecting a lower tendency towards yellow or bluish tones. This behavior was maintained in the other proportions (50/50 and 90/10), where lower L and higher a values were recorded in drinks with red fruits (Table 2). This pattern resulted from the high concentration of anthocyanins, water-soluble pigments present in red fruits (such as strawberries, raspberries, and blueberries), which generate dark purple tones and produce a decrease in luminosity and an increase in absorbance [38]. Furthermore, the relationship between anthocyanin intensity and CIELAB values has been documented as inversely proportional to L and directly proportional to a [39]. On the other hand, pear-flavored powders exhibited high L and low a values, reflecting their light color and absence of intense pigments, such as anthocyanins [38]. This behavior is characteristic of fruits with low phenolic pigment content and has been reported in studies involving clear juices and flavored dairy products.

2.5. Structural and Molecular Characterization of Reconstituted Beverages by Raman Spectroscopy

Regarding molecular characterization and interactions, Raman spectroscopy enabled the evaluation of the chemical composition of the reconstituted beverage from their respective powders. Figure 4A,B (pear and red fruit flavors) show the relative intensity of the Raman signal (Y axis) as a function of the wavenumber (X axis, in cm^{-1}) for three samples labeled C1, C2, and C3. Several characteristic bands associated with the molecular vibrations of the compounds present in the samples were observed, allowing the identification of possible structural changes resulting from the addition of different ingredients and whey concentrations.

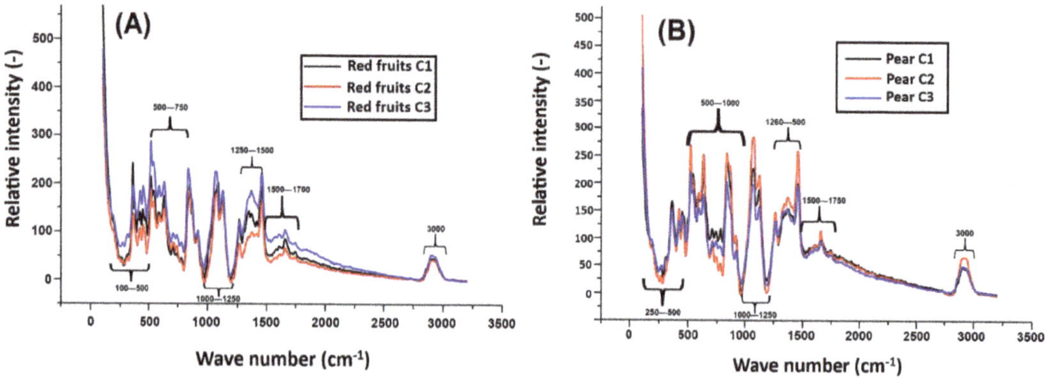

Figure 4. Raman spectra of functional beverages prepared with different concentrations of whey and milk. (**A**) Red fruit-flavored beverage: C1 (10% whey, 90% milk), C2 (50% whey, 50% milk), and C3 (90% whey, 10% milk). (**B**) Pear-flavored beverage: C1 (10% whey, 90% milk), C2 (50% whey, 50% milk), and C3 (90% whey, 10% milk). All formulations contain 20 g of stevia and 102 g of the corresponding fruit.

In the case of the pear-flavored beverage (Figure 4A), the peaks around 200–499 cm^{-1} are attributed to bond vibrations and heavy atom movements, such as those present in polymeric compounds or crystalline structures. These peaks are also associated with antioxidants and fruit-derived pectins. In contrast, in Figure 4B (red fruits), in this same region, fat-soluble fibers, antioxidants, and pectins present in the fruits (especially strawberries and blueberries) are highlighted, which contribute to the functional properties of the beverage and its structural stability. These results are consistent with those reported by Schulz and Baranska (2007) [40], who identified similar bands in fruits with high carbohydrate content, structural fibers, and pectins in fresh fruits.

On the other hand, in the pear drink, the vibrations observed in the 500–1000 cm^{-1} region are associated with the presence of iron, casein, and fat-soluble fibers, possibly originating from milk and whey. In this sense, for the red fruit powder, these peaks indicate the presence of iron, proteins, and casein, in addition to structural carbohydrates. This combination is typical of fruit-enriched dairy drinks. In the 1000–1250 cm^{-1} region in Figure 4A, pectins and key components of fruit cell walls, such as C-O-C groups and esters, are identified. These findings highlight the structural role of pectins in pear drinks. In contrast, in red fruit drinks, some compounds such as anthocyanins, sugars, and fats are more prominent in this spectral region. In this regard, Tadapaneni et al. (2012) [41] demonstrated that strawberry anthocyanins interact with dairy matrices, contributing to antioxidant functionality and influencing the stability of fruit-based dairy beverages.

In the 1250–1500 cm^{-1} region, the peaks observed in the pear-flavored beverage are associated with C–H bonds in methyl and methylene groups, as well as dairy proteins such as casein. Etzion et al. (2004) [42] reported that FTIR spectra exhibit characteristic amide I and II bands (1500–1700 cm^{-1}) related to milk proteins, and that lipid-related vibrations (C–H stretching) appear in the 1400–1477 cm^{-1} region. Similarly, this region also indicates the presence of casein and milk fat. In the 1500–1750 cm^{-1} region, C = C and C = O stretching vibrations are associated with phenolic compounds, methylated pectins, and carboxylic acids. However, although these signals are present in the pear drink, a lower intensity is observed compared to the red fruit drink, since red fruits have a notably high content of phenolic compounds, especially anthocyanins and vitamin C, which are recognized for their antioxidant potential.

Finally, in the 2800–3000 cm^{-1} region, peaks related to O-H and C-H stretching vibrations were detected, which are characteristic of hydroxyl and methyl groups present in carbohydrates, lipids, and water. In the pear-flavored beverages, the higher intensity of these peaks was associated with the presence of total solids and a slight lipid contribution, although the fat content did not exceed 1%. This behavior is likely due to structural water retention and the possible formation of hydrogel networks resulting from interactions between proteins and pectins present in the beverage.

These structures are relevant for the development of gel-based systems capable of modulating the release of bioactive compounds and enhancing the physical stability of the product. These observations are consistent with the O-H and C-H group vibrations commonly found in carbohydrates and lipids, as reported by Schulz and Baranska (2007) [40]. In contrast, the red fruit-flavored beverages exhibited lower intensity peaks in this region, reflecting a compositional profile with a lower solids and fats content and a reduced ability to form hydrogel-like structures.

3. Conclusions

This study demonstrated that electrohydrodynamic spraying and dripping techniques are effective for encapsulating *Lactobacillus reuteri DSM 17938* using sodium alginate and chia seed mucilage as wall materials, achieving encapsulation efficiencies above 99%. The physicochemical characterization of functional powders made with different ratios of whey and milk revealed that matrix composition significantly influences moisture content, macronutrient distribution, and molecular interaction. Raman spectroscopy analysis revealed structural differences between pear- and berry-flavored beverages, highlighting the presence of pectin, casein, and specific phenolic compounds in the formulation. The highest intensity of the carbohydrate- and lipid-associated bands in the 3000 cm^{-1} region suggests the possible formation of hydrogel-like networks in the re-constituted beverages, particularly in the pear-flavored formulations. These hydrogels could promote water retention, structural stability, and controlled release of bioactive compounds.

These findings support the potential of combining dairy by-products with encapsulated fruit and probiotic ingredients to develop reconstituted functional beverages with enhanced structural, nutritional, and gelling properties. Future studies should investigate the rheological behavior, gel-forming capacity, and in vitro gastrointestinal release to validate the functionality and consumer acceptance of these formulations.

From a theoretical standpoint, this study contributes to the understanding of how hydrogel-based microencapsulation with chia mucilage and sodium alginate can enhance the stability and functional integration of probiotics into food matrices, particularly under freeze-drying conditions. Methodologically, it supports the use of electrohydrodinamic atomization (EHDA) as a non-thermal encapsulation technique suitable for sensitive probiotic strains, and it highlights the relevance of Raman spectroscopy and multiscale analysis (SEM, CLSM) for structural and functional evaluation of encapsulated systems.

4. Materials and Methods

4.1. Materials

Chia seeds (*Salvia hispanica*) and strawberries (*Fragaria ananassa*) were purchased from the local market in San Felipe del Progreso, State of Mexico. *Lactobacillus reuteri DSM 17938* (BioGaia, Stockholm, Sweden), De Man Rogosa and Sharpe (MRS) broth (Merck, 04053252452864, Rahway, NJ, USA), MRS agar (Merck, 110660), monobasic potassium phosphate (Sigma-Aldrich, 7778-77-0, St. Louis, MO, USA), sodium citrate (Merck, 6132-04-3), magnesium sulfate (Sigma-Aldrich, 7487-88-9), and glycerol (Merck, 104057) were used. Peptone water (Merck, 107228), sodium alginate (Deiman S.A. de C.V., 8060), $CaCl_2$

(JT Baker, 1313-01, Phillipsburg, NJ, USA), propidium iodide (5 mg/mL; Sigma-Aldrich, 81845), and Calcofluor White stain (Sigma-Aldrich, 18909) were also utilized. Additionally, pasteurized whey and whole milk (approx. 3.5% fat content) were included as components of the beverage matrices. All reagents were of analytical grade and used as received, unless otherwise stated.

4.2. Methods

4.2.1. Chia Seed Mucilage Extraction

Chia mucilage (CM) was extracted following the method previously described by Rentería-Ortega et al. (2021) [43]. Briefly, 15 g of *Salvia hispanica* seeds were mixed with 1000 mL of distilled water and stirred at 6000 rpm for 60 min at 40 °C. The mucilage was separated by pressing the hydrated seeds against a 40-mesh screen, and the resulting filtrate was precipitated using an ethanol–acetone solution (70:30, v/v). The extracted mucilage was dried at 40 °C for 4 h and it reached a final moisture content of 6.0%; then, it was milled and stored in sealed plastic containers for further use.

4.2.2. Activation and Growth of *Lactobacillus reuteri DSM 17938*

The activation of the microorganism (*Lactobacillus reuteri DSM 17938*) was carried out in saline solution through serial dilutions according to that reported by Castro (2020) [44]. In the first tube, 100 µL of the commercial preparation *Lactobacillus reuteri DSM 17938* and 900 µL of saline solution were added and vortexed for 30 s. Then, in the second tube, 100 µL of the previous tube and 900 µL of saline solution were added, and it was vortexed again for 30 s. This process was repeated until eight serial dilutions were prepared, which were then incubated for 45 min at 25 °C. To determine the growth of *lactobacillus reuteri*, 1 mL of each dilution was transferred to Falcon tubes containing 9 mL of Man, Rogosa, and Sharpe (MRS) broth, resulting in a total of eight tubes. These tubes were incubated for 24 h at 37 °C, allowing for optimal development of the strain.

4.2.3. Solution Preparation

For the preparation of the chia mucilage (CM), sodium alginate (AlgNa), and *Lactobacillus reuteri DSM 17938* (LR) solution, the methodology proposed by Sadeghi-Varkani et al. (2018) [45] was used with some modifications. The CM-AlgNa solution (0.5% w/v) was stirred at 6000 rpm (Thermo Scientific, Shanghai, China) for 15 min at room temperature (25 °C) until a homogeneous mixture was achieved. Then, *Lactobacillus reuteri* (6.98×10^{10} CFU/mL) was added by stirring until a uniform distribution was obtained.

The bacterial concentration was selected based on previous scientific recommendations and regulatory guidelines to ensure probiotic effectiveness. Burgain et al. (2011) [46] report that doses above 10^8 CFU/mL are required to achieve gastrointestinal benefits. Additionally, Hill et al. (2014) [47] indicate that probiotic foods or supplements must contain at least 1×10^9 viable cells per daily dose to be considered functional and properly labeled.

4.2.4. Preparation of Encapsulates by Electrohydrodynamic Atomization (EHDA) and Drip Mode (DM)

The encapsulates were obtained using the electrohydrodynamic atomization equipment. The equipment was fed with the MC-AlgNa-LR solution, which was placed in a 10 mL syringe. The syringe was connected to the positive electrode of the current unit, with a voltage of 17 kV, a distance of 5.5 cm, and a flow rate of 2.0 mL/h. The collector was used to ground the system. To determine the preparation of the capsules made by DM, the same process conditions were used, except for the voltage. Finally, during the atomization and drip process, the encapsulates were collected from the $CaCl_2$ solution and stored for later analysis [7].

4.2.5. Encapsulation Efficiency (EE)

Regarding the determination of EE, the capsules were broken using the procedure described by Chávarri et al. (2010) [48], which involved combining 9 mL of sterile 100 mM sodium citrate solution, pH 7, with 1 g of capsules and vortexing until they were completely dissolved. Serial dilutions were then performed to obtain a countable number of cells. The dilutions were spread on Petri dishes containing MRS agar, and the dishes were incubated for 24 h at 37 °C for subsequent counting. The seedings were performed in triplicate. MRS agar (de Man, Rogosa, and Sharpe, Merk, Darmstadt, Germany; 110660) was used as the selective medium for the growth of *Lactobacillus reuteri*. This medium contains peptone, meat extract, yeast extract, dextrose, sodium acetate, triammonium citrate, polysorbate 80, dipotassium phosphate, magnesium sulfate, and manganese sulfate. The pH of the medium was adjusted to 6.5 ± 0.1 prior to autoclaving. After pouring the agar into Petri dishes, they were incubated in a humidity-controlled incubator at 37 °C for 24 h in aerobic conditions. Triplicate seedings were performed, and encapsulation efficiency was calculated as the ratio of viable cells recovered from broken capsules to the initial cell concentration. The colony-forming unit (CFU) was counted, and the encapsulation efficiency was determined according to the methodology reported by Mohammad et al. (2021) [49], which calculated the percentage of cells trapped in the capsule relative to the initial number of viable cells in the sample.

4.2.6. Lyophilization of Encapsulates

The encapsulates were distributed in a Petri dish, frozen at −80 °C, and lyophilized using a 75,034 Bench Top Freeze Dryer (Labconco, Kansas City, MO, USA) for 24 h. The obtained powders were stored in amber bottles wrapped in aluminum foil until further analysis. Three individually prepared replicates were analyzed for each formulation.

4.2.7. Microstructure Analysis

Confocal Laser Scanning Microscopy (CLSM) Analysis

The microstructure of chia seed and sodium alginate (MC-AlgNa) capsules loaded with *Lactobacillus reuteri DSM 17938* (LR) was evaluated by multiphoton confocal microscopy using an LSM 710 NLO (Carl Zeiss, Oberkochen, Germany). Staining of the probiotic and the coating material was performed before encapsulation.

The bacterial pellet (6.98×10^{10} CFU/mL) was resuspended in 1.5 mL of sterile distilled water, 500 µL of propidium iodide (5 mg/mL; Sigma-Aldrich, 81845) was added, and the pellet was sonicated in an ultrasonic bath (Branson-1510, Danbury, CT, USA) for 60 s at 40 kHz. It was then mixed with 16 mL of a solution containing 0.5 mg of chia mucilage and 0.5 mg of sodium alginate. Further, 2 mL of Calcofluor White (Sigma-Aldrich, 18909) was added to this mixture as a wall fluorochrome. The solution was transferred to 10 mL syringes (BD Plastipak, Dickson and Company, Franklin Lakes, NJ, USA) fitted with blunt-tip hypodermic needles (0.80 mm inner diameter) and subjected to electrohydrodynamic spraying (electrospraying), as described in Section 4.2.4.

Three-dimensional images of 100 sections were obtained using the LSM 710 NLO microscope, equipped with excitation lasers at 405 nm (for calcofluor) and 594 nm (for propidium iodide), and employing a Plan-Apochromat 63×/1.40 Oil DIC M27 objective. Image analysis was performed with NIS-Elements software (version 5.30.02; Nikon, Melville, NY, USA).

Scanning Electron Microscopy (SEM)

The capsules were characterized using micrographs obtained with a Field Emission Scanning Electron Microscope (JEOL, JSM-7800F, Akishima, Japan), with an acceleration voltage of 25 kV and a back-scatter electron detector (BSE). The capsules were lyophilized in a Freeze Dryer (ECO-FD10PT, LabTech, Sorisole, Bergamo, Sorisole, Italy) and sprinkled onto aluminum trays provided with conductive double-sided carbon tape and gold coating (Sputter coater, SPI-module, West Chester, PA, USA).

4.3. Beverage Preparation

The beverages were prepared following the methodology reported by Cuellas & Wagner et al. (2010) [50], with some modifications. Different concentrations of whey (10, 50, and 90%) and raw milk (10, 50, and 90%) were used, combined with fruit and a sugar substitute, stevia (20 g). The selected fruit proportions—98 g strawberry, 60 g raspberry, 45 g blueberry, and 203 g pear—were established through pilot-scale sensory screening to determine the most acceptable formulation when combined with the dairy matrix. These exact quantities were retained to ensure consistency and reflect the outcome of the optimization process. The mixtures were frozen at −80 °C and freeze-dried for 24 h (Labconco, model 75034, Kansas City, MO, USA). Afterwards, 10 g of *Lactobacillus reuteri DSM 17938* encapsulates were homogeneously mixed with 100 g of the lyophilized beverage powder, ensuring uniform distribution of the probiotic cells within the MC-AlgNa hydrogel matrix, as confirmed by CLSM imaging.

4.4. Physicochemical Characterization

4.4.1. Protein

The crude protein content was determined using the Kjeldahl digestion method, with the standard nitrogen-to-protein conversion factor (N × 6.25). In detail, 15 g of each sample was weighed, and 0.4 g of copper sulfate pentahydrate, 6 g of potassium sulfate, and 20 mL of concentrated sulfuric acid were added as digestion reagents. The mixture was subjected to sequential digestion in a graphite digester (Hanon SH220, Hanon Instrument Co., Ltd., Jinan, China) at different temperatures and times: 120 °C for 30 min, 240 °C for 30 min, 360 °C for 1 h, and 420 °C for 1 h. The samples were then allowed to cool to room temperature. The determination of total nitrogen content was performed using an automatic Kjeldahl analyzer (ZDDN-II, Tuopu Instrument Co., Ltd., Hangzhou, China), according to the methodology reported by Lai Jing (2022) [51].

4.4.2. Total Lipids

To determine total lipid content, 7 mL of each sample was added, followed by 90 mL of chloroform–methanol solution (2:1 *v*/*v*) at 45 °C for 2 h. A 30 mL NaCl solution (0.9%, *w*/*w*) was immediately added and mixed thoroughly. The chloroform was removed using a rotary evaporator at 40 °C (RE-52AA, Shanghai Yarong Biochemistry Instrument Factory, Shanghai, China). The remaining material was weighed using an electronic analytical balance (CP114, Ohaus International Trading Co., Ltd., Changzhou, China). It was expressed as a percentage of total lipid [51].

4.4.3. Fiber

The fiber content was determined using the methodology reported by Guaman Rivera et al. (2023) [52], which involves sequential acid and alkaline digestion to remove proteins, lipids, and soluble carbohydrates. Briefly, 1 g of defatted sample was digested in 1.25% H_2SO_4 followed by 1.25% NaOH, with intermediate filtration and washing steps. The dried residue was weighed and then incinerated at 550 °C to obtain the ash content. The fiber percentage was calculated using Equation (1):

$$Crude\ fiber\ (\%) = \frac{(W2 - W1) - (W4 - W3)}{W5} \times 100 \quad (1)$$

where

W1 = Weight of filter paper (g);
W2 = Weight of filter paper + dried residue (g);
W3 = Weight of empty crucible (g);
W4 = Weight of crucible after incineration (g);
W5 = Weight of defatted sample (g).

4.4.4. Ash Percentage

Ash content was determined using AOAC Method 923.03. First, the sample was weighed on an analytical balance (Precisa, model XT220A, Dietikon, Switzerland). After drying, the sample was incinerated at 550 °C in a muffle furnace (Selecta, model Select-furnace, Murten, Switzerland), and the incineration residue was determined by calculating the weight difference. The results are expressed as the percentage of ash calculated according to Equation (2).

$$Ash\ (\%) = \left(\frac{W\ ash}{W\ sample}\right) \times 100 \tag{2}$$

where

W ash = Weight of ash after incineration (g);
W sample = Initial weight of the dried sample (g).

4.4.5. Percentage of Total Carbohydrates

The total carbohydrate content was determined by difference, using the proximate analysis data. The percentage was calculated according to Equation (3):

$$\text{Carbohydrates (\%)} = 100 - (\text{Moisture (\%)} + \text{Protein (\%)} + \text{Lipids (\%)} + \text{Ash (\%)} + \text{Crude fiber (\%)}) \tag{3}$$

4.5. Color

For colorimetric analysis, 0.3 g of freeze-dried beverage powder was taken from the total formulation weight (110 g: 100 g beverage base + 10 g encapsulates). This sample size was selected according to the specifications of the colorimeter and allowed for consistent evaluation across all replicates. The reflection spectrum readings of the powders were performed as described by Rentería-Ortega et al. (2023) [53], using a Konica Minolta CR-400 colorimeter (Konica Minolta, NJ, USA), which had been previously calibrated with a standard reference plate (Y = 93.7, x = 0.3159, y = 0.3324). Measurements were taken at five different points on each sample, and the results represented the average of these readings. The coordinates obtained in the CIELab color space were reported, where L* indicates lightness (values between 0 and 100), a* represents the chromatic component from green (−) to red (+), and b* corresponds to the chromatic component from blue (−) to yellow (+). All measurements were taken on a standard white plate background to minimize interference. The results were analyzed statistically using one-way ANOVA, with a significance level of $p < 0.05$, using SigmaPlot 12.5 software. The reported values are the average of three independent measurements taken for each experimental condition.

4.6. Humidity

The moisture content was determined using an Ohaus MB120 moisture analyzer (Ohaus Corporation, Pine Brook, NJ, USA), according to the manufacturer's specifications. For each measurement, 0.500 g of sample was placed directly onto the equipment's weighing pan. Humidity was determined at a constant temperature of 105 °C for a 10 min analysis time, and the final value was obtained as a percentage of mass loss.

4.7. Raman Spectroscopy

For Raman analysis, 0.2 g of the powdered sample was used, extracted from the same 110 g total formulation. This amount complies with the standard sample volume required for instrument calibration and spectral stability. The molecular and structural changes in the powdered beverage added with whey-based LR encapsulants under different times, ingredients, and methods were described using Raman spectroscopy (LabRAM model HR800, Horiba Jobin Yvon, Kyoto, Japan) equipped with an Nd: YAG laser (785 nm) at room temperature (25 °C) and an integration time of 6 s. Three spectra were obtained for each sample and averaged using Origen Pro 8 software (v8.0724, OriginLab, Northampton, MA, USA). A baseline was then created and smoothed using the Savitzky–Golay method with a 2nd-order polynomial and 12 reference points in each spectrum, as reported by Rojas-Candelas et al. (2023) [54].

4.8. Experimental Design

A face-centered rotatable composite statistical design was used for each whey and milk concentration, with five replicates at the center point and duplicates at the other points. Data were processed using multiple regression analysis, and statistical significance ($p \leq 0.005$) was found for the model and its variables. Design Expert V9 software (Stat-Ease, Inc., Minneapolis, MN, USA) was used to independently analyze the effects of the variables on the responses of protein, fat, moisture, fiber, and carbohydrates.

Author Contributions: M.R.-O. and A.Y.C.-C.; methodology: M.R.-O., M.d.J.P.-F., A.P.-B., G.C.-D. and R.B.-F.; software: G.C.-D., M.R.-O. and R.B.-F.; validation: M.R.-O. and G.C.-D.; formal analysis: A.Y.C.-C. and F.S.S.-V.; investigation: A.Y.C.-C. and G.C.-D.; resources: M.R.-O., M.G.-V. and G.C.-D.; original draft preparation: A.Y.C.-C. and M.R.-O.; writing—reviewing and editing: M.R.-O., A.P.-B., M.d.J.P.-F. and G.C.-D.; visualization: F.S.S.-V. and M.R.-O.; supervision: M.R.-O.; project administration: M.R.-O. All authors have read and agreed to the published version of the manuscript.

Funding: This research received no external funding.

Institutional Review Board Statement: Not applicable.

Informed Consent Statement: Not applicable.

Data Availability Statement: The original contributions presented in the study are included in the article, further inquiries can be directed to the corresponding author.

Acknowledgments: Minerva Rentería-Ortega thanks Tecnológico Nacional de México, Tecnológico de Estudios Superiores de San Felipe del Progreso, Escuela Nacional de Ciencias Biológicas, and Centro de Nanociencias y Micro y Nano Tecnologías del Instituto Politécnico Nacional for the experimental support to prepare electrospraying capsules and to analyze the samples by CSLM and SEM. All individuals mentioned in this section have given their consent to be acknowledged.

Conflicts of Interest: The authors declare no conflicts of interest.

Abbreviations

The following abbreviations are used in this manuscript:

CFU	Colony Forming Units
CLSM	Confocal Laser Scanning Microscopy
CM	Chia Mucilage
DM	Drip Mode
EHDA	Electrohydrodynamic Atomization
FTIR	Fourier Transform Infrared Spectroscopy
LR	*Lactobacillus reuteri*
SEM	Scanning Electron Microscopy

References

1. Mosquera-Vivas, E.; Ayala-Aponte, A.; Serna-Cock, L.; Torres-León, C.; Tirado, D.F. Viability of *Lactobacillus reuteri* DSM 17938 Encapsulated by Ionic Gelation during Refractance Window® Drying of a Strawberry Snack. *Foods* **2024**, *13*, 823. [CrossRef] [PubMed]
2. Bagchi, D. (Ed.) *Nutraceutical and Funcional Food Regulations in the United States and Around the Word*, 2nd ed.; Elsevier: Houston, TX, USA, 2014; ISBN 978-0-12-405870-5.
3. Niu, X.; Yin, X.; Wu, X.; Zhang, Q.; Jiang, Y.; He, J.; Zhao, Y.; Zhang, C.; Ren, Y.; Lai, M.; et al. Heat-killed Bifidobacterium longum BBMN68 in pasteurized yogurt alleviates mugwort pollen-induced allergic airway responses through gut microbiota modulation in a Murine model. *Foods* **2023**, *12*, 2049. [CrossRef]
4. Rentería-Ortega, M.; Salgado-Cruz, M.D.L.P.; Morales-Sánchez, E.; Alamilla-Beltrán, L.; Valdespino-León, M.; Calderón-Domínguez, G. Liberación de glucosa oxidasa de cápsulas de mucílago de chía-alginato de sodio estresadas preparadas por electropulverización. *Rev. Proces. Conserv. Aliment.* **2021**, *45*, e15484.
5. Coghetto, C.C.; Webster, K.; Silva, M.L.; Leite, B.; Castell-Palou, M.A. Electrospraying microencapsulation of Lactobacillus plantarum enhances cell viability under refrigeration storage and simulated gastric and intestinal fluids. *Colloids Surf. B Biointerfaces* **2016**, *146*, 677–685. [CrossRef]
6. Yang, S.; Wei, S.; Wu, Y.; Fang, Y.; Deng, Z.; Xu, J.; Zhang, H. Encapsulation techniques, action mechanisms, and evaluation models of probiotics: Recent advances and future prospects. *Food Front.* **2024**, *5*, 1212–1239. [CrossRef]
7. Rentería-Ortega, M.; Salgado-Cruz, M.D.L.P.; Morales-Sánchez, E.; Alamilla-Beltrán, L.; Farrera-Rebollo, R.R.; Valdespino León, M.; Calderón-Domínguez, G. Effect of electrohydrodynamic atomization conditions on morphometric characteristics and mechanical resistance of chia mucilage-alginate particles. *CyTA-J. Food* **2020**, *18*, 461–471. [CrossRef]
8. Jia, H.; Li, Y.; Tian, Y.; Li, N.; Zheng, M.; Zhang, W.; Jiang, Y.; Zhao, Q.; Man, C. Recent advances in electrospray encapsulation of probiotics: Influencing factors, natural polymers and emerging technologies. *Crit. Rev. Food Sci. Nutr.* **2025**, 1–18. [CrossRef]
9. Librán, C.M.; Castro, S.; Lagaron, J.M. Encapsulation by electrospray coating atomization of probiotic strains. *Innov. Food Sci. Emerg. Technol.* **2017**, *39*, 216–222. [CrossRef]
10. Premjit, Y.; Mitra, J. Synthesis, characterization, and in vitro digestion of electrosprayed and freeze-dried probiotics encapsulated in soy protein isolate-sunflower oil emulsions. *Food Biosci.* **2023**, *53*, 102532. [CrossRef]
11. Hua, C.; Yang, F.; Jia, X.; Lu, Y.; Li, X.; Zhao, P.; Xing, M.; Lyu, G. Multi-comparted microgels delivering human derived probiotics and deferoxamine for multidrug-resistant infection and healing. *Chem. Eng. J.* **2024**, *483*, 148432. [CrossRef]
12. Perea-Flores, M.D.J.; Aguilar-Morán, H.F.; Calderón-Domínguez, G.; García-Hernández, A.B.; Díaz-Ramírez, M.; Romero-Campos, H.E.; Cortés-Sánchez, A.D.J.; Salgado-Cruz, M.d.l.P. Entrapment efficiency (EE) and release mechanism of rhodamine B encapsulated in a mixture of chia seed mucilage and sodium alginate. *Appl. Sci.* **2020**, *13*, 1213. [CrossRef]
13. Silva e Alves, A.T.; Spadoti, L.M.; Zacarchenco, P.B.; Trento, F.K. Probiotic functional carbonated whey beverages: Development and quality evaluation. *Beverages* **2018**, *4*, 49. [CrossRef]
14. Malos, I.G.; Ghizdareanu, A.I.; Vidu, L.; Matei, C.B.; Pasarin, D. The role of whey in functional microorganism growth and metabolite generation: A biotechnological perspective. *Foods* **2025**, *14*, 1488. [CrossRef]
15. Lee, D.W.; Hwang, S.J.; Park, J.B.; Park, H.J. Preparation and release characteristics of polymer-coated and blended alginate microspheres. *J. Microencapsul.* **2003**, *20*, 179–192. [CrossRef]
16. Popović, M.; Stojanović, M.; Veličković, Z.; Kovačević, A.; Miljković, R.; Mirković, N.; Marinković, A. Characterization of potential probiotic strain, *L. reuteri* B2, and its microencapsulation using alginate-based biopolymers. *Int. J. Biol. Macromol.* **2021**, *183*, 423–434. [CrossRef]
17. Dragoni-Rosado, J.J. Micro-encapsulación de Lactobacillus Casei y el Efecto Sobre la Supervivencia Durante el Procesamiento del Yogurt y Bajo Condiciones Simuladas del Estómago Humano. Ph.D. Thesis, Universidad De Puerto Rico, San Juan, Puerto Rico, 2014.
18. Huang, H.Y.; Tang, Y.J.; King, V.A.E.; Chou, J.W.; Tsen, J.H. Properties of *Lactobacillus reuteri* chitosan-calcium-alginate encapsulation under simulated gastrointestinal conditions. *Int. Microbiol.* **2015**, *18*, 61–69. [CrossRef]
19. Gowda, H.; Ivanisevic, J.; Johnson, C.H.; Kurczy, M.E.; Benton, H.P.; Rinehart, D.; Siuzdak, G. Interactive XCMS Online: Simplifying Advanced Metabolomic Data Processing and Subsequent Statistical Analyses. *Anal. Chem.* **2014**, *86*, 6931–6939. [CrossRef]
20. Nasiri, H.; Golestan, L.; Shahidi, S.A.; Darjani, P. Encapsulation of *Lactobacillus casei* in sodium alginate microcapsules: Improvement of the bacterial viability under simulated gastrointestinal conditions using wild sage seed mucilage. *J. Food Meas. Charact.* **2021**, *15*, 4726–4734. [CrossRef]
21. Hernández San José, C. Análisis de Distribuciones de Carga Espacial Emitidas por Atomización Electrohidrodinámica (Electrospray) en Vacío. Ph.D. Thesis, Universidad Nacional de Educación a Distancia, Madrid, Spain, 2018. Portal Científico UNED. Available online: https://portalcientifico.uned.es/documentos/5f63fc8d29995274fc8e8d01 (accessed on 1 July 2025).

22. Bustamante, M.; Oomah, B.D.; Burgos-Díaz, C.; Vergara, D.; Flores, L.; Shene, C. Viability of Microencapsulated Probiotics in Crosslinked Alginate Matrices and Chia Seed Mucilage during Spray-Drying and Storage. *Microorganisms* **2025**, *13*, 1457. [CrossRef]
23. Gómez-Mascaraque, L.G.; López-Rubio, A. Optimization of electrospraying conditions for the microencapsulation of probiotics and evaluation of their resistance during storage and in-vitro digestion. *LWT-Food Sci. Technol.* **2016**, *69*, 438–446. [CrossRef]
24. Rajam, R.; Karthik, P.; Parthasarathi, S.; Joseph, G.S.; Anandharamakrishnan, C. Retention of Encapsulated Lactobacillus plantarum in Spray-Dried Milk–Whey Powder during Storage and Simulated Gastrointestinal Conditions. *J. Funct. Foods* **2012**, *4*, 891–898. [CrossRef]
25. Chacón-Gurrola, L.R.; Chávez, A.; Rentería-Monterrubio, A.L.; Rodríguez-Figueroa, J.C. Proteínas del lactosuero: Usos, relación con la salud y bioactividades. *Interciencia* **2017**, *42*, 712–718.
26. Shiby, V.K.; Radhakrishna, K.; Bawa, A.S. Development of whey-fruit-based energy drink mixes using D-optimal mixture design. *Int. J. Food Sci. Technol.* **2013**, *48*, 742–748. [CrossRef]
27. Sepúlveda Valencia, J.U.; Flórez Flórez, L.E.; Peña Álvarez, C.M. Utilización de lactosuero de queso fresco en la elaboración de una bebida fermentada con adición de pulpa maracuyá (passiflora edulis) variedad púrpura y carbóximetil celulosa (cmc), enriquecida con vitaminas ay d. *Rev. Fac. Nac. Agron. Medellín* **2002**, *2*, 1633–1674.
28. Cosme, F.; Pinto, T.; Aires, A.; Morais, M.C.; Bacelar, E.; Anjos, R.; Ferreira-Cardoso, J.; Oliveira, I.; Vilela, A.; Gonçalves, B. Red fruits composition and their health benefits—A review. *Foods* **2022**, *11*, 644. [CrossRef]
29. Aziz, M.G.; Yusof, Y.A.; Blanchard, C.; Saifullah, M.; Farahnaky, A.; Scheiling, G. Material properties and tableting of fruit powders. *Food Eng. Rev.* **2018**, *10*, 66–80. [CrossRef]
30. Zamani, H.; Zamani, S.; Zhang, Z.; Abbaspourrad, A. Exceptional colloidal stability of acidified whey protein beverages stabilized by soybean soluble polysaccharide. *J. Food Sci.* **2020**, *85*, 989–997. [CrossRef] [PubMed]
31. Gómez-Mascaraque, L.G.; Miralles, B.; Recio, I.; López-Rubio, A. Microencapsulation of a whey protein hydrolysate within micro-hydrogels: Impact on gastrointestinal stability and potential for functional yoghurt development. *J. Funct. Foods* **2016**, *26*, 290–300. [CrossRef]
32. Liang, L.I.; Luo, Y. Casein and pectin: Structures, interactions, and applications. *Trends Food Sci. Technol.* **2020**, *97*, 391–403. [CrossRef]
33. Sabater, C.; Ramírez, J.; Blanco, L.; Torres, A. Effect of Low-Esterified Pectin Interactions with Milk Proteins in Reduced-Syneresis Yogurts: Consequences on Carbohydrate Retention and Compositional Profile. *J. Dairy Sci.* **2020**, *103*, 2342–2353.
34. Arcila, N.; Mendoza, Y. Elaboración de una bebida instantánea a base de semillas de amaranto (*Amaranthus cruentus*) y su uso potencial en la alimentación humana. *Rev. Fac. Agron.* **2006**, *23*, 114–124.
35. Hernández-Rojas, M.; Vélez-Ruiz, J.F. Suero de leche y su aplicación en la elaboración de alimentos funcionales. *Temas Sel. Ing. Aliment.* **2014**, *8*, 13–22.
36. Laurent, M.A.; Boulenguer, P. Stabilization mechanism of acid dairy drinks (ADD) induced by pectin. *Food Hydrocoll.* **2003**, *17*, 445–454. [CrossRef]
37. Smithers, G.W. Whey and whey proteins—From 'gutter-to-gold'. *Int. Dairy J.* **2008**, *18*, 695–704. [CrossRef]
38. Ren, S.; Jiménez-Flores, R.; Giusti, M.M. The interactions between anthocyanin and whey protein: A review. *Compr. Rev. Food Sci. Food Saf.* **2021**, *20*, 5992–6011. [CrossRef]
39. Lee, J.; Durst, R.W.; Wrolstad, R.E. Determination of total monomeric anthocyanin pigment concentration of fruit juices, beverages, natural colorants, and wines by the pH differential method: Collaborative study. *J. AOAC Int.* **2005**, *88*, 1269–1278. [CrossRef]
40. Schulz, H.; Baranska, M. Identification and quantification of valuable plant substances by IR and Raman spectroscopy. *Vib. Spectrosc.* **2007**, *43*, 13–25. [CrossRef]
41. Tadapaneni, R.K.; Banaszewski, K.; Patazca, E.; Edirisinghe, I.; Cappozzo, J.; Jackson, L.; Burton-Freeman, B. Effect of high-pressure processing and milk on the anthocyanin composition and antioxidant capacity of strawberry-based beverages. *J. Agric. Food Chem.* **2012**, *60*, 5795–5802. [CrossRef]
42. Etzion, Y.; Linker, R.; Cogan, U.; Shmulevich, I. Determination of protein concentration in raw milk by mid-infrared Fourier transform infrared/attenuated total reflectance spectroscopy. *J. Dairy Sci.* **2004**, *87*, 2779–2788. [CrossRef]
43. Rentería-Ortega, M.; Salgado-Cruz, M.P.; Morales-Sánchez, E.; Alamilla-Beltrán, L.; Valdespino-León, M.; Calderón-Domínguez, G. Glucose oxidase release of stressed chia mucilage-sodium alginate capsules prepared by electrospraying. *J. Food Process Preserv.* **2021**, *45*, e15484. [CrossRef]
44. Castro Crespo, A. Cultivo de *Lactobacillus Reuteri* en Solitario y en Cocultivo con *Escherichia Coli* a 37 °C. Bachelor's Thesis, Facultad de Ciencias, Universidad Nacional de Colombia, Bogotá, Colombia, 2020.
45. Sadeghi-Varkani, A.; Emam-Djomeh, Z.; Askari, G. Physicochemical and microstructural properties of a novel edible film synthesized from Balangu seed mucilage. *Int. J. Biol. Macromol.* **2018**, *108*, 1110–1119. [CrossRef]
46. Burgain, J.; Gaiani, C.; Linder, M.; Scher, J. Encapsulation of probiotic living cells: From laboratory scale to industrial applications. *J. Food Eng.* **2011**, *104*, 467–483. [CrossRef]

47. Hill, C.; Guarner, F.; Reid, G.; Gibson, G.R.; Merenstein, D.J.; Pot, B.; Morelli, L.; Canani, R.B.; Flint, H.J.; Salminen, S.; et al. Activity of cecropin P1 and FA-LL-37 against urogenital microflora. *Nat. Rev. Gastroenterol. Hepatol.* **2014**, *11*, 506. [CrossRef] [PubMed]
48. Chávarri, M.; Marañón, I.; Ares, R.; Ibáñez, F.C.; Marzo, F.; del Carmen Villarán, M. Microencapsulation of a probiotic and prebiotic in alginate-chitosan capsules improves survival in simulated gastro-intestinal conditions. *Int. J. Food Microbiol.* **2010**, *142*, 185–189. [CrossRef] [PubMed]
49. Muhammad, Z.; Ramzan, R.; Zhang, R.; Zhang, M. Resistant starch-based edible coating composites for spray-dried microencapsulation of *Lactobacillus acidophilus*: Thermal protection, in vitro digestion, and physicochemical characteristics. *Coatings* **2021**, *11*, 587. [CrossRef]
50. Cuellas, A.V.; Wagner, J.R. Elaboración de bebida energizante a partir de suero de quesería. *INNOTEC* **2010**, *5*, 54–57.
51. Lai, J.; Wu, R.; Wang, J.; Wang, Y.; Zhang, X.; Zhou, L.; Zhu, Y. Effect of cooking modes on quality and flavor characteristic in Clitocybe squamulose chicken soup. *Front. Nutr.* **2022**, *9*, 1048352. [CrossRef]
52. Guaman Rivera, S.A.; Guerrero-Pincay, A.E.; Ortiz-Naveda, N.R.; González-Marcillo, R.L. Prediction of the nutritional values by INRA in grass forage: Crude fiber was analyzed according to the Weende method by acid hydrolysis with 1.25% H_2SO_4, followed by alkaline hydrolysis with 1.25% NaOH. *J. Agric. Environ. Int. Dev.* **2023**, *117*, 117–140. [CrossRef]
53. Rentería-Ortega, M.; Colín-Alvarez, M.D.L.; Gaona-Sánchez, V.A.; Chalapud, M.C.; García-Hernández, A.B.; León-Espinosa, E.B.; Valdespino-León, M.; Serrano-Villa, F.S.; Calderón-Domínguez, G. Characterization and Applications of the Pectin Extracted from the Peel of *Passiflora tripartita* var. mollissima. *Membranes* **2023**, *13*, 797. [CrossRef]
54. Rojas-Candelas, L.E.; Díaz-Ramírez, M.; Rayas-Amor, A.A.; Cruz-Monterrosa, R.G.; Méndez-Méndez, J.V.; Salgado-Cruz, M.D.L.P.; Calderón-Domínguez, G.; Cortes-Sanchez, A.D.J.; González-Vázquez, M. Development of Biodegradable Films Produced from Residues of Nixtamalization of Popcorn. *Appl. Sci.* **2023**, *13*, 8436. [CrossRef]

gels

MDPI

Article

Improvement of Gel Properties of *Nemipterus virgatus* Myofibrillar Protein Emulsion Gels by Curdlan: Development and Application to Emulsified Surimi

Zhiqin Wu [1], Yongyan Qu [1], Ouhongyi Li [1], Soottawat Benjakul [2] and Aimei Zhou [1,*]

[1] College of Food Science, South China Agricultural University, Guangzhou 510642, China; wuzhiqin1983@163.com (Z.W.); quyongyan0919@163.com (Y.Q.); 13626909168@163.com (O.L.)
[2] Department of Food Technology, Faculty of Agro-Industry, Prince of Songkla University, Hat Yai, Songkhla 90112, Thailand; soottawat.b@psu.ac.th
* Correspondence: zhouam@scau.edu.cn

Abstract: This study aims to improve the gel properties of *Nemipterus virgatus* myofibrillar protein (MP) emulsion gels by Curdlan (Cur) and investigate the effect of the emulsion gels on the quality of emulsified surimi gels. The effects of different concentrations of Cur on the gel properties of MP emulsion gels were investigated. Fourier transform infrared (FTIR) results indicated that intermolecular interactions between Cur and MP were primarily hydrogen bonds. Cur enhanced the adsorption capacity of MP at the oil/water interface, inducing the formation of a more uniform and dense composite network structure in Cur/MP emulsion gels. Adding 6% (*w*/*v*) of Cur significantly increased the hardness, gel strength, water-holding capacity (WHC) and rheological properties of the gel. In addition, microstructural images showed that MP formed a complex interpenetrating network with Cur, thus enhancing the gel network skeleton. Low-field NMR confirmed that the addition of Cur decreased water mobility in the emulsion gel system. Compared to the direct addition of oil, the application of Cur/MP emulsion gels to surimi significantly improved the texture, gel strength, and WHC of the surimi gel. These findings provide a reference for the development of myofibrillar protein emulsion gels and broaden their potential application in the food industry.

Keywords: emulsion gel; composite network; textural properties; surimi

1. Introduction

Dietary fat is an important nutrient for maintaining human physiological functions and provides energy for human metabolism. Saturated fatty acids are the most common type of lipid in daily consumed meat products, but excessive intake may induce cardiovascular diseases, coronary heart disease, type 2 diabetes, and obesity [1]. Replacing saturated fats with plant oils or marine liquid oils in surimi products confers health benefits [2]. However, their liquid state results in a softer surimi texture. Furthermore, direct oil addition hinders protein cross-linking, thereby disrupting the surimi gel network structure [3]. Therefore, developing healthy and sustainable fat substitutes with fat-mimetic textures has become urgent to meet growing consumer demand for natural and health-promoting alternatives [4]. In recent years, emulsion gels have been widely recognized as a promising fat substitute due to their unique structural and functional properties [5]. Emulsion gels possess dual emulsion and gel characteristics, which reduce oil droplet size while helping maintain fat-mimicking sensory properties [6]. Emulsion gels formulated with vegetable oils and natural materials (proteins, polysaccharides, or oligosaccharides) offer advantages including safety, health benefits, low calorie content, and richness in unsaturated fatty acids [7]. Among them, proteins are ideal components for the development of emulsion gels due to their dual capabilities, excellent ability to stabilize the water/oil interfaces, and gel-forming properties [8]. Animal proteins, especially fish proteins, have garnered increasing attention due to their high nutritional value and functional properties. As a high-quality protein resource, fish protein contains essential amino acids (lysine, etc.) and exhibits advantages such as easy digestibility and hypoallergenicity, making it suitable for diverse populations requiring nutritional supplementation [9]. However, its high-value applications in emulsification and delivery systems remain insufficient. Myofibrillar protein (MP), the primary structural protein of fish muscle, demonstrates excellent gel-forming ability and emulsification properties. MP not only forms intermolecular cross-linked networks, but also its abundant hydrophobic groups anchors on the surface of oil droplets to form interfacial protein films that prevent oil droplet aggregation [10]. However, pure MP stabilized emulsion gels are susceptible to thermal denaturation and aggregation of MP molecules during heating, leading to unsatisfactory product quality [11]. Additionally, the texture of pure MP emulsion exhibits insufficient texture firmness. Pei et al. [12] reported that while the addition of MP emulsion gels improved the water/oil retention of fish sausage, they reduced the gel strength of surimi gel, potentially diminishing consumer sensory satisfaction. Previous studies demonstrate that polysaccharides can enhance the functional properties of proteins by strengthening the protein gel network through covalent or non-covalent interactions [13]. Zhao et al. [5] demonstrated that the addition of *Konjac glucomannan* improved the gel properties of soybean isolate protein emulsion gels through physical bonding. In order to enhance the thermal stability, emulsification, and gel strength of MP emulsion gels, polysaccharide incorporation is a natural and safe strategy.

Curdlan (Cur), a linear β-1,3-glucan produced by microbial fermentation, which exhibits the advantages of food safety, water-holding capacity, thermal stability, and gel-forming ability [14]. Cur has been widely used to improve the gelation properties of protein gels by hydration and filling effects, including water-holding capacity and gel strength [15]. Cur is insoluble in cold water, but heating Cur aqueous suspension to 55–60 °C or above 80 °C yields thermo-reversible and thermo-irreversible gels, respectively. At temperatures above 80 °C, the disordered coil-like molecular conformation of Cur transforms into a stable and ordered triple-helix structure and undergoes hydration [16]. Upon cooling, this triple-helix molecular structure forms a thermally irreversible three-dimensional network through strong intermolecular hydrogen bonding interactions that encapsulate water in the network [17]. Cur after high temperature heating exhibits excellent gel strength and mechanical stiffness [18]. It had been shown that strategically mixing Cur with soybean isolate proteins yields a desirable texture comparable to animal fat, which prevented emulsion gels from forming an imbalanced texture (excessive rigidity or flexibility) [19]. Furthermore, unfolded Cur is able to enhance protein gel strength through hydrogen bonding and hydrophobic interactions with MP [20]. Although the effect of Cur on the properties of MP gels has been extensively studied, there is a lack of reports on emulsion gels prepared by mixing fish MP with Cur as the aqueous phase. Meanwhile, the effect of replacing unemulsified vegetable oils with MP/Cur emulsion gels on the gel properties of surimi products unknown. *Nemipterus virgatus* has become one of the major raw materials in surimi processing due to its high nutritional value, rapid growth rate, and high yield, ranking as the second most important resource for surimi production after Alaska pollock [21]. As an abundant and low-value marine fish species, the development and utilization of *Nemipterus virgatus* for high-value applications is of significant research importance. Therefore, this study used *Nemipterus virgatus* as the raw material for the extraction of MP and the preparation of surimi gel.

Accordingly, this study aimed to investigate the formation and gel properties of emulsion gels prepared by blending MP with Cur, as well as the effect of such emulsion gels on the gel characteristics of surimi products. Cur/MP emulsion gel was prepared using a one-step method and characterized for their gel properties such as structure, texture and water retention capacity. Subsequently, surimi gels were prepared by mixing Cur/MP emulsion gels containing different Cur contents with surimi, and their effects on the properties of surimi gels were evaluated compared to the direct addition of soybean oil. This study provides an efficient strategy for improving the physical properties of fish MP emulsion gels. The results are expected to provide theoretical references to advance the development of MP emulsion gels and their application in surimi products.

2. Results and Discussion

2.1. Characterization of Cur/MP Mixtures

2.1.1. FTIR

The intermolecular interactions of Cur/MP complexes were investigated by FTIR spectroscopy. As shown in Figure 1a, the characteristic bands of the samples at about 3000–3500 cm^{-1} corresponded to O–H or N–H stretching vibrations, while the absorption bands at 2990–2890 cm^{-1} were attributed to C–H stretching vibrations [22]. Typical protein bands were observed: the absorption band at 1680–1630 cm^{-1} corresponded to C–O stretching vibration (amide I band), the absorption band at 1560–1530 cm^{-1} corresponded to N–H vibration (amide II band), and the absorption band at 1450–1240 cm^{-1} corresponded to N-H deformation and C–H stretching vibration (amide III band) [19]. In addition, the absorption peak at 1038.34 cm^{-1} in the spectrum of MP represented C–O or C–C stretching vibration [23]. For Cur, the absorption peaks at 1077.93 cm^{-1} (C–O glycosidic bond) and 890.04 cm^{-1} (β-1, 3-D-glucan) were observed [24]. Notably, no new absorption peaks were observed in the FTIR spectra of Cur/MP complexes with different Cur concentrations compared to pure MP (Figure 1a). This suggested that no new functional groups were formed between Cur and MP molecules and that no covalent interactions occurred between the compounds. In the Cur/MP complexes profile, the signals of the unique absorption peaks of Cur (3433 cm^{-1}, 2891.68 cm^{-1}, 1077.93 cm^{-1} and 890.04 cm^{-1}) were intensified and the signals of the MP absorption peaks (3191.98 cm^{-1}, 2989.77 cm^{-1}, 1631.91 cm^{-1}, 1553 cm^{-1} and 1038.34 cm^{-1}) were diminished as the concentration of Cur was increased. The absorption bands of Cur/MP complexes gradually changed from being similar to pure MP (Cur $\leq$ 3%) to being similar to pure Cur (Cur > 3%), indicating that Cur gradually became the dominant component of the complexes, and intermolecular interactions were enhanced with the increase in Cur concentration. Blue shifts in hydroxyl absorption peaks of the Cur/MP complexes were observed in comparison to the spectrum of Cur and red shifts compared to the spectrum of MP [24]. These wavelength shifts confirmed hydrogen bonding between Cur and MP. Similar findings were found by Zhu et al. [20] in a study of TGase and curdlan to improve the gel strength of surimi gels.

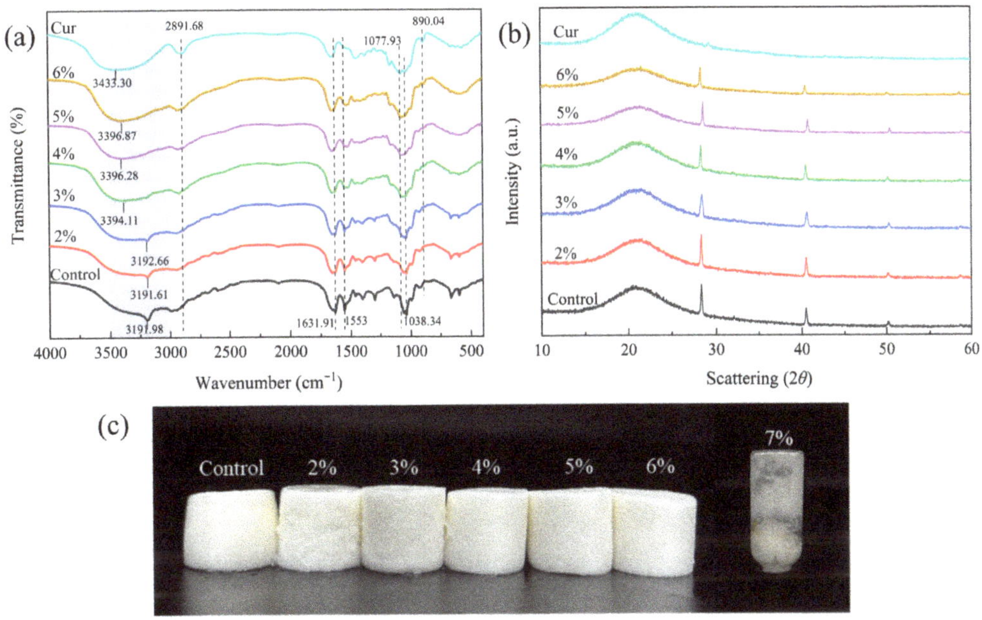

Figure 1. FTIR spectrum (**a**), XRD spectrum (**b**) appearance and (**c**) of Cur/MP emulsion gels prepared with different concentrations of Cur.

2.1.2. XRD

The crystal structures of Cur/MP complexes with different Cur contents were characterized by XRD to gather information related to the interaction between Cur and MP (Figure 1b). Pure Cur exhibited a broad diffraction peak at about $2\theta = 21^\circ$ and a weak peak at $2\theta = 29^\circ$, confirming the amorphous structure of Cur [25]. Pure MP displayed an amorphous band at $2\theta = 21^\circ$. The presence of differential amorphous peaks in Cur and MP maps also indicated the inconsistent arrangement of atoms in proteins or polysaccharides [26]. Notably, the disappearance of the $2\theta = 29^\circ$ diffraction peaks in the profile of Cur-MP complexes compared to pure Cur might be caused by the interaction of Cur with MP disrupting the original crystal structure domains of Cur [24]. The profile of Cur/MP complexes was similar to that of pure MP and the signal intensity of all the diffraction peaks of Cur/MP polymer varied with the difference in the concentration of Cur, which suggested that Cur may have altered the crystalline properties of the complexes, and that Cur-MP interactions resulted in conformational changes in MP [27]. The intensity of the diffraction peaks of Cur/MP complexes with a Cur concentration of 2% tended to increase but tended to decrease slightly with continued increase in Cur concentration. This may be attributed to the interaction between excess Cur [28]. The strong interactions formed by high concentrations of Cur prevent spontaneous interactions between Cur-MP and MP molecules [28]. Collectively, the experimental results of XRD were in agreement with the existence of natural interactions in Cur/MP complexes as evidenced by FTIR spectroscopy. Furthermore, XRD analysis confirmed that the addition of Cur enhanced the intermolecular interactions within the emulsion gel system, thereby contributing to improved gel strength. This enhancement was further reflected in TPA (Table 1) and rheological properties.

Table 1. Effects of different Cur concentrations on gel strength and TPA of Cur/MP emulsion gels.

Concentration of Cur	Gel Strength (g·mm)	Hardness (g)	Springiness (cm)	Gumminess (g)	Chewiness (mJ)	Cohesiveness
Control	105.85 ± 6.34 [e]	125.06 ± 5.78 [f]	7.60 ± 0.02 [c]	75.24 ± 10.25 [e]	57.29 ± 9.28 [e]	0.6 ± 0.06 [c]
2%	241.06 ± 20.08 [d]	247.82 ± 18.77 [e]	8.06 ± 0.00 [b]	136.73 ± 21.44 [d]	110.12 ± 16.96 [d]	0.55 ± 0.06 [cd]
3%	316.36 ± 23.73 [c]	306.39 ± 24.46 [d]	7.96 ± 0.02 [b]	158.20 ± 11.6 [d]	125.86 ± 7.18 [d]	0.52 ± 0.00 [e]
4%	378.28 ± 65.87 [c]	424.04 ± 36.6 [c]	8.76 ± 0.02 [a]	290.45 ± 22.33 [c]	254.59 ± 20.58 [c]	0.69 ± 0.04 [b]
5%	481.04 ± 72.29 [b]	521.05 ± 5.76 [b]	8.65 ± 0.01 [a]	384.47 ± 9.19 [b]	332.52 ± 14.26 [b]	0.74 ± 0.02 [ab]
6%	734.89 ± 64.97 [a]	736.45 ± 30.76 [a]	8.66 ± 0.05 [a]	568.19 ± 21.57 [a]	492.06 ± 21.41 [a]	0.77 ± 0.00 [a]

Different letters within the same column indicate statistically significant differences ($p < 0.05$).

2.2. Cur/MP Emulsion Gels Property Determination

2.2.1. TPA and Gel Strength

The texture of the food affects the acceptance of consumer, and TPA can present a comprehensive picture of the texture condition of emulsion gels from a mechanical point of view. As shown in Figure 1c, with increasing Cur concentration, Cur/MP emulsion gels exhibited a harder solid morphology compared to the pure MP emulsion gel. In addition, the emulsification system collapsed when the Cur concentration reached 7% and subsequent experiments could not be performed, necessitating the maximum concentration of Cur was set at 6% in this study. Hardness is determined by the ability of the sample to resist external pressure, chewiness and springiness respond to the combined texture properties of the food when it is chewed [25]. Gumminess reflects the force of the sample to resist external separation, and cohesion responds to the strength of the internal interactions of the sample when it resists damage from external forces. As reflected in Table 1, the hardness, chewiness, springiness, gumminess and cohesiveness of Cur/MP emulsion gels were significantly increased ($p < 0.05$) with increasing Cur content. At the maximum concentration of Cur (6%), the chewiness, springiness, gumminess and cohesiveness of Cur/MP emulsion gel reached the maximum value and increased by 488.86%, 758.85%, 13.95%, 286.01% and 28.67%, respectively, over the pure MP emulsion gel. The results indicated that Cur could significantly improve the texture and mouthfeel of Cur/MP emulsion gels and enhance their mechanical properties ($p < 0.05$). The improvement in textural properties could be attributed to the strong hydrogen bonding interactions between Cur molecules in the continuous phase of the emulsion gel to form an elastic three-dimensional network induced at high temperatures, enhancing the strength of the MP gel network structure. This is consistent with the findings from Figure 1a FTIR, where the hydrogen bond signal peak gradually intensifies with increasing Cur concentration, indicating that Cur addition enhances the hydrogen bonding interactions between Cur and MP molecules. The heating treatment induced the gradual stretching of the Cur molecular conformation from the curled state, and the resulting unfolded Cur and MP molecules generated noncovalent interactions and formed a dense microstructure, which enhanced the textural properties of emulsion gels [29]. A large increase in the Cur content led to a significant increase in the intermolecular forces between Cur molecules and the formation of a Cur-dominated composite gel, and a high concentration of Cur reduced the interference of MP on the intermolecular interactions between Cur molecules, as evidenced by the FITR spectrum.

Gel strength is an important index to assess the performance of gels and reflects the density of the gel structure. The gel strength of all Cur/MP emulsion gels were significantly higher than that of the control ($p < 0.05$), with a positive correlation between gel strength and Cur concentration. The gel strength reached a maximum value of 734.89 g·mm at a Cur concentration of 6%, representing a 594.26% increase over pure MP emulsion gels. The textural properties and the enhancement of gel strength of Cur/MP emulsion gels indicated that the density of the gel structure was enhanced by Cur. The results were consistent with the TPA. The enhancement of textural properties and gel strength of Cur/MP emulsion gels suggests that Cur is able to improve the network structure of the gels. This is due to the thermogenic gel properties of Cur. Cur can form thermally irreversible highly solidifying gels and enhance the mechanical properties of emulsion gels through hydration and filling.

2.2.2. Water-Holding Capacity (WHC)

The water holding capacity of emulsion gels represents the ability of emulsion gels to immobilize and retain water. The ability of the gel network to retain water also indirectly reflects the state of the gel network structure [30]. As shown in Figure 2, the addition of Cur significantly increased the WHC of pure MP emulsion gels ($p < 0.05$). At 6% Cur, WHC reached a maximum value of 97.56%, representing a 38.11% increase compared to pure MP emulsion gel (70.64%). Cur retained water by exposing a large number of hydroxyl groups that combine with water molecules to form hydrogen bonds during gel formation [31]. The addition of Cur enhanced hydrogen bonding interactions with MP will improve the water binding capacity of emulsion gels [20]. In addition, Cur produced a three-dimensional network structure when forming gels. The results from Figure 1a FTIR and Figure 1b XRD collectively demonstrated that the addition of Cur enhanced the intermolecular interactions between Cur and MP within the emulsion gel, which was a key factor contributing to the improved gel network structure. Furthermore, such a dense and compact network played a critical role in enhancing the WHC of the gel [32]. As the concentration of Cur increases, the resultant denser and multi-hierarchical network structure from the interaction of Cur and MP contributed to generating stronger capillary forces to trap water.

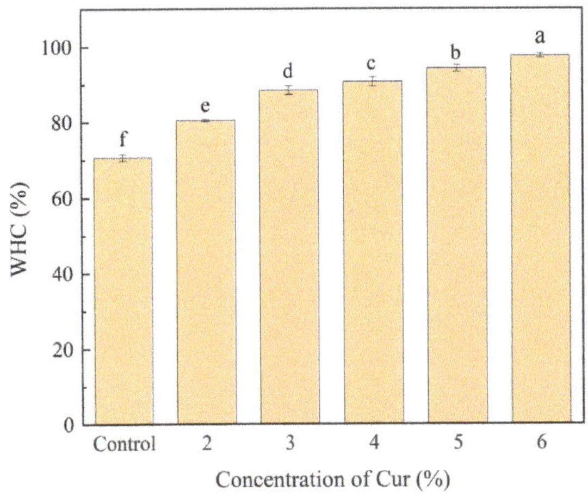

Figure 2. WHC of Cur/MP emulsion gels prepared with different concentrations of Cur. Different letters indicate significant differences ($p < 0.05$).

2.2.3. Rheological Behavior

Information on the dynamic rheological properties of emulsion gels at increasing strain at a fixed frequency was obtained by strain scanning. The energy storage modulus (G′) and loss modulus (G″) reflected the elastic and viscous properties of the samples, respectively. The results are shown in Figure 3a, where the energy storage modulus (G′) is higher than the loss modulus (G″) in the linear viscoelastic region (LVR) for all experimental groups, indicative of their elastic behavior at low strains. Both G′ and G″ of the emulsion gels progressively increased with Cur concentration. This trend indicated that Cur enhanced the solid-like characteristics of MP emulsion gels. It was observed that all samples exhibited relatively stable G′ values and G″ values in the 1% strain range, due to the structural integrity of the emulsion gel. Consequently, a strain of 1% was selected for subsequent frequency and temperature scans within the LVR.

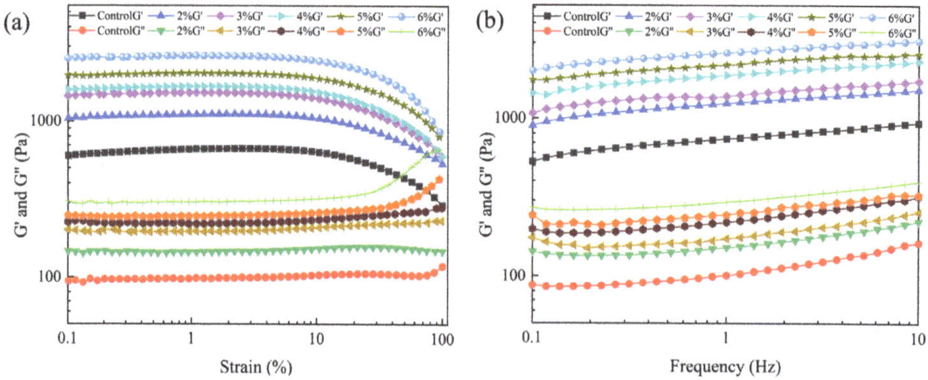

Figure 3. Strain scans (**a**) and frequency scans (**b**) of Cur/MP emulsion gels prepared with different concentrations of Cur.

The frequency scans of the emulsion gels were able to respond to the effect of the available gel on the viscoelasticity of the MP emulsion gel network structure. As shown in Figure 3b, the G′ values of all experimental groups were higher than the G″ values throughout the frequency range, which indicated that all samples exhibited elastic gel properties. The G′ values of the emulsion gels increased with increasing Cur concentration, peaking at a Cur addition of 6%. This trend was consistent with the strain scan results, indicating that Cur improved the gel strength of the MP emulsion gels. The enhancements of gel properties of gel systems were associated with the improvement of gel network structures. Cur formed a dense three-dimensional network with MP in the emulsion gel matrix, thus enhancing the viscoelastic behavior [33]. Cur molecules were linked to the protein network to enhance the stability of the emulsion gel network.

2.2.4. LF-NMR Relaxation

Transverse relaxation time (T_2) can characterize the binding of hydrogen protons in samples and thus reflects the strength of water mobility in the sample. The T_2 of emulsion gels was determined by low field nuclear magnetic resonance (LF-NMR) to analyze the distribution and migration of water. A shorter relaxation time represents a smaller degree of freedom for the hydrogen proton. Figure 4 shows the relaxation peaks of three different hydrogen proton signals for emulsion gel samples: the T_{21} (0.5–12 ms) was associated with water/oil tightly bound by the macromolecules, the T_{22} (12–900 ms) corresponded to immobilized water/oil in the emulsion gel network, and the T_{23} (>900 ms) represented free water/oil [26]. With the increase in Cur content, the relaxation times of T_{22} and T_{23} of Cur/MP emulsion gels were gradually shortened compared to the control group. In particular, the relaxation time of peaks in the emulsion gel samples did not exceed 1000 ms at Cur concentrations ≥5%, which was closely related to the Cur induced unique stability of MP emulsion gels. The results suggested that the addition of Cur to the MP emulsion gel system decreased the degree of freedom of hydrogen protons and enhanced the binding capacity to water/oil. To further analyze the water distribution in each experimental group, the relative contents of water in different states are shown in Table 2. A_{21}, A_{22} and A_{23} represented the relaxation peak relative areas of T_{21}, T_{22} and T_{23}, respectively. The A_{21} of all groups had no significant difference ($p > 0.5$) and was much lower than the other relaxation peaks (not more than 1% of the total content), indicating that the content of Cur had no significant effect on the bound water/oil of MP emulsion gels. The addition of Cur resulted in a significant increase in A_{22} along with a significant decrease in A_{23} compared to the control ($p < 0.5$). The higher the concentration of Cur the higher the proportion of immobilized water in the system, and no free water was detected at higher concentrations of Cur (5–6%). These changes indicated that the addition of Cur resulted in the migration of free water to immobilized water in MP emulsion gels, which was consistent with the results of WHC. This may be because the dense gel network induced by Cur was helpful to improve the water retention of MP emulsion gels, a finding further supported by CLSM (Figure 5) and Cryo-SEM (Figure 6) images. Combined with the microstructural imaging results, it was demonstrated that Cur/MP emulsion gels formed a denser network, which increased capillary forces and thereby enhanced the retention of both water and oil. The Cur molecule was rich in hydroxyl groups capable of binding more free water through hydrogen bonding interactions. In addition, the complex network system formed by the interpenetration of Cur and MP networks also reduced the water/oil mobility [20].

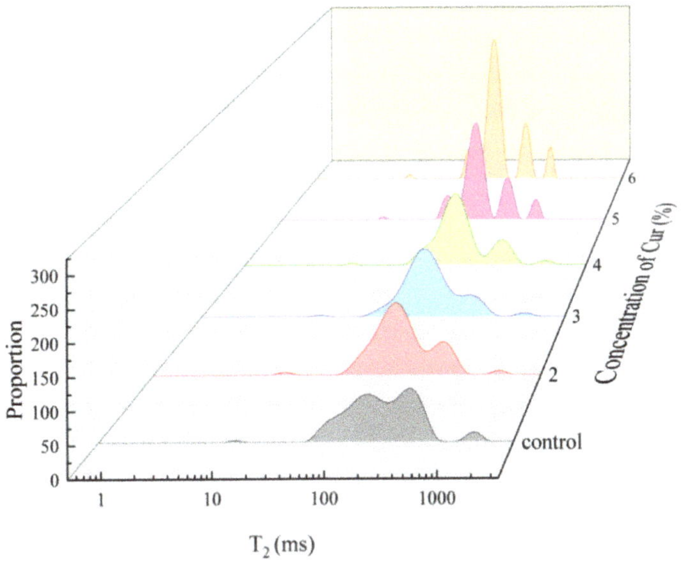

Figure 4. Spin-spin relaxation spectra (T_2) of Cur/MP emulsion gels prepared with different concentrations of Cur.

Table 2. Effects of different Cur concentrations on T_2 relaxation peak relative area of Cur/MP emulsion gels.

Concentration of Cur	T_2 Relaxation Peak Relative Area (%)		
	A_{21}	A_{22}	A_{23}
Control	0.03 ± 0.02 b	79.87 ± 0.71 d	20.10 ± 0.68 a
2%	0.04 ± 0.01 ab	84.11 ± 1.99 c	14.64 ± 1.04 b
3%	0.03 ± 0.01 b	86.83 ± 1.41 bc	13.14 ± 1.41 b
4%	0.05 ± 0.02 ab	87.67 ± 3.31 b	12.28 ± 3.3 b
5%	0.07 ± 0.02 a	99.93 ± 0.02 a	-
6%	0.07 ± 0.03 a	99.93 ± 0.03 a	-

Different letters within the same column indicate statistically significant differences ($p < 0.05$).

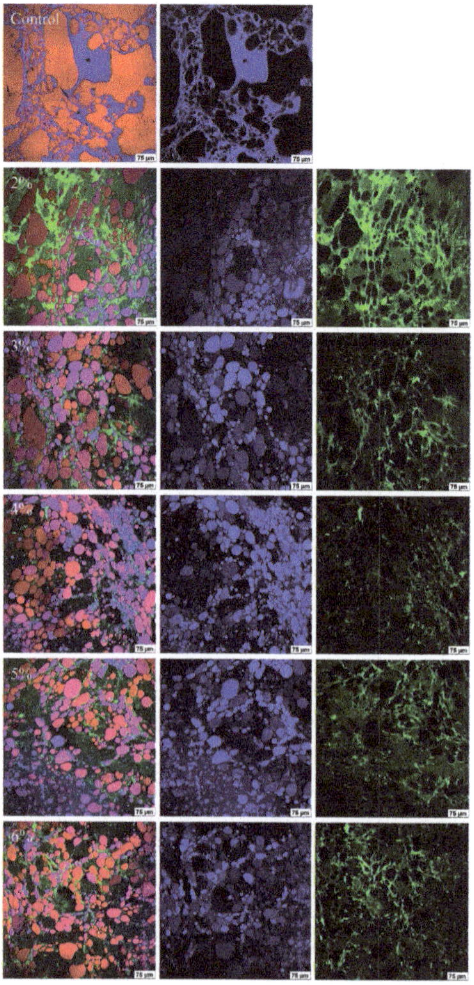

Figure 5. CLSM images of Cur/MP emulsion gels prepared with different concentrations of Cur.

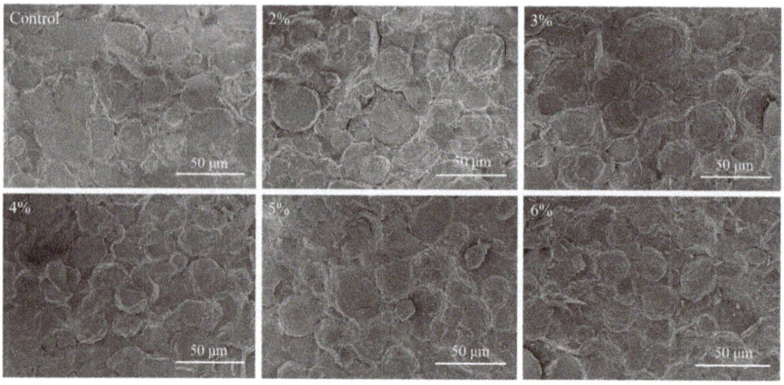

Figure 6. Cryo-SEM images of Cur/MP emulsion gels prepared with different concentrations of Cur.

2.2.5. CLSM

The influence of Cur on the morphological conformation of MP emulsion gels was observed by CLSM. As shown in Figure 5, Cur labeled green, MP labeled blue and oil droplets labeled red can be observed to determine the spatial distribution of these three substances. The image of pure MP emulsion gel showed that the particle size and distribution of oil droplets in the gel network structure were severely inhomogeneous, and there were some larger irregularly shaped oil droplets, which impeded the cross-linking between MP molecules and led to the creation of large voids in the microstructure of the pure MP emulsion gels, thus decreasing their ability to retain water and oil. With the addition of Cur (2–6%), Cur/MP emulsion gels achieved a more uniform microstructure, the shape of the oil droplets showed a more regular spherical shape, and the particle size of the oil droplets were smaller compared with that of the pure MP emulsion gel. In terms of the interfacial state, from CLSM images of Cur/MP emulsion gels, it could be observed that part of the blue fluorescence signals overlapped with the red color, and MP was adsorbed on the surface of the oil droplets to form an interfacial layer. Moreover, the oil droplet was surrounded by strong blue and green fluorescence, and part of the blue and green fluorescence signals overlapped with red to form purple and orange fluorescence, indicating that the interface around the oil droplets were totally encapsulated by Cur and MP. In addition to these, more Cur and MP filled in the voids among the oil droplets and connected with the Cur and MP at the droplet interface, which helped to form a spatial barrier to stabilize the droplets. Cur has been shown to have certain hydrophobic and emulsifying properties [34]. Incorporation of an appropriate amount of Cur in the continuous phase contributed to enhancing the tight adsorption of MP on the surface of oil droplets, intertwining with Cur at droplet-to-droplet interfaces. This interpenetrating network structure effectively restricted the movement of oil droplets, preventing aggregation and thereby improving the stability of the emulsion gel [35]. Specifically, molecular chains of Cur were interconnected with MP in the continuous phase of emulsion gels through hydrophobic interactions, hydrogen bonding and van der Waals forces to form a more stable three-dimensional network structure [19]. At the same time, Cur also formed aggregates to fill in the emulsion gel networks. The synergistic effect of bridging and flocculation structures contributed to the formation of a more stable emulsion system [20]. In summary, in addition to the adsorption of Cur and MP at the oil/water interfacial layer (MP predominantly), another part of Cur and MP filled the space between the oil droplets and acted as the emulsion gel skeleton, thus improving the stability of the emulsion gel.

2.2.6. Cryo-SEM

Cryo-SEM was used to analyze the reasons for the changes in the gel properties of emulsion gels from a microscopic point of view. Figure 6 shows Cryo-SEM images of emulsion gels with different Cur contents. Some irregular, large and unevenly distributed oil droplets could be observed in the microstructure of pure MP emulsion gels, the protein interfacial film could not be uniformly and tightly wrapped around the surface of the oil droplets, and the structure of the gel network was relatively loose and disordered. This phenomenon may be due to the fact that the amount of MP was not sufficient to completely cover all the oil droplets formed during the homogenization process [36]. As the Cur concentration increased (2–6%), the oil droplet size was gradually reduced and tightly wrapped by the interfacial film. The Cur/MP emulsion gels showed a more homogeneous and compact network structure. The possible reasons affecting the emulsification effect of Cur/MP emulsion gels are as follows. On the one hand, Cur had a thickening effect on the emulsion gel system, which prevented oil droplets from agglomerating by strengthening the spatial site resistance during high-speed shear. It has been reported that the higher viscosity aqueous phase in the emulsified system contributes to the formation of smaller oil droplets [37]. On the other hand, Cur increased the network structure strength of Cur/MP emulsion gels by self-assembling from a mixture of single helices and loosely interconnected triple helices to form a more cohesive rod-like triple-helical structure during thermal treatment [6]. These would significantly alter the microstructure of emulsion gels, thereby affecting the water- and oil-holding capacity and textural properties of emulsion gels. Similarly, Li et al. [38] reported that appropriate addition of *Konjac glucomannan* improved the emulsification properties of MP and contributed to the formation of network structure. As shown in Figures 5 and 6, the observed trend in the microstructural changes in Cur/MP emulsion gels was consistent with Figure 2 WHC and Figure 3 rheological behavior results. This reinforced gel network structure contributed to enhanced WHC and exhibited favorable gel elasticity.

2.3. Emulsified Surimi Gels Property Determination

2.3.1. TPA and Gel Strength

TPA and gel strength are critical indicators for assessing surimi gel quality. As shown in Table 3, pure MP emulsion gel did not significantly improve the gel strength or TPA properties of surimi gels compared to those with added soybean oil ($p > 0.05$). Pei et al. [12] similarly observed that incorporating pure tilapia MP emulsion gel weakened surimi gel strength due to the aqueous phase in the MP emulsion gel. In contrast, adding Cur/MP emulsion gel significantly enhanced gel strength, hardness, chewiness, and gumminess of surimi gels at 4–6% Cur concentrations ($p < 0.05$), with optimal values at 6% Cur. Numerous studies have shown that pre-emulsification reduces the particle size of oil droplets and thus decreases their damage to the protein gel network [27]. Figures 5 and 6 demonstrated that increasing Cur concentration reduced oil droplet size and improved uniformity in Cur/MP emulsion gels. Additionally, the texture of the surimi gel was hardened due to the thickening and gelling properties of Cur.

Table 3. Gel strength, TPA, WHC and whiteness of emulsified surimi gels.

Sample Name	Gel Strength (g·mm)	Hardness (g)	Springiness (cm)	Gumminess (g)	Chewiness (mJ)	Cohesiveness	WHC	Whiteness
Control	2276.94 ± 121.62 ^{d}	3363.11 ± 45.82 ^{d}	8.99 ± 0.00 ^{a}	2778.56 ± 5.21 ^{d}	2498.68 ± 8.19 ^{d}	0.83 ± 0.01 ^{a}	82.39 ± 1.85 ^{f}	81.45 ± 0.31 ^{a}
C0	2327.28 ± 141.69 ^{d}	3439.76 ± 41.41 ^{d}	9.08 ± 0.01 ^{a}	2864.74 ± 29.92 ^{d}	2601.66 ± 35.69 ^{cd}	0.83 ± 0.00 ^{a}	84.40 ± 0.47 ^{e}	76.85 ± 0.31 ^{e}
C2	2354.61 ± 107.67 ^{d}	3608.62 ± 64.26 ^{c}	9.03 ± 0.03 ^{a}	3012.90 ± 23.68 ^{d}	2720.11 ± 106.41 ^{bc}	0.84 ± 0.01 ^{a}	85.20 ± 0.33 ^{de}	77.76 ± 0.63 ^{d}
C3	2470.32 ± 125.07 ^{cd}	3775.74 ± 14.28 ^{b}	9.05 ± 0.01 ^{a}	3137.31 ± 53.47 ^{d}	2837.85 ± 77.31 ^{b}	0.83 ± 0.01 ^{a}	86.28 ± 0.41 ^{cd}	78.36 ± 0.11 ^{d}
C4	2654.41 ± 107.10 ^{bc}	3837.46 ± 30.98 ^{b}	9.11 ± 0.01 ^{a}	3149.79 ± 60.39 ^{c}	2867.74 ± 39.98 ^{b}	0.82 ± 0.02 ^{a}	87.22 ± 0.55 ^{bc}	78.46 ± 0.44 ^{d}
C5	2766.86 ± 111.26 ^{ab}	4111.54 ± 106.42 ^{a}	9.14 ± 0.00 ^{a}	3455.46 ± 152.53 ^{b}	3157.24 ± 151.49 ^{a}	0.84 ± 0.02 ^{a}	88.31 ± 0.45 ^{b}	79.35 ± 0.31 ^{c}
C6	2916.58 ± 56.45 ^{a}	4163.11 ± 89.49 ^{a}	9.18 ± 0.02 ^{a}	3481.12 ± 116.91 ^{a}	3196.54 ± 133.86 ^{a}	0.84 ± 0.01 ^{a}	90.14 ± 0.40 ^{a}	80.10 ± 0.17 ^{b}

Different letters within the same column indicate statistically significant differences ($p < 0.05$).

2.3.2. WHC and Whiteness

Surimi gels form elastic colloids by trapping water through a protein network. Compact and homogeneous gel network exhibits excellent WHC [39]. As shown in Table 3, emulsion gel addition significantly improved surimi gel WHC compared to soybean oil incorporation ($p < 0.05$). At 6% Cur concentration, WHC increased by 9.41% relative to the control group. This enhancement can be attributed to the smaller oil droplets in the emulsion gel minimizing the interference of cross-linking between surimi proteins [27].

As shown in Table 3, whiteness was significantly higher in the control group than in the emulsion gel group ($p < 0.05$). This difference was attributed to soybean oil enhancing light scattering in surimi gels [40]. This may also be attributed to fading caused by increased light scattering from smaller droplets, as light waves penetrate less deeply into the emulsion before being backscattered, resulting in reduced light absorption [41]. Furthermore, whiteness increased with higher Cur concentrations in emulsion gel formulations. This may stem from improved light scattering properties due to structural modifications induced by Cur/MP emulsion gels in the surimi matrix [32]. The exceptional ability of Cur to bind water molecules and the denser three-dimensional gel network structure induced by Cur addition both contributed to the retention of more water and oil droplets, which ultimately enhanced light scattering [11].

2.3.3. LF-NMR

Table 4 and Figure 7 show the effect of soybean oil and Cur/MP emulsion gel with varying Cur contents on the water distribution of surimi gels. Compared with the control group, the addition of emulsion gel promoted the migration of free water to immobilized and bound water ($p < 0.05$), which was consistent with the results of WHC. This suggested that pre-emulsification facilitates the reduction in network structure disruption by oil droplets. As shown in Table 4, Cur/MP emulsion gel with high Cur content (4–6%) significantly reduced the water mobility in surimi gels, with C6 group exhibiting the highest proportion of immobilized and bound water ($p < 0.05$). These findings collectively demonstrate that Cur/MP emulsion gels enhanced structural integrity in emulsified surimi gels.

Table 4. T_2 relaxation peak relative area of emulsified surimi gels.

Samples	T_2 Relaxation Peak Relative Area (%)			
	A_{21}	A_{22}	A_{23}	
			A_{23a}	A_{23b}
Control	0.08 ± 0.01 ^{c}	65.34 ± 1.50 ^{d}	18.82 ± 1.28 ^{b}	15.77 ± 2.56
C0	0.07 ± 0.01 ^{c}	67.34 ± 0.22 ^{c}	16.71 ± 0.34 ^{c}	15.87 ± 0.40
C2	0.09 ± 0.01 ^{bc}	77.78 ± 0.49 ^{b}	22.13 ± 0.50 ^{a}	-
C3	0.09 ± 0.01 ^{bc}	79.57 ± 1.49 ^{ab}	20.34 ± 1.50 ^{ab}	-
C4	0.11 ± 0.02 ^{ab}	80.08 ± 0.99 ^{a}	19.80 ± 0.99 ^{b}	-
C5	0.11 ± 0.02 ^{a}	81.00 ± 0.64 ^{a}	18.89 ± 0.63 ^{b}	-
C6	0.13 ± 0.01 ^{a}	80.99 ± 0.57 ^{a}	18.88 ± 0.58 ^{b}	-

Different letters within the same column indicate statistically significant differences ($p < 0.05$).

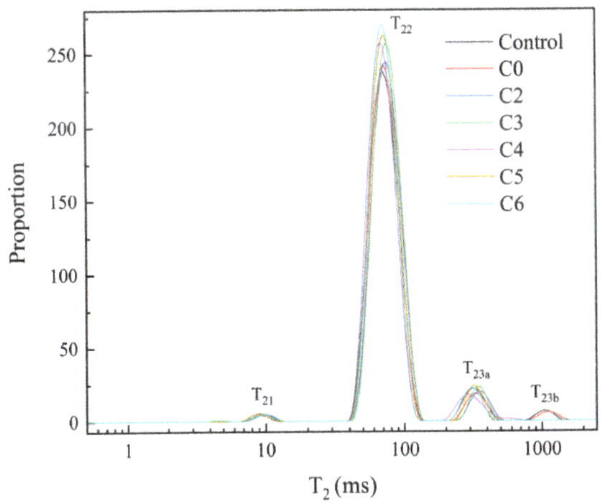

Figure 7. Spin-spin relaxation spectra (T_2) of emulsified surimi gels.

3. Conclusions

In this research, the Cur/MP emulsion gels were prepared by adding different proportions of Cur, and the feasibility of improving the gel properties of MP emulsion gels by Cur was confirmed. The results indicated that Cur had a filling role and enhanced the strength of the network structure in Cur/MP emulsion gels. The high concentration (6%) of Cur dominated the continuous phase of the emulsion gel, resulting in a substantial enhancement of the hardness of the MP emulsion gel. Cur formed a composite network structure with MP molecules mainly through hydrogen bonding interactions, thus promoting the migration of free water to immobilized water in the emulsification system, ultimately exhibiting better gel elasticity. Moreover, the addition of Cur enhanced the adsorption of MP at oil/water interface and improved the emulsion stability. Compared with the pure MP emulsion gel, the Cur/MP emulsion gel had superior water-holding properties, gel texture, rheological properties and microstructure. In conclusion, this study presents a simple and efficient method for the construction of Cur/MP emulsion gels and elucidates the effect of Cur on the performance of Cur/MP emulsion gels. This study provides a reference for improving the utilization of low-value fish protein and meeting public demand for dietary health. Future research should further investigate the mechanism by which the Cur/MP emulsion gel improves the gel properties of emulsified surimi gels and evaluate the effects of its addition on the sensory properties and shelf life of surimi products.

4. Material and Methods

4.1. Materials and Chemicals

Frozen surimi of *Nemipterus virgatus* used in this study was purchased from Qingdao Tengbenwei Foods Co., Ltd. (Qingdao, China). Cur (food grade), supplied by Jiangsu Yiming Technology Co., Ltd. (Taixing, China), consists of molecules with 300–500 glucose residues, an average polymerization degree of 450, and a molecular weight of approximately 74,000. Soybean oil was procured from Yonghui Fresh Food Supermarket (Guangzhou, China). Nile Red and Nile Blue were purchased from Shanghai Yuanye Biotechnology Co., Ltd. (Shanghai, China). Calcofluor White was available from Sigma-Aldrich Co., Ltd. (Shanghai, China). All other chemicals were of analytical grade and purchased from Sinopharm Chemical Reagent Co., Ltd. (Shanghai, China).

4.2. Extraction of MP

MP was extracted from frozen surimi of *Nemipterus virgatus* following the method of Lv et al. [42]. The extraction process was maintained at a low temperature of 4 °C throughout, and the concentration of MP was determined using the Biuret method [43].

4.3. Preparation of Cur/MP Mixtures

The MP concentration was diluted to 20 mg/mL with phosphate-buffer solution (0.6 M NaCI, pH 7.4). Subsequently, different proportions of Cur powder were added to the MP suspension and continuously stirred for 2 h to form a mixed homogeneous dispersion. The concentrations of Cur added into MP suspension were 0, 2, 3, 4, 5 and 6% (*w*/*v*), respectively. After being heated at 40 °C for 20 min, the samples were immediately transferred to a water bath preheated to 90 °C and held for 20 min. Following heating, the gels were rapidly cooled in crushed ice for 1 h. Finally, they were stored in a 4 °C refrigerator until further testing.

4.4. Characterization of Cur/MP Mixtures

4.4.1. Fourier Transform Infrared Spectrometer (FTIR)

The Cur/MP mixture was lyophilized for 48 h and homogenized into a powder for testing. The powder was mixed with KBr at a 1.5:100 (*w*/*w*) ratio, ground and pressed into a thin sheet. Subsequently, spectra were collected in the 4000–400 cm^{-1} range in 64 scans by a Vertex-70 spectrophotometer (Bruker Technologies Ltd., Billerica, MA, USA).

4.4.2. X-Ray Diffraction (XRD)

X-ray diffraction analyses of pure Cur, pure MP, and Cur/MP mixtures with varying Cur concentrations were carried out using an X-ray diffractometer (Ultima IV; Kuraray Co., Ltd., Tokyo, Japan) in reflection mode at 40 kV and 40 mA. The 2θ range was scanned from 10° to 60° with a step size of 0.02° [27].

4.5. Preparation of Cur/MP Emulsion Gels

Cur powder was added to MP suspensions at designated ratios then uniformly dispersed (as described in Section 2.3). The Cur/MP suspension and soybean oil were emulsified at a high speed of 12,000 rpm for 2 min to form emulsified systems, and the volume of the oil phase φ = 50%. Subsequently, emulsions were incubated at 40 °C for 30 min and then heated to 90 °C for 20 min to obtain Cur/MP emulsion gels, rapidly cooled, and stored overnight at 4 °C for subsequent characterization.

4.6. Cur/MP Emulsion Gels Property Determination

4.6.1. Texture Profile Analysis (TPA) and Gel Strength

TPA and gel strength of the samples were analyzed by TA-XT texture analyzer (TA-XT plus, Stable Micro Systems Ltd., Godalming, UK). After equilibrating to room temperature, the emulsion gel samples were cut into cylinders with a height of 20 mm for subsequent analysis. TPA was performed according to the method described by Yang et al. [44] with slight modifications. The TA/50 probe was used under the following parameters: constant test speed of 1 mm/s, trigger force of 5 g, and engineering strain of 50%. Gel strength was determined based on the method of Jiang et al. [17] with minor adjustments. Measurements were carried out using a TA/0.5 probe, which penetrated the gel axially to a depth of 15 mm at a speed of 1.5 mm/s. All measurements were performed in triplicate.

4.6.2. Water-Holding Capacity (WHC)

The WHC of the sample was measured by referring to the method of Lv et al. [45] with minor modifications. Briefly, emulsion gel samples (5 g) were centrifuged for 10 min (10,000× *g*, 4 °C). Surface moisture was removed using filter paper. The WHC was calculated as the percentage ratio of post-centrifugation to pre-centrifugation sample weight.

4.6.3. Rheological Properties

The rheological properties of emulsion gels were analyzed using a rheometer (HR20, TA Corporation, Santa Fe Springs, CA, USA) equipped with T20 parallel plates at a gap distance of 1 mm with reference to previous methods [34]. The parameters of the strain scan are shown below: frequency 1 Hz, shear strain range 0.01–100%, temperature 25 °C. A strain of 0.1–1% in the viscoelastic region (LVE) was determined from the strain scan. The parameters for the frequency scan were as follows: shear strain 1%, frequency 0.1–10 Hz, temperature 25 °C.

4.6.4. Low-Field Nuclear Magnetic Resonance (LF-NMR)

In order to determine the effect of Cur on the stability of MP emulsion gels, the water and oil distributions of the emulsion gel samples were assessed by a nuclear magnetic resonance (NMR) analyzer (Niumag Co., Ltd., Shanghai, China). The setup parameters were consistent with those of Zhu et al. [20].

4.6.5. Confocal Laser Scanning Microscopy (CLSM)

Samples were sectioned stained and imaged using a CLSM (TCS SP8 STED 3×, Leica Microsystems Inc., Wetzlar, Germany) at 20× magnification. The emulsion gel samples were stained after cutting into thin slices. Oil, protein and polysaccharide were labeled with Nile Red (1 mg/mL, excitation wavelength 488 nm), Nile Blue (1 mg/mL, excitation wavelength 633 nm) and Calcium Fluorescent White stain (1 mg/mL, excitation wavelength 405 nm), respectively.

4.6.6. Cryo-Scanning Electron Microscope (Cryo-SEM)

Referring to the previous method [46], the emulsion gel samples were sublimated with liquid nitrogen at −80 °C for 15 min and sputter plated with gold. The microstructures of the emulsion gel samples were examined using a scanning electron microscope (S4800, Hitachi, Ltd., Tokyo, Japan) equipped with a Cryo-SEM preparation system (PP3010T, Quorum Technologies Ltd., Lewes, UK) at 700× magnification.

4.7. Preparation of Emulsified Surimi Gels

Frozen surimi thawed overnight at 4 °C, was mixed with 2.5% NaCl for 1 min. Subsequently, 10% unheated Cur/MP emulsion gel (prepared as described in Section 4.5) with varying Cur contents was added to the surimi. These mixtures were designated C0, C2, C3, C4, C5, and C6, respectively. The control consisted of surimi supplemented with 10% soybean oil. The moisture content of all surimi mixtures was adjusted to 78% using crushed ice, followed by continuous chopping for 4 min. The surimi mixtures were then stuffed into polyvinylidene chloride casings (inner diameter 30 mm), incubated at 40 °C for 30 min followed by 90 °C for 20 min, rapidly cooled. After heat treatment, all samples were stored overnight at 4 °C.

4.8. Emulsified Surimi Gels Property Determination

4.8.1. TPA and Gel Strength

TPA and gel strength of the samples were analyzed using a TA-XT plus texture analyzer (Stable Micro Systems Ltd., Godalming, UK). After equilibrating to room temperature, the surimi gel samples were cut into cylinders with a height of 20 mm for subsequent analysis. The TPA and gel strength of the surimi gel samples were measured using the same test parameters as those described in Section 4.6.1. Each sample was analyzed with three replicates.

4.8.2. WHC and Whiteness

WHC and whiteness were measured according to the previous method [8] with slight modifications. For WHC determination, a surimi gel sample (3 g) was wrapped in filter paper and centrifuged at 10,000× *g* for 10 min at 4 °C. WHC was calculated as the percentage of the sample weight retained after centrifugation relative to its initial weight.

A colorimeter (CR-410, Konica Minolta Camera, Co, Tokyo, Japan) was used to determine the lightness (L*), red-green value (a*), and yellow-bulb value (b*) of surimi gels with a height of 20 mm. All measurements were performed in triplicate. Whiteness was calculated as follows:

$$\text{Whiteness} = 100 - [(100 - L^*)^2 + a^{*2} + b^{*2}]^{1/2}$$

4.8.3. LF-NMR

The moisture distribution of surimi gel was determined according to the method of Zhou et al. [47] using low-field NMR (MesoMR23-040V-I; Niumag Analytical Instrument Corporation, Shanghai, China). Cylindrical surimi gel samples (30 mm thickness) were prepared and placed in NMR tubes. Spin-spin relaxation time (T_2) was measured at 32 °C. Each sample was analyzed with three replicates.

4.9. Statistical Analysis

All experiments were repeated three times and data were expressed as mean ± standard deviation. Significance analysis was performed using SPSS 22.0 software (IBM Corp., Armonk, NY, USA), with and one-way analysis of variance (ANOVA) applied to determine the data for significant differences. Significantly different groups ($p < 0.05$) were denoted by distinct superscript letters. Tukey's Honestly Significant Difference test was performed on all data following analysis of ANOVA.

Author Contributions: Conceptualization: Z.W. and A.Z.; methodology: Z.W. and Y.Q.; investigation: Z.W.; formal analysis: Z.W. and Y.Q.; software: Z.W., Y.Q. and O.L.; data curation: Z.W., Y.Q. and O.L.; writing—original draft: Z.W.; writing—review and editing: S.B.; project administration: A.Z.; funding acquisition: A.Z. All authors have read and agreed to the published version of the manuscript.

Funding: This research was funded by the National Natural Science Foundation of China (No. 31972103).

Institutional Review Board Statement: Not applicable.

Informed Consent Statement: Not applicable.

Data Availability Statement: The original contributions presented in this study are included in the article. Further inquiries can be directed to the corresponding author.

Acknowledgments: The authors are grateful to the National Natural Science Foundation of China (No. 31972103).

Conflicts of Interest: The authors declare no conflicts of interest.

References

1. Zembyla, M.; Murray, B.S.; Sarkar, A. Water-in-oil emulsions stabilized by surfactants, biopolymers and/or particles: A review. *Trends Food Sci. Technol.* **2020**, *104*, 49–59. [CrossRef]
2. Yu, J.; Song, L.; Xiao, H.; Xue, Y.; Xue, C. Structuring emulsion gels with peanut protein isolate and fish oil and analyzing the mechanical and microstructural characteristics of surimi gel. *LWT* **2022**, *154*, 112555. [CrossRef]
3. Zhang, X.; Xie, W.; Liang, Q.; Jiang, X.; Zhang, Z.; Shi, W. High inner phase emulsion of fish oil stabilized with rutin-grass carp (*Ctenopharyngodon idella*) myofibrillar protein: Application as a fat substitute in surimi gel. *Food Hydrocoll.* **2023**, *145*, 109115. [CrossRef]
4. Salminen, H.; Herrmann, K.; Weiss, J. Oil-in-water emulsions as a delivery system for n-3 fatty acids in meat products. *Meat Sci.* **2013**, *93*, 659–667. [CrossRef]
5. Zhao, J.; Chang, B.; Wen, J.; Fu, Y.; Luo, Y.; Wang, J.; Zhang, Y.; Sui, X. Fabrication of soy protein isolate-konjac glucomannan emulsion gels to mimic the texture, rheological behavior and in vitro digestion of pork fat. *Food Chem.* **2025**, *468*, 142462. [CrossRef] [PubMed]
6. Zhao, D.; Sun, L.; Wang, Y.; Liu, S.; Cao, J.; Li, H.; Liu, X. Salt ions improve soybean protein isolate/curdlan complex fat substitutes: Effect of molecular interactions on freeze-thaw stability. *Int. J. Biol. Macromol.* **2024**, *272*, 132774. [CrossRef]
7. Li, J.; Huang, G.; Qian, H.; Pi, F. Fabrication of soy protein isolate—High methoxyl pectin composite emulsions for improving the stability and bioavailability of carotenoids. *Food Biosci.* **2023**, *53*, 102738. [CrossRef]
8. Li, M.; Feng, L.; Xu, Y.; Nie, M.; Li, D.; Zhou, C.; Dai, Z.; Zhang, Z.; Zhang, M. Rheological property, β-carotene stability and 3D printing characteristic of whey protein isolate emulsion gels by adding different polysaccharides. *Food Chem.* **2023**, *414*, 135702. [CrossRef]

9. Zhang, X.; Liu, Z.; Shi, W. Pickering emulsion stabilized by grass carp myofibrillar protein via one-step: Study on microstructure, processing stability and stabilization mechanism. *Food Chem.* **2024**, *447*, 139014. [CrossRef]
10. Lin, L.; Xiong, Y.L. Competitive adsorption and dilatational rheology of pork myofibrillar and sarcoplasmic proteins at the O/W emulsion interface. *Food Hydrocoll.* **2021**, *118*, 106816. [CrossRef]
11. Wu, Y.; Zhao, H.; Lv, Y.; Xu, Y.; Yi, S.; Li, X.; Li, J. Improved gel properties of *Nemipterus virgatus* myofibrillar protein emulsion gel by Konjac glucomannan incorporation: Insight into the modification of protein conformation. *Int. J. Biol. Macromol.* **2024**, *282*, 136833. [CrossRef] [PubMed]
12. Pei, Z.; Wang, H.; Xia, G.; Hu, Y.; Xue, C.; Lu, S.; Li, C.; Shen, X. Emulsion gel stabilized by tilapia myofibrillar protein: Application in lipid-enhanced surimi preparation. *Food Chem.* **2023**, *403*, 134424. [CrossRef] [PubMed]
13. Shao, T.; Zhou, Y.; Dai, H.; Ma, L.; Feng, X.; Wang, H.; Zhang, Y. Regulation mechanism of myofibrillar protein emulsification mode by adding psyllium (*Plantago ovata*) husk. *Food Chem.* **2022**, *376*, 131939. [CrossRef] [PubMed]
14. Li, X.-M.; Meng, R.; Xu, B.-C.; Zhang, B. Investigation of the fabrication, characterization, protective effect and digestive mechanism of a novel Pickering emulsion gels. *Food Hydrocoll.* **2021**, *117*, 106708. [CrossRef]
15. Hu, Y.; Liu, W.; Yuan, C.; Morioka, K.; Chen, S.; Liu, D.; Ye, X. Enhancement of the gelation properties of hairtail (*Trichiurus haumela*) muscle protein with curdlan and transglutaminase. *Food Chem.* **2015**, *176*, 115–122. [CrossRef]
16. Jiang, S.; Mo, F.; Liu, Q.; Jiang, L. Insights into the in vitro digestibility and rheology properties of myofibrillar protein with different incorporation types of curdlan. *Food Chem.* **2024**, *459*, 140255. [CrossRef]
17. Jiang, S.; Zhao, S.; Jia, X.; Wang, H.; Zhang, H.; Liu, Q.; Kong, B. Thermal gelling properties and structural properties of myofibrillar protein including thermo-reversible and thermo-irreversible curdlan gels. *Food Chem.* **2020**, *311*, 126018. [CrossRef]
18. Zhao, J.; Wu, P.; He, J.; Zhao, Y.; Fang, Y. Microstructure and mechanical behavior of curdlan hydrogels: The role of thermal pre-treatment temperature. *Carbohydr. Polym.* **2025**, *367*, 123982. [CrossRef]
19. Choi, M.; Choi, H.W.; Jo, M.; Hahn, J.; Choi, Y.J. High-set curdlan emulsion gel fortified by transglutaminase: A promising animal fat substitute with precisely simulated texture and thermal stability of animal fat. *Food Hydrocoll.* **2024**, *154*, 110063. [CrossRef]
20. Zhu, S.; Wang, Y.; Ding, Y.; Xiang, X.; Yang, Q.; Wei, Z.; Song, H.; Liu, S.; Zhou, X. Improved texture properties and toughening mechanisms of surimi gels by double network strategies. *Food Hydrocoll.* **2024**, *152*, 109900. [CrossRef]
21. Mi, H.; Su, Q.; Chen, J.; Yi, S.; Li, X.; Li, J. Starch-fatty acid complexes improve the gel properties and enhance the fatty acid content of *Nemipterus virgatus* surimi under high-temperature treatment. *Food Chem.* **2021**, *362*, 130253. [CrossRef]
22. Cando, D.; Borderías, A.J.; Moreno, H.M. Combined effect of aminoacids and microbial transglutaminase on gelation of low salt surimi content under high pressure processing. *Innov. Food Sci. Emerg. Technol.* **2016**, *36*, 10–17. [CrossRef]
23. Hong, Z.; Kong, Y.; Chen, J.; Guo, R.; Huang, Q. Collaborative stabilizing effect of trehalose and myofibrillar protein on high internal phase emulsions: Improved freeze-thaw stability and 3D printability. *Food Chem.* **2025**, *469*, 142564. [CrossRef]
24. Wang, F.; Li, J.; Wang, Y.; Liu, H.; Yu, B.; Zhao, H.; Zhang, R.; Tao, H.; Ren, X.; Cui, B. The dispersibility of biphasic stabilized oil-in-water emulsions improved by the interaction between curdlan and soy protein isolate. *Food Chem.* **2024**, *457*, 140101. [CrossRef]
25. Tao, H.; Guo, L.; Qin, Z.; Yu, B.; Wang, Y.; Li, J.; Wang, Z.; Shao, X.; Dou, G.; Cui, B. Textural characteristics of mixed gels improved by structural recombination and the formation of hydrogen bonds between curdlan and carrageenan. *Food Hydrocoll.* **2022**, *129*, 107678. [CrossRef]
26. Liu, C.; Wang, L.; Chen, H.; Gao, P.; Xu, Y.; Xia, W.; Liu, S.-Q. Interfacial structures and processing stability of surimi particles-konjac glucomannan complexes stabilized pickering emulsions via one-step and layer-by-layer. *Food Hydrocoll.* **2024**, *147*, 109349. [CrossRef]
27. Zhang, N.; Han, J.; Chen, F.; Gao, C.; Tang, X. Chitosan/gum arabic complexes to stabilize Pickering emulsions: Relationship between the preparation, structure and oil-water interfacial activity. *Food Hydrocoll.* **2022**, *129*, 107532. [CrossRef]
28. Liu, S.-Y.; Lei, H.; Li, L.-Q.; Liu, F.; Li, L.; Yan, J.-K. Effects of direct addition of curdlan on the gelling characteristics of thermally induced soy protein isolate gels. *Int. J. Biol. Macromol.* **2023**, *253*, 127092. [CrossRef] [PubMed]
29. Wu, C.; Yuan, C.; Chen, S.; Liu, D.; Ye, X.; Hu, Y. The effect of curdlan on the rheological properties of restructured ribbonfish (*Trichiurus* spp.) meat gel. *Food Chem.* **2015**, *179*, 222–231. [CrossRef]
30. Jiang, X.; Chen, Q.; Xiao, N.; Du, Y.; Feng, Q.; Shi, W. Changes in Gel Structure and Chemical Interactions of *Hypophthalmichthys molitrix* Surimi Gels: Effect of Setting Process and Different Starch Addition. *Foods* **2022**, *11*, 9. [CrossRef]
31. Hatakeyama, T.; Iijima, M.; Hatakeyama, H. Role of bound water on structural change of water insoluble polysaccharides. *Food Hydrocoll.* **2016**, *53*, 62–68. [CrossRef]
32. Chen, H.; Zhou, A.; Benjakul, S.; Zou, Y.; Liu, X.; Xiao, S. The mechanism of low-level pressure coupled with heat treatment on water migration and gel properties of *Nemipterus virgatus* surimi. *LWT* **2021**, *150*, 112086. [CrossRef]
33. Cui, B.; Mao, Y.; Liang, H.; Li, Y.; Li, J.; Ye, S.; Chen, W.; Li, B. Properties of soybean protein isolate/curdlan based emulsion gel for fat analogue: Comparison with pork backfat. *Int. J. Biol. Macromol.* **2022**, *206*, 481–488. [CrossRef] [PubMed]

34. Zhang, L.; Han, X.; Guo, K.-J.; Ren, Y.-P.; Chen, Y.; Yang, J.; Qian, J.-Y. Pickering emulsion gels with curdlan as both the emulsifier and the gelling agent: Emulsifying mechanism, gelling performance and gel properties. *Food Chem.* **2025**, *465*, 141971. [CrossRef]
35. Yu, H.; Zhang, J. Emulsion co-stabilized with high methoxyl pectin and myofibrillar protein: Used to enhance the application in emulsified gel. *Food Chem.* **2025**, *475*, 143359. [CrossRef]
36. Tcholakova, S.S.; Denkov, N.; Sidzhakova, D.; Ivanov, I.B.; Campbell, B.E.J.L. Interrelation between Drop Size and Protein Adsorption at Various Emulsification Conditions. *Langmuir* **2003**, *19*, 5640–5649. [CrossRef]
37. Hu, X.; McClements, D.J. Construction of plant-based adipose tissue using high internal phase emulsions and emulsion gels. *Innov. Food Sci. Emerg. Technol.* **2022**, *78*, 103016. [CrossRef]
38. Li, D.; Li, N.; Wang, Y.; Zhang, K.; Tan, Z.; Liu, H.; Liu, X.; Wu, Q.; Zhou, D. Effect of konjac glucomannan on gelling and digestive properties of myofibrillar protein in *Litopenaeus vannamei* based on molecular docking. *Food Hydrocoll.* **2024**, *149*, 109595. [CrossRef]
39. Hu, W.; Xu, X.; Wang, X.; Ma, T.; Li, Y.; Qin, X.; Wei, J.; Chen, S. Effect of curdlan on the gel properties and interactions of whey protein isolate gels. *Int. J. Biol. Macromol.* **2024**, *277*, 134161. [CrossRef]
40. Yao, W.; Huang, X.; Li, C.; Kong, B.; Xia, X.; Sun, F.; Liu, Q.; Cao, C. Underlying the effect of soybean oil concentration on the gelling properties of myofibrillar protein-based emulsion gels: Perspective on interfacial adsorption, rheological properties and protein conformation. *Food Hydrocoll.* **2025**, *162*, 110935. [CrossRef]
41. Hu, X.; Xiang, X.; Ju, Q.; Li, S.; McClements, D.J. Impact of lipid droplet characteristics on the rheology of plant protein emulsion gels: Droplet size, concentration, and interfacial properties. *Food Res. Int.* **2024**, *191*, 114734. [CrossRef]
42. Lv, Y.; Zhao, H.; Xu, Y.; Yi, S.; Li, X.; Li, J. Properties and microstructures of golden thread fish myofibrillar proteins gel filled with diacylglycerol emulsion: Effects of emulsifier type and dose. *Food Hydrocoll.* **2023**, *144*, 108935. [CrossRef]
43. Jain, B.P.; Pandey, S.; Goswami, S.K. Chapter 36—Estimation of Proteins by the Biuret Method. In *Protocols in Biochemistry and Clinical Biochemistry*, 2nd ed.; Jain, B.P., Pandey, S., Goswami, S.K., Eds.; Academic Press: Cambridge, MA, USA, 2025; pp. 85–86. [CrossRef]
44. Yang, Z.; He, X.; Song, Y.; Zhang, W.; Chen, L.; Jiang, L.; Huang, Z.; Tian, T. Fabrication and characterization of novel curcumin-loaded thermoreversible high amylose maize starch emulsion gel. *Int. J. Biol. Macromol.* **2024**, *280*, 136173. [CrossRef] [PubMed]
45. Lv, D.; Chen, F.; Yin, L.; Zhang, P.; Rashid, M.T.; Yu, J. Wheat bran arabinoxylan-soybean protein isolate emulsion-filled gels as a β-carotene delivery carrier: Effect of polysaccharide content on textural and rheological properties. *Int. J. Biol. Macromol.* **2023**, *253*, 126465. [CrossRef]
46. Lv, Y.; Sun, X.; Jia, H.; Hao, R.; Jan, M.; Xu, X.; Li, S.; Dong, X.; Pan, J. Antarctic krill (*Euphausia superba*) oil high internal phase emulsions improved the lipid quality and gel properties of surimi gel. *Food Chem.* **2023**, *423*, 136352. [CrossRef]
47. Zhou, A.; Chen, H.; Zou, Y.; Liu, X.; Benjakul, S. Insight into the mechanism of optimal low-level pressure coupled with heat treatment to improve the gel properties of *Nemipterus virgatus* surimi combined with water migration. *Food Res. Int.* **2022**, *157*, 111230. [CrossRef] [PubMed]

gels

MDPI

Article

A Comparative Study of Soy Protein Isolate-κ-Carrageenan Emulsion Gels and Bigels for the Encapsulation, Protection, and Delivery of Curcumin

Emmanueline T Gray [1], Weining Huang [1], Zhongkai Zhou [2], Hao Cheng [1,*] and Li Liang [1]

[1] State Key Laboratory of Food Science and Resources, School of Food Science and Technology, Jiangnan University, Wuxi 214122, China; 6230112907@stu.jiangnan.edu.cn (E.T.G.); wnhuang@jiagnan.edu.cn (W.H.); liliang@jiangnan.edu.cn (L.L.)

[2] School of Food Science and Technology, Shihezi University, Shihezi 832000, China; zkzhou@shzu.edu.cn

* Correspondence: haocheng@jiangnan.edu.cn

Abstract: Protein-based emulsion gels and bigels serve as ideal delivery systems owing to their distinctive structural properties, high encapsulation efficiency, and adjustable digestive behavior. However, limited research has examined the differences between emulsion gels and bigels as polyphenol delivery systems. In this study, oil-in-water (O/W)-type emulsion gels formulated with soy protein isolate (SPI) and κ-carrageenan (κ-CG) were fabricated using a cold-set gelation method, and then the bigels were prepared through further oil gelation by the addition of glycerol monostearate (GMS). Both SPI-κ-CG emulsion gels and bigels were mainly stabilized by electrostatic and hydrophobic interactions, exhibiting high gel strength, varying from 940 g to 1304 g, and high water holding capacity (~84%). Both the SPI-κ-CG emulsion gels and bigels demonstrated high curcumin encapsulation efficiency, reaching 98~99%. Stability testing revealed that bigels prepared with 15% and 20% GMS exhibited the highest curcumin retention ratios, with a value of around 78% after storage for 21 days at 25 °C, suggesting that denser network structures more effectively prevent the degradation of the encapsulated compound. During the in vitro simulated gastric digestion, higher GMS content significantly delayed curcumin release by over 7%. Increasing GMS concentration from 0% to 20% elevated lipolysis by over 8% and concurrently improved the release of curcumin by more than 18% during the in vitro simulated intestinal digestion. This study provides comparative insights into polyphenol delivery performance between emulsion gels and bigels, offering valuable guidance for developing functional foods based on gel delivery systems.

Keywords: bigels; emulsion gels; soy protein isolate; κ-carrageenan; curcumin; delivery

1. Introduction

Curcumin, a natural polyphenol characterized by its bis-feruloylmethane structure, demonstrates multiple pharmacological activities including antioxidant, anti-inflammatory, anticancer, hepatoprotective, and anti-atherosclerotic effects [1]. Recent studies have demonstrated that the health benefits of curcumin result from its numerous active metabolites transformed by the host and microbial enzymes [2,3]. However, its poor aqueous solubility, sensitivity to environmental factors, and gastrointestinal tract conditions (e.g., oxygen, light, heat, enzymes) limit its application in functional foods [4]. Therefore, developing edible carriers is crucial to enhancing curcumin's physicochemical stability and targeted release properties.

To date, food gels have gained considerable interest owing to their distinctive thermodynamic stability, tunable rheological properties, and controlled-release capabilities [5]. Among these gel systems, emulsion-filled gels consisting of emulsified oil droplets entrapped within a hydrogel have been developed for the encapsulation, protection, and delivery of both hydrophobic and hydrophilic bioactives because they combine the advantages of both emulsion and gel network characteristics [6–10]. Within emulsion gel matrices, protein-bound oil droplets functionally integrate bioactive encapsulation with structural modulation of the 3D network [11]. This configuration enhances protection efficacy for oil-phase bioactives through compartmentalization effects and achieves spatiotemporally programmed release during gastrointestinal transit [12,13].

Bigels possess a novel biphasic system integrating both hydrogel and oleogel phases, making them serve as novel carriers, 3D printing materials, and fat replacers [14]. The bigels with an oleogel-in-hydrogel microstructure can serve as a special form of emulsion-filled gels, exhibiting higher stability due to both gelled phases in the gel network [15]. The gelled oil phase in bigels not only protects encapsulated ingredients against environmental stressors (e.g., oxidation, humidity) but also critically governs microstructure and mechanical properties. By modulating the oleogel's composition, such as crystalline network density or oleogelator concentration, the entire bigel system can be engineered to optimize loading capacity and achieve stimuli-responsive release kinetics for incorporated bioactives [14]. Lu and co-workers fabricated oleogel-in-hydrogel bigels for the co-delivery of epigallocatechin gallate and curcumin [16]. It was found that the increase in oleogelator concentration resulted in more compact gel networks, leading to a lower release rate of curcumin in gastric digestion but a higher release rate in the intestinal stage. In another study, bigels with konjac glucomannan-gelatin hydrogel phase and stearic acid oleogel phase were fabricated for the delivery of curcumin, and found that the increase in oleogel/hydrogel volume ratio could achieve a sustained free fatty acid release but decrease the bioaccessibility of curcumin [17]. Bigels have been also used for the encapsulation and delivery of other polyphenols, such as quercetin, catechin [18], and epigallocatechin gallate [19]. These bioactive-loaded bigel systems show great promise as functional ingredients for use as fat replacers, in food packaging applications, and in 3D printing foods [20]. Recently, Kaimal and co-workers compared the difference in the encapsulation of ascorbic acid within emulsion gels and bigels formulated with xanthan-guar gum and ethyl cellulose; they found that bigels and emulsion gels had the same bioaccessibility of ascorbic acid with a value of 87% [21]. While oleogel-in-hydrogel bigels, with their structured oil phase, are theoretically expected to provide enhanced protection and tunable release profiles for bioactives compared to emulsion-filled gels, direct comparative studies between these two systems remain scarce.

Proteins and polysaccharides are widely used in the formation of food gels. The combination of these two biomacromolecules can lead to distinctive structural and functional properties, owing to the interactions between proteins and polysaccharides [22]. Soy protein isolate (SPI), a commercially significant food protein, is widely recognized for its nutritional value, functional properties, and health benefits. κ-Carrageenan (κ-CG), a linear sulfated polysaccharide (15~40% ester sulfate), consists of alternating α-(1→3)-D-galactopyranose and β-(1→4)-3,6-anhydro-D-galactopyranose units. Our previous studies demonstrated that protein-κ-CG hydrogels sequentially crosslinked by KCl and glucono-δ-lactone (GDL) showed enhanced mechanical strength and decreased swelling ratio. In this study, SPI-κ-CG emulsion-filled gels were fabricated using the double crosslinking gelation method. Glycerol monostearate (GMS) with various concentrations was further used as an oleogelator to prepare SPI-κ-CG bigels. This study systematically investigated structural evolution mechanisms, microstructural architectures, textural profiles, and in vitro digestive behaviors to elucidate functional divergences between emulsion gels and bigels in encapsulating, protecting, and delivering curcumin. Our research establishes a theoretical framework for selecting optimal gel carriers to enhance hydrophobic polyphenol delivery in nutraceutical applications.

2. Results and Discussion

2.1. Size Distribution and ζ-Potential of Emulsions

As shown in Figure 1A, all SPI-stabilized emulsions with various GMS concentrations in the absence and presence of κ-CG exhibited a unimodal size distribution. SPI-stabilized emulsions without GMS exhibited the largest droplet size with a peak of around 300 nm. The addition of GMS resulted in the gradual decrease in oil droplet size of emulsions, with a peak of around 150 nm at 20% GMS. These results suggested that the addition of GMS facilitates smaller oil droplet formation. This was attributed to the high surface activity of GMS with low molecular weight absorbing faster than SPI into the water–oil interface during the homogenization [23]. Additionally, unlike sunflower oil, which consists largely of triacylglycerides, the GMS molecule lacks esterified fatty acyl groups and is more hydrophilic. This property causes a substantial reduction in the interfacial tension of the oil phase, thereby decreasing the size of emulsified oil droplets [24,25]. The addition of 0.25% κ-CG had a slight impact on the size distribution of emulsions, which was observed in the pea protein-κ-CG sunflower oil emulsions and whey protein-gum Arabic sunflower oil emulsions due to the weak interaction between the protein and anion polysaccharide in the neutral environment [13,26].

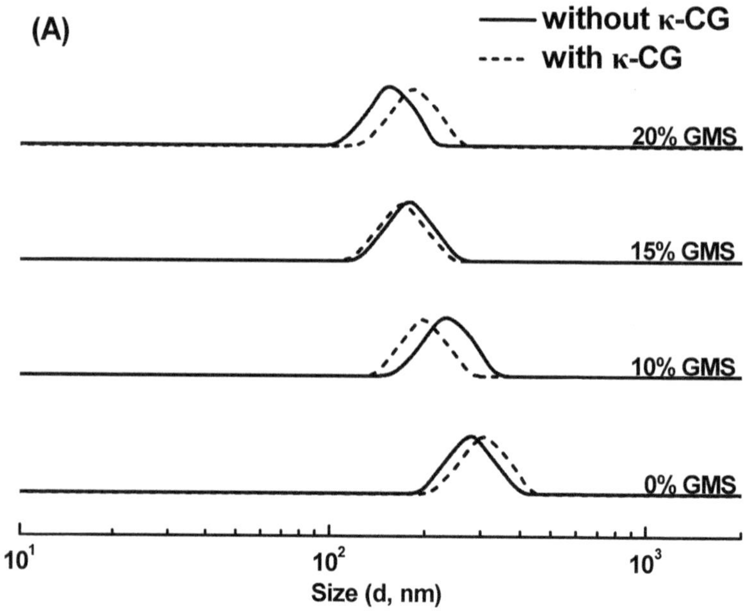

Figure 1. *Cont.*

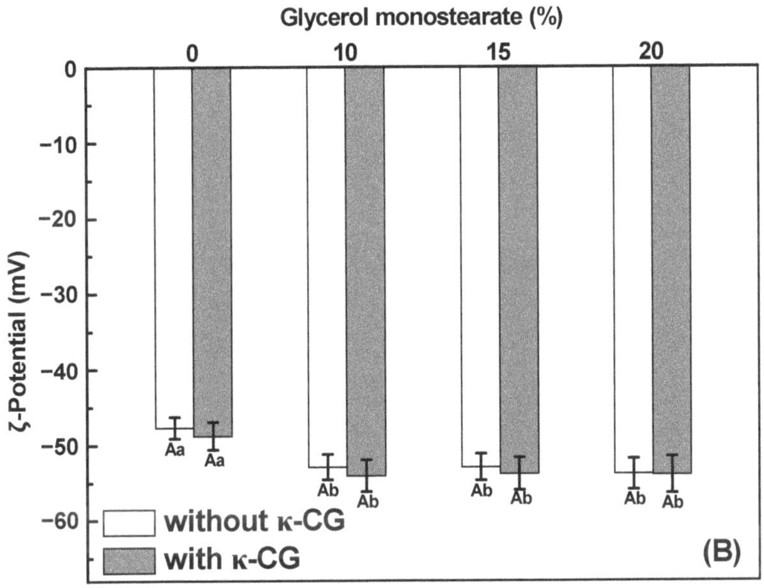

Figure 1. The size distribution (**A**) and ζ-potential (**B**) of soy protein isolate-stabilized emulsions with various glycerol monostearate (GMS) concentrations in the absence and presence of 0.25% κ-carrageenan (κ-CG). Different letters (uppercase letters for samples with or without κ-CG, lowercase letters for GMS concentration) indicate statistically significant differences ($p < 0.05$).

SPI-stabilized emulsions exhibited a ζ-potential of approximately −48 mV (Figure 1B). Since the emulsion pH (~7.0) exceeded the isoelectric point of SPI (~4.5) [27], the emulsified oil droplets possessed negatively charged surfaces. Adding GMS slightly increased the ζ-potential of the droplets (e.g., −54 mV at 20% GMS), possibly due to altered protein adsorption at the oil–water interfaces [24]. However, incorporating κ-CG into the SPI-stabilized emulsion had no significant effect ($p < 0.05$) on ζ-potential. This suggests that the negatively charged κ-CG molecules remained primarily in the aqueous phase and did not adsorb onto the oil droplet surfaces.

2.2. Gel Formation and Microstructure

Figure 2 shows the visual appearance and microstructure of the curcumin-loaded SPI-κ-CG emulsion gels and bigels with various GMS concentrations. All formulations could form self-supporting gels with a yellow color. Generally, the emulsifier oil droplets were almost dispersed within the protein network in the absence and presence of GMS, suggesting that both oil droplets and proteins were co-localized during the gel formation. It should be noted that larger oil droplets could be observed in the SPI-κ-CG emulsion gel, indicating that the aggregation or coalescence of oil droplets occurred during the gelation. There was no significant oil coalescence in SPI-κ-CG bigels, which is consistent with the DLS results (Figure 1A).

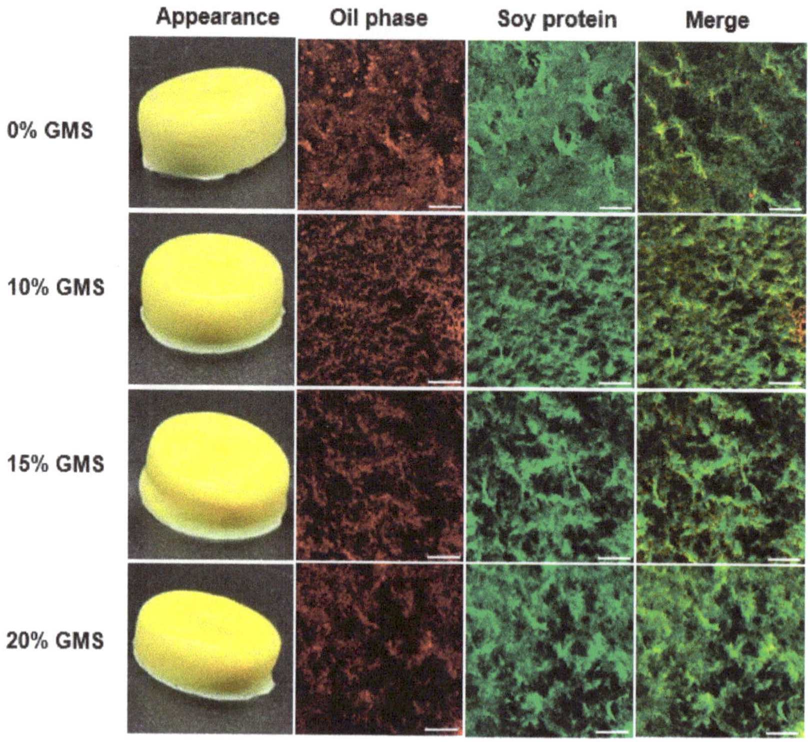

Figure 2. Appearance images and confocal laser scanning microscopy images of curcumin-loaded soy protein isolate-κ-carrageenan emulsion gels and bigels with various glycerol monostearate (GMS) concentrations. The red and green represent the oil phase and protein phase, respectively. The scale bar is 75 μm.

2.3. Encapsulation Efficiency of Curcumin

Curcumin possesses very low hydrosolubility of 11 ng/mL, which can be overcome by encapsulating with protein particles and emulsions [4]. As shown in Table 1, the encapsulation efficiency of curcumin in SPI-κ-CG emulsions was around 93%. This result was consistent with goat whey protein-gum Arabic complex-stabilized emulsion (96%) [28], glutelin-stabilized emulsions (~94%) [29], and mussel protein-stabilized sunflower oil emulsions (~95%) [7]. GMS addition slightly increased the encapsulation efficiency of curcumin, as the oil gelation prevented the release and diffusion of curcumin from the oil to the aqueous phase [30]. The encapsulation efficiency of curcumin in emulsion gels and bigels was around 100% (Table 1), suggesting that curcumin was completely encapsulated in emulsified oil droplets or colloidal particles in the aqueous phase during the gel formation [31]. That is to say, curcumin molecules did not release from the gel system during the KCl and GDL double crosslinked gelation process, resulting from the low hydrosolubility of curcumin and complexation of curcumin with soy protein particles dispersed in the aqueous phase [32].

Table 1. Encapsulation efficiency of curcumin within soy protein isolate-κ-carrageenan emulsions, emulsion gels, and bigels with various glycerol monostearate (GMS) concentrations.

GMS (%)	Encapsulation Efficiency (%)	
	Oil Droplet of Emulsion	Gel Sample
0	93.2 ± 0.9 [a]	98.2 ± 0.7 [a]
10	95.5 ± 1.1 [b]	99.1 ± 0.9 [a]
15	96.4 ± 0.8 [b]	98.8 ± 0.4 [a]
20	97.1 ± 1.0 [b]	99.3 ± 0.3 [a]

Different letters in the same column indicate statistically significant differences ($p < 0.05$).

2.4. Textural Properties

The textural characteristics of SPI-κ-CG emulsion gels and bigels containing different GMS concentrations are shown in Table 2. Overall, the gel matrix structure, emulsified oil droplets, and their interactions primarily influenced the textural profile of emulsion-filled gels and bigels [33,34]. The gel hardness and chewiness of SPI-κ-CG emulsion gel were 1304 g and 392 g, respectively. The incorporation of 10% GMS initially led to a slight reduction in the mechanical strength of the bigels. This may be explained by the ability of GMS to partially displace proteins from the oil–water interface, thereby weakening the interactions between emulsified oil droplets and the gel matrix and resulting in lower overall hardness [11]. However, with further increase in GMS concentration, a moderate enhancement in gel strength was observed. This strengthening effect is likely attributable to the role of GMS as an oleogelator, which contributes to the formation of a more structured and robust oleogel network within the bigel system, thereby increasing its overall mechanical properties [16]. It should be noted that the double crosslinking SPI-κ-CG bigels still showed higher gel strength than SPI-agar-alginate double crosslinking emulsion gels (gel hardness ~360 g) [35] and whey protein-gellan gum double crosslinking emulsion gels (gel hardness ~123.5 g) [36] at similar protein and polysaccharide concentrations. Therefore, the KCl-GDL sequential double crosslinking is an effective approach for the improvement of both SPI-κ-CG emulsion gel and bigel strength, mainly attributed to the higher degree of crosslinking and intertwined double protein-polysaccharide networks. The addition of GMS slightly decreased the gel springiness and cohesiveness of SPI-κ-CG emulsion gels and bigels (Table 2). This was primarily due to oil gelation weakening interactions between emulsified oil droplets and the gel matrix, thereby reducing gel elasticity and cohesiveness [37].

Table 2. Textural characteristics of soy protein isolate-κ-carrageenan emulsion gels and bigels with various glycerol monostearate (GMS) concentrations.

GMS (%)	Hardness (g)	Springiness	Cohesiveness	Chewiness (g)
0	1304.6 ± 24.7 [a]	0.594 ± 0.025 [a]	0.506 ± 0.027 [a]	392.9 ± 32.3 [a]
10	940.5 ± 38.9 [b]	0.592 ± 0.014 [a]	0.465 ± 0.001 [ab]	259.1 ± 21.1 [b]
15	1117.8 ± 30.3 [c]	0.573 ± 0.022 [ab]	0.429 ± 0.021 [bc]	274.7 ± 6.7 [b]
20	1285.8 ± 15.8 [c]	0.544 ± 0.003 [b]	0.414 ± 0.037 [c]	289.6 ± 30.3 [b]

Different letters in the same column indicate statistically significant differences ($p < 0.05$).

2.5. Water Holding Capacity

Water holding capacity of SPI-κ-CG emulsion gels and bigels was around 84%, which was independent of the GMS concentration (Figure 3). The double crosslinked SPI-κ-CG emulsion gels and bigels showed higher water holding capacity than Ca^{2+}-induced pea protein-κ-CG emulsion gels (~75%) [13]. The water holding capacity of gels is affected by the crosslinking methods due to the diverse gel microstructure [38,39]. The double crosslinking emulsion gels and bigels had denser networks and higher gel strength (Table 2) against water loss during centrifugation.

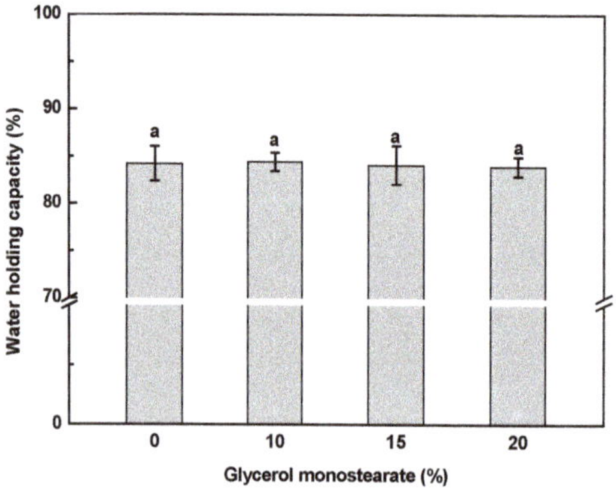

Figure 3. Water holding capacity of soy protein isolate-κ-carrageenan emulsion gels and bigels with various glycerol monostearate concentrations. Different letters indicate the statistically significant differences ($p < 0.05$).

2.6. FTIR

In Figure 4A, the broad absorption peaks at 3200~3600 cm^{-1} could be observed in curcumin (~3510 cm^{-1}), GMS (~3312 cm^{-1}), κ-CG (~3388 cm^{-1}), and SPI (~3293 cm^{-1}), indicating the presence of O-H stretching vibrations participating in hydrogen bonding. For the sunflower oil and GMS, peaks at 2800~3000 cm^{-1} corresponded to the stretching of C-H groups of triglycerides and monoglyceride [40]. SPI displayed two diagnostic peaks of 1659 cm^{-1} (amide I, C=O stretch) and 1539 cm^{-1} (amide II, N-H bend/C-N stretch). κ-CG exhibits characteristic peaks at 1644 cm^{-1} (carbonyl group stretching), 1240 cm^{-1} (ester sulfate group), 848 cm^{-1} (galactose-4-sulfate), 916 cm^{-1} (3,6-anhydro-D-galactose), and 1070 cm^{-1} (glycosidic bond) [41]. The FTIR spectrum of curcumin exhibited characteristic peaks in the regions 800~1627 cm^{-1}, while no significant peaks were observed in the carbonyl region (1650~1800 cm^{-1}), confirming that curcumin adopts the keto-enol tautomeric form [42].

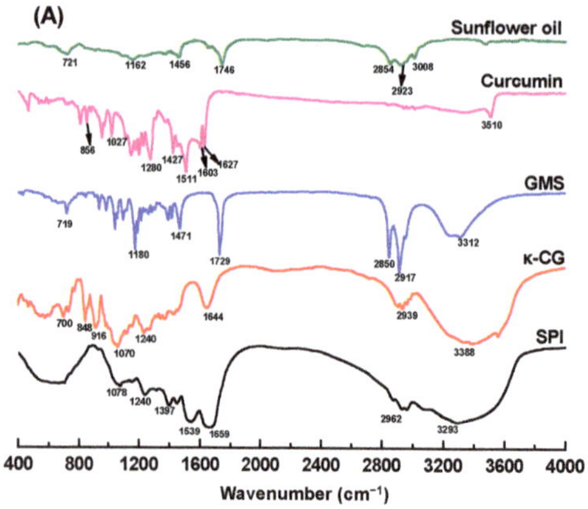

Figure 4. *Cont.*

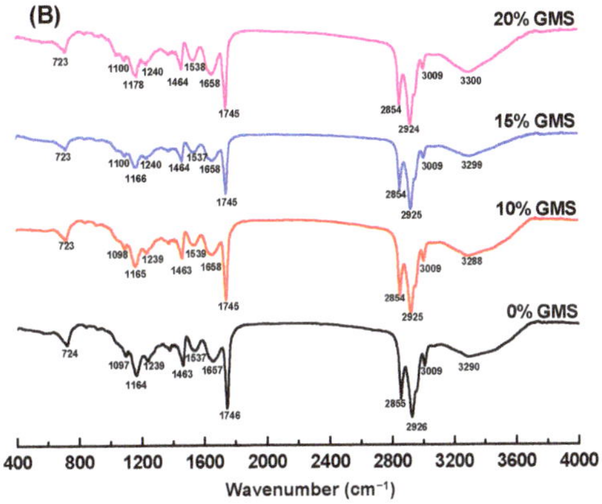

Figure 4. Fourier-transform infrared (FTIR) spectra of soy protein isolate (SPI), κ-carrageenan (κ-CG), curcumin, glycerol monostearate (GMS), and sunflower oil (**A**), as well as emulsion gels and bigels with various concentrations of GMS (**B**).

FTIR analysis of curcumin-loaded emulsion gels and bigels (Figure 4B) revealed substantial attenuation of characteristic curcumin peaks, indicating successful compartmentalized encapsulation within emulsified oil droplets and protein particles during gel network formation [43]. The FTIR spectra of emulsion gels and bigels were similar to those of sunflower oil and GMS, attributing to the high proportion of the oil phase in the freeze-dried gel powders. Furthermore, the characteristic peaks of SPI and κ-CG were included in the spectrum of emulsion gels and bigels. No new peaks and drastic shifting were observed in the spectra of emulsion gels and bigels. It is therefore speculated that the SPI-κ-CG emulsion gels and bigels were mainly stabilized by the noncovalent bonds.

2.7. Intermolecular Force

The significant gel disintegration (Figure 5A) and protein dissolution (Figure 5B) of all gel samples were observed in EDTA and SDS solutions, while no significant gel dissolution was observed in water, β-mercaptoethanol solution, and urea, indicating that electrostatic and hydrophobic interactions are the primary forces for maintaining SPI-κ-CG emulsion gels' and bigels' integrity. During the double crosslinking emulsion gel and bigel formation, the presence of K^+ would induce the aggregation of the helices and κ-CG gel network formation through the hydrogen bonds and hydrophobic interactions by shielding the electrostatic repulsion of the κ-CG sulfate groups [44]. Meanwhile, GLD-induced pH decrease facilitates the electrostatic interactions between SPI and κ-CG to form the secondary networks. Therefore, the electrostatic and hydrophobic interactions play an important role in the SPI-κ-CG emulsion gel and bigel formation.

2.8. Chemical Stability of Curcumin in Emulsion Gels and Bigels

Curcumin is sensitive to environmental conditions (e.g., pH, light, and heat) and rapidly degraded in the aqueous phase [45]. As shown in Figure 6A,B, curcumin retention in SPI-κ-CG emulsion gels decreased progressively during storage at both 25 °C and 45 °C, with 45% and 8% remaining after 21 days, respectively. This high initial retention was due to curcumin encapsulation within the inner oil phase, which provided substantial protection from the external environment [46]. Furthermore, the pH decrease during GDL-induced gel formation enhanced curcumin's physicochemical stability [45]. GMS incorporation further reduced curcumin degradation (Figure 6), demonstrating that the bigel structure offers enhanced protection. This effect intensified at higher GMS concentrations. At

20% GMS, curcumin retention reached ~78% at 25 °C and 50% at 45 °C after storage for 21 days (Figure 6A,B). These findings align with studies showing that entrapment within gel networks improves bioactive compounds' stability [5]. In bigels, the concurrent gelation of both aqueous and lipid phases synergistically impedes ingredient degradation. The hydrogel matrix acts as a physical barrier against pro-oxidants and free radicals [47], while oleogel networks restrict molecular mobility and limit contact with environmental degradation factors [14]. Therefore, the encapsulation of curcumin in SPI-κ-CG bigels with high GMS concentration could effectively improve its chemical stability during storage.

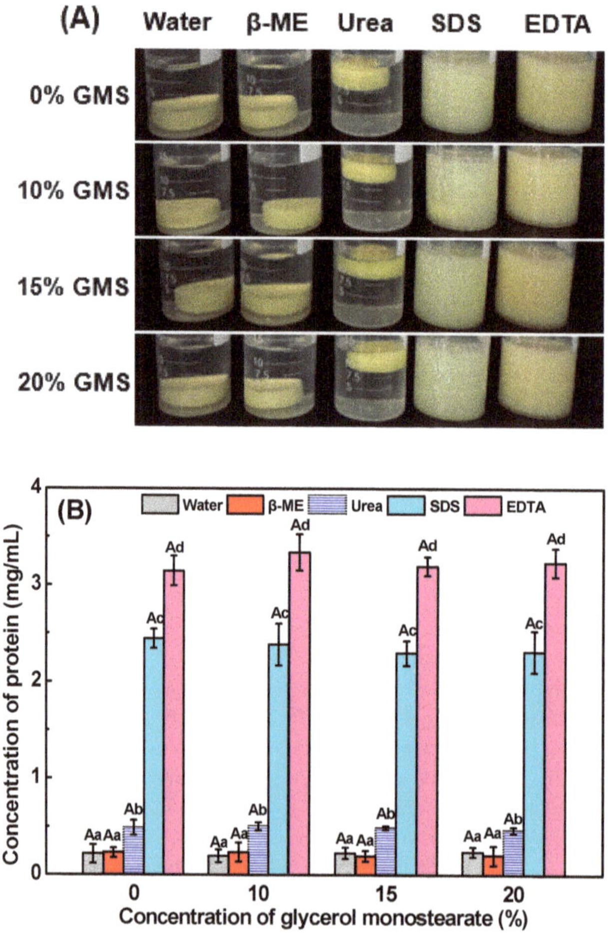

Figure 5. Appearance images (**A**) and protein solubility (**B**) of the emulsion gels and bigels with various concentrations of glycerol monostearate (GMS) in water, β-mercaptoethanol (β-ME), urea, sodium dodecyl sulfate (SDS), and ethylene diamine tetraacetic acid•2Na (EDTA). Different letters (uppercase letters for the samples with different GMS concentration in the same solvent, lowercase letters for the samples at the same GMS concentration in different solvents) indicate statistically significant differences ($p < 0.05$).

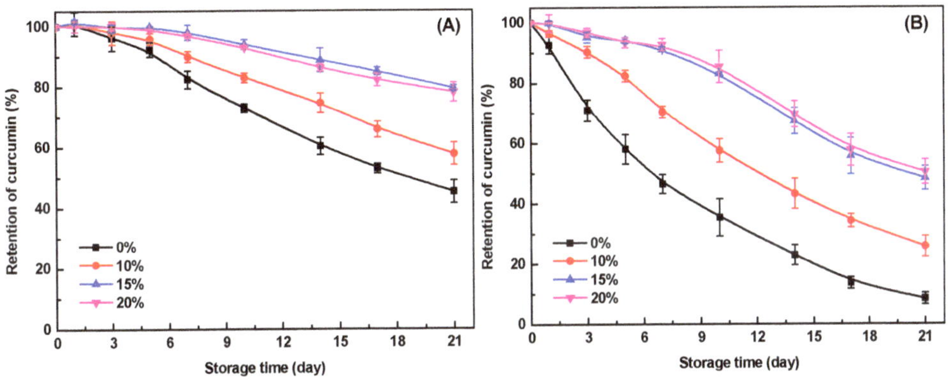

Figure 6. Retention of curcumin encapsulated within soy protein isolate-κ-carrageenan emulsion gels and bigels with various glycerol monostearate (GMS) concentrations during storage for 21 days at 25 °C (**A**) and 45 °C (**B**).

2.9. *In Vitro Digestion*

2.9.1. Free Fatty Acids Release

Figure 7A depicts free fatty acid (FFA) release profiles during intestinal digestion for SPI-κ-CG emulsion gels and bigels with varying GMS concentrations. All systems exhibited rapid initial FFA liberation followed by progressively slower release kinetics. Without GMS, about 15% FFAs were released within 30 min, increasing to just 26% by the end of intestinal digestion. Incorporation of 10% GMS slightly enhanced both release rate and extent, with values of 17% and 28% after 30 min and 120 min, respectively (Figure 7A). Further increase in GMS concentration resulted in a slight increase in lipid digestion rate and extent. As the GMS concentration reached 20%, there was a further increase in lipid digestion, with 23% and 34% of lipids being digested after 30 min and 120 min, respectively. These results suggested that the addition of GMS could improve the lipid digestion of bigels during in vitro intestinal digestion. The oelogel with three-dimensional networks has been expected to delay lipid digestion through inhibiting the liquid oil diffusion and the access of the lipase to oils, depending on the oil composition, oleogelator types, gelation mechanism, and the gel structure and strength [17,48,49]. However, it has been reported that GMS presented significantly higher digestibility than glyceryl tripalmitate, because pancreatic lipase could selectively hydrolyze GMS with fatty acids located at sn-1 or 3 positions [50]. Therefore, the increase in GMS concentration might contribute to the higher overall fatty acids release values.

2.9.2. In Vitro Release Profile of Curcumin

Figure 7B displays curcumin release profiles from emulsion gels and bigels during simulated gastrointestinal digestion. Generally, a small fraction of curcumin was released (<11%) after SGF digestion, as curcumin was mainly present in the oil phase or oleogel phase. Minimal curcumin release (<4%) in bigels with 20% GMS occurred during the simulated gastric fluid (SGF), as most curcumin remained encapsulated within the oleogel phases. Bigels with higher GMS concentrations exhibited significantly slower release rates than emulsion gels, demonstrating enhanced gastric protection against curcumin degradation. This aligns with findings in whey protein-based [51], gelatin-based [16], and alginate-based emulsion gel and bigel systems [19].

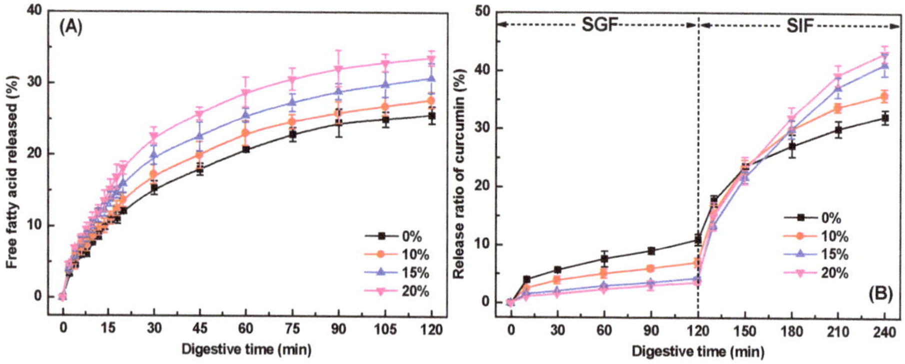

Figure 7. (**A**) The release of free fatty acids from soy protein isolate-κ-carrageenan emulsion gels and bigels with various glycerol monostearate concentrations during in vitro simulated intestinal digestion. (**B**) The release of curcumin from emulsion gels and bigels with various glycerol monostearate concentrations during simulated gastrointestinal digestion.

Upon transition to simulated intestinal fluids, both emulsion gels and bigels showed rapid curcumin release (Figure 7B), primarily driven by protein and lipid hydrolysis. Notably, bigels with higher GMS concentrations ultimately released more curcumin than emulsion gels, which correlates with their higher FFA release (Figure 7A). The increased availability of FFAs promotes the formation of mixed micelles, thereby enhancing the solubilization and bioaccessibility of curcumin [50]. This improvement in bioaccessibility suggests a corresponding increase in curcumin bioavailability, demonstrating that modulating GMS content enables tailored control over nutraceutical release kinetics in bigel-based delivery systems.

The bigel system demonstrates excellent protection and controlled release of curcumin, making it suitable for enhancing the stability of hydrophobic bioactive compounds in functional food products, such as fortified beverages, yogurts, or health bars. Furthermore, the ability to suppress premature release in gastric conditions while enabling intestinal-specific liberation suggests potential for designing targeted nutrient delivery systems that improve absorption and efficacy. However, it should be noted that this study has certain limitations, including the absence of in vivo tests to evaluate the bioavailability and physiological efficacy of the released curcumin, as well as a lack of assessment regarding the sensory properties and stability of the bigels when incorporated into real food matrices. Future studies should therefore focus on validating the in vivo performance of curcumin-loaded SPI-κ-CG bigels and evaluating their sensory compatibility and stability in actual food applications.

3. Conclusions

This study successfully fabricated soy protein isolate-κ-carrageenan emulsion gels and bigels, demonstrating excellent curcumin encapsulation efficiency, protective capacity, and targeted delivery performance. Incorporating GMS to form hydrogel-in-oleogel bigel structures slightly decreased gel mechanical strength but significantly improved curcumin's chemical stability during storage. Crucially, GMS-enriched bigels suppressed premature curcumin release in gastric conditions while enhancing intestinal-phase liberation. Comparative analysis revealed that the bigels, with their embedded oleogel phase, outperformed emulsion gels in encapsulating hydrophobic polyphenols like curcumin. These structures notably improved the compound's resistance to degradation and enabled more efficient intestinal-specific release. The findings underscore the potential of SPI-κ-CG bigels as effective oral delivery systems for lipophilic bioactive compounds, offering promising applications in functional foods, nutraceuticals, and targeted nutrient delivery platforms.

4. Materials and Methods

4.1. Materials

Soy protein isolate (BR, 99%) was purchased from Shandong Xiya Reagent Co., Ltd. (Linyi, Shandong, China). Sunflower oil (Brand Duoli) was obtained from a local retailer (Wuxi, Jiangsu, China). κ-Carrageenan (molecular weight 300,000 g/mol), bile salts, porcine pancreatin (4 × USP specifications), Nile Red, Nile Blue, β-mercaptoethanol, and porcine pepsin (≥500 U/mg) were purchased from Sigma-Aldrich Co., Ltd. (St. Louis, MO, USA). Glycerol monostearate, sodium dodecyl sulfate (SDS), HCl, ethylene diamine tetraacetic acid (EDTA), and NaOH were of analytical grade and obtained from SinoPharm CNCM Ltd. (Shanghai, China). All samples were formulated with deionized water purified using a Milli-Q purification system (Millipore, Billerica, MA, USA).

4.2. Emulsion Preparation

A 5 wt% soy protein isolate (SPI) solution was subjected to thermal denaturation at 90 °C for 30 min, followed by controlled cooling to room temperature. Separately, a 0.5 wt% κ-carrageenan (κ-CG) stock solution was formulated by dissolving κ-carrageenan powder in pure water at 55 °C with continuous stirring for 2 h. Glycerol monostearate (0, 10, 15, or 20 wt% based on oil weight) was incorporated into sunflower oil at 75 °C under agitation. Curcumin (0.5 mg/g oil phase) was dispersed in the lipid phase. Primary emulsions were formulated via homogenization of a 90 wt% SPI dispersion with a 10 wt% oil phase using a high-speed homogenizer (Ultra-Turrax T25, IKA, Staufen, Germany) at 12,000 rpm for 2 min, followed by high-pressure homogenization (AH-2010, ATS Engineering, Brampton, ON, Canada) under 50 MPa pressure and at 50 °C for 2 min. The resulting emulsion was combined with an equal-volume κ-CG solution under continuous stirring (55 °C, 30 min). Final compositions contained 2.5 wt% SPI, 5 wt% oil phase, and 0.25 wt% κ-CG.

4.3. Emulsion Size Distribution and ζ-Potential Measurement

The particle size distribution and ζ-potential of emulsion samples were measured via dynamic light scattering (DLS) using a Brookhaven Instruments Co., Ltd. (New York, NY, USA) analyzer at 25 °C after suitable dilution. For particle size analysis, intensity distributions were derived from measurements conducted at a 90° scattering angle. ζ-Potential values were computed using the Smoluchowski theory.

4.4. Fabrication of Emulsion Gels and Bigels

The freshly prepared emulsion was mixed with 1 M KCl to achieve a final gel concentration of 10 mM KCl. After overnight storage at 4 °C for gelation, KCl-induced emulsion gels/bigels were immersed in twice the volume of 1 wt% GDL solution and incubated at 4 °C. This yielded dual-crosslinked gels/bigels through successive KCl and GDL crosslinking mechanisms.

4.5. Encapsulation Efficiency of Curcumin in the Emulsions and Gels

Curcumin content in emulsified oil droplets, emulsion gels, and bigels was quantified following established protocols [52]. Emulsions were centrifuged at 20,000× *g* for 30 min. The supernatant (0.5 mL) and GDL solution (0.5 mL) used for the gel preparation were mixed with 9.5 mL ethanol to extract curcumin, followed by centrifugation at 10,000× *g* for 10 min. Curcumin concentration in ethanol extracts was determined at 425 nm using a

Shimadzu UV-Vis spectrophotometer (Tokyo, Japan) against a standard curve ($y = 137.73x - 0.0017$, $r^2 = 0.9996$). Encapsulation efficiency was calculated using Equation (1):

$$\text{Encapsulation efficiency } (\%) = \left(1 - \frac{C_a}{C_t}\right) \times 100\% \quad (1)$$

where C_t represents total curcumin content in whole emulsion, emulsion gels, and bigels, C_a represents curcumin content in the supernatant and GDL solution during gelation.

4.6. Water Holding Capacity (WHC)

WHC was determined using an established method [53]. Approximately 3 g gel samples were centrifuged (8000× *g*, 20 min) in 50 mL tubes. Excess water was removed with filter paper, and WHC was calculated using Equation (2):

$$\text{WHC } (\%) = \left(1 - \frac{W_t - W_1}{W_t}\right) \times 100\% \quad (2)$$

where W_t represents the initial gel mass (g), W_1 represents gel mass after centrifugation and water removal (g).

4.7. Texture Profile Analysis (TPA)

TPA was conducted using a TA.XT Plus texture analyzer (Stable Micro Systems, Surrey, UK) equipped with a P-36R cylindrical probe [35]. Cylindrical samples (2 cm diameter × 1.5 cm height) were compressed twice to 50% strain at 1.0 mm/s with a 3 g trigger force. Textural parameters were derived from the resultant force-time curves using Exponent software (Version 8).

4.8. Microstructure Characterization

The microstructure of samples was observed using a TCS SP8 confocal laser scanning microscope (Leica Microsystems GmbH, Heidelberg, Germany). Nile Red (2 mg/mL, ethanol, dyes in oil phase) and Nile Blue (2 mg/mL, distilled water, dyes in protein phase) were added to the emulsion at ratios of 1:50 and 1:100 (*v*/*v*, dye: emulsion), respectively, before the gel preparation. The dyeing gel samples were placed on a slide and covered with a coverslip. At excitation spectra of 633 nm for Nile Blue and 552 nm for Nile Red, images of the regions represented by each sample were taken with 40× (objective lens) magnification using an He-Ne laser.

4.9. Fourier Infrared Spectroscopy (FTIR)

Freeze-dried gel samples were ground with KBr in a 1:100 (*w*/*w*) mass ratio and homogenized to form a uniform powder. FTIR spectral data were collected using a Nicolet iS10 Fourier-transform infrared (FTIR) spectrometer (Thermo Fisher Scientific, Waltham, MA, USA). Spectra were recorded from 4000~400 cm^{-1} at 4 cm^{-1} resolution (32 scans co-added), with background subtraction using pure KBr.

4.10. Intermolecular Force Analysis

Intermolecular forces stabilizing emulsion gels and bigels were characterized using an established solvent-disruption protocol [54]. Gel samples (2.5 g) were immersed in 10 mL of target solvents and continuously dissolved at 25 °C for 3 h. Post dissolution, residual gels were filtered to obtain solubilized fractions. Four solvent systems were employed: pure water, Tris buffer containing 4 mmol/L EDTA·2Na (0.086 mol/L Tris, 0.09 mol/L glycine, pH 8.0; disrupting electrostatic interactions), 2% (*w*/*v*) SDS (disrupting hydrophobic interactions), 8 mol/L urea (disrupting hydrogen bonding), and 1% (*v*/*v*) β-mercaptoethanol (disrupting disulfide bonds). Soy protein concentrations in each solubilized fraction were quantified via the Kjeldahl method (N × 6.25 conversion factor).

4.11. Stability of Curcumin in Emulsion Gels and Bigels

All samples underwent accelerated stability testing at 25 °C and 45 °C for 21 days. The curcumin content in gel samples was determined using previously published protocols [52,55]. Briefly, 0.5 g gel samples were homogenized with 9.5 mL ethanol to extract curcumin, then centrifuged at 20,000× *g* for 30 min. The supernatant was analyzed at 425 nm using UV-Vis spectrophotometry. The stability of curcumin in emulsion gels and bigels was assessed by its retention throughout storage.

4.12. In Vitro Digestion of Emulsion Gels and Bigels

4.12.1. Simulated Gastrointestinal Tract Digestion

Gel samples underwent in vitro digestion following a modified INFOGEST 2.0 protocol [56]. Briefly, 10 g of gel samples were minced and dispersed in 10 mL of PBS (pH 7.0) for 3 min. Subsequently, the resulting mixture was incubated with simulated gastric fluid (SGF) containing 2000 U/mL pepsin at pH 3.0 over a 2 h period. Following SGF incubation, the digesta were combined with 40 mL of simulated intestinal fluid (SIF) containing 100 U/mL pancreatin and 10 mM bile salts, and mixed for 2 h. The pH of the digestive fluid was maintained at 7.0 by incremental addition of 0.1 M NaOH throughout the entire intestinal digestion phase. Quantification of free fatty acid (FFA) release was performed using Equation (3) [57,58]. All digestions occurred at 37 °C with constant agitation (100 rpm).

$$\text{FFA } (\%) = \frac{V_{NaOH} \times C_{NaOH} \times M_{Lipid}}{2 \times W_{Lipid}} \times 100\% \qquad (3)$$

where V_{NaOH} represents volume of NaOH titrant consumed (mL), C_{NaOH} represents molarity of NaOH solution (0.1 M), M_{Lipid} represents average molecular weight of sunflower oil triglycerides (880 g/mol), and W_{Lipid} represents mass of lipids in digested samples (g). The stoichiometric relationship arises from lipase-catalyzed hydrolysis, where each triacylglycerol molecule releases two free fatty acids.

4.12.2. Release Profile of Curcumin During the Digestion

To evaluate curcumin release kinetics from emulsion gels and bigels, digestive aliquots were collected every 30 min during gastrointestinal simulation. Each sample was immediately quenched in ice-water to halt enzymatic activity. Curcumin content in quenched aliquots was quantified per Section 2.5 methodology. Cumulative release was calculated using Equation (4):

$$\text{Release ratio of curcumin } (\%) = \frac{C_t}{C_i} \times 100\% \qquad (4)$$

where C_i represents the initial content of curcumin in emulsion gels or bigels, C_t represents the content of curcumin at time t during digestion.

4.13. Statistical Analysis

Each experiment was conducted in at least triplicate replicates. The results are expressed as mean ± standard deviation (SD). Statistical analysis was performed using one-way analysis of variance (ANOVA) combined with Dunnett's post hoc test, executed via SPSS 20.0 software (IBM Corporation, New York, NY, USA). A p-value less than 0.05 was considered statistically significant.

Author Contributions: Conceptualization, H.C.; methodology, E.T.G.; investigation, E.T.G.; writing—original draft preparation, H.C.; writing—review and editing, W.H., H.C., Z.Z., and L.L.; supervision, H.C. and L.L.; funding acquisition, H.C. and L.L. All authors have read and agreed to the published version of the manuscript.

Funding: This work was supported by Key project of Xinjiang Production and Construction Corps: Quality improvement and efficiency promotion of soybean deep processing and demonstration of high value-added products development (No. 2024AB052) and the National Natural Science Foundation of China (grant number, 32101939).

Data Availability Statement: The original contributions presented in this study are included in the article. Further inquiries can be directed to the corresponding author.

Conflicts of Interest: The authors declare no conflict of interest.

References

1. Kotha, R.R.; Luthria, D.L. Curcumin: Biological, pharmaceutical, nutraceutical, and analytical aspects. *Molecules* **2019**, *24*, 2930. [CrossRef] [PubMed]
2. Luo, M.N.; Han, Y.H.; Chen, Y.L.; Du, H.J.; Chen, B.; Gao, Z.L.; Wang, Q.; Cao, Y.; Xiao, H. Unveiling the role of gut microbiota in curcumin metabolism using antibiotic-treated mice. *Food Chem.* **2024**, *460*, 140706. [CrossRef] [PubMed]
3. Scazzocchio, B.; Minghetti, L.; D'Archivio, M. Interaction between gut microbiota and curcumin: A new key of understanding for the health effects of curcumin. *Nutrients* **2020**, *12*, 2499. [CrossRef]
4. Araiza-Calahorra, A.; Akhtar, M.; Sarkar, A. Recent advances in emulsion-based delivery approaches for curcumin: From encapsulation to bioaccessibility. *Trends Food Sci. Technol.* **2018**, *71*, 155–169. [CrossRef]
5. Mao, L.K.; Lu, Y.; Cui, M.N.; Miao, S.; Gao, Y.X. Design of gel structures in water and oil phases for improved delivery of bioactive food ingredients. *Crit. Rev. Food Sci. Nutr.* **2020**, *60*, 1651–1666. [CrossRef]
6. Zhang, B.; Meng, R.; Li, X.L.; Liu, W.J.; Cheng, J.S.; Wang, W. Preparation of Pickering emulsion gels based on κ-carrageenan and covalent crosslinking with EDC: Gelation mechanism and bioaccessibility of curcumin. *Food Chem.* **2021**, *357*, 129726. [CrossRef] [PubMed]
7. Chen, W.W.; Jin, W.; Ma, X.Y.; Wen, H.B.; Xu, G.C.; Xu, P.; Cheng, H. Impact of κ-carrageenan on the freshwater mussel (Solenaia oleivora) protein emulsion gels: Gel formation, stability, and curcumin delivery. *Gels* **2024**, *10*, 659. [CrossRef]
8. Zhang, L.M.; Zheng, J.Q.; Wang, Y.; Ye, X.Q.; Chen, S.G.; Pan, H.B.; Chen, J.L. Fabrication of rhamnogalacturonan-I enriched pectin-based emulsion gels for protection and sustained release of curcumin. *Food Hydrocoll.* **2022**, *128*, 107592. [CrossRef]
9. Su, J.Q.; Wang, L.L.; Dong, W.X.; Wei, J.; Liu, X.; Yan, J.X.; Ren, F.Z.; Yuan, F.; Wang, P.J. Fabrication and characterization of ultra-high-pressure (UHP)-induced whey protein isolate/κ-carrageenan composite emulsion gels for the delivery of curcumin. *Front. Nutr.* **2022**, *9*, 839761. [CrossRef]
10. Cheng, H.; Chen, W.W.; Jiang, J.; Khan, M.A.; Wusigale; Liang, L. A comprehensive review of protein-based carriers with simple structures for the co-encapsulation of bioactive agents. *Compr. Rev. Food. Sci. Food Saf.* **2023**, *22*, 2017–2042. [CrossRef]
11. Farjami, T.; Madadlou, A. An overview on preparation of emulsion-filled gels and emulsion particulate gels. *Trends Food Sci. Technol.* **2019**, *86*, 85–94. [CrossRef]
12. Guo, Q.; Bellissimo, N.; Rousseau, D. Role of gel structure in controlling in vitro intestinal lipid digestion in whey protein emulsion gels. *Food Hydrocoll.* **2017**, *69*, 264–272. [CrossRef]
13. Li, X.J.; Chen, X.; Cheng, H. Impact of κ-carrageenan on the cold-set pea protein isolate emulsion-filled gels: Mechanical property, microstructure, and in vitro digestive behavior. *Foods* **2024**, *13*, 483. [CrossRef]
14. Hashemi, B.; Assadpour, E.; Jafari, S.M. Bigels as novel carriers of bioactive compounds: Applications and research trends. *Food Hydrocoll.* **2024**, *147*, 109427. [CrossRef]
15. Shakeel, A.; Farooq, U.; Iqbal, T.; Yasin, S.; Lupi, F.R.; Gabriele, D. Key characteristics and modelling of bigels systems: A review. *Mater. Sci. Eng. C-Mater. Biol. Appl.* **2019**, *97*, 932–953. [CrossRef]
16. Lu, Y.; Zhong, Y.; Guo, X.; Zhang, J.; Gao, Y.; Mao, L. Structural modification of O/W bigels by glycerol monostearate for improved co-delivery of curcumin and epigallocatechin gallate. *ACS Food Sci. Technol.* **2022**, *2*, 975–983. [CrossRef]
17. Tian, W.; Huang, Y.; Liu, L.; Yu, Y.; Cao, Y.; Xiao, J. Tailoring the oral sensation and digestive behavior of konjac glucomannan-gelatin binary hydrogel based bigel: Effects of composition and ratio. *Int. J. Biol. Macromol.* **2024**, *256*, 127963. [CrossRef]
18. Xie, D.; Hu, H.; Huang, Q.; Lu, X. Influence of oleogel/hydrogel ratios and emulsifiers on structural and digestion properties of food-grade 3D printed bigels as carriers for quercetin and catechin. *Food Hydrocoll.* **2023**, *144*, 108948. [CrossRef]
19. Liu, G.; Wang, Y.; Yang, J.; Wang, Y.; He, H.; Mao, L. Roles of different polysaccharides on the structures of alginate-based Bigel beads and co-delivery of bioactives. *Food Chem.-X* **2025**, *27*, 102359. [CrossRef]

20. Chao, E.; Li, J.; Duan, Z.; Fan, L. Bigels as emerging biphasic systems: Properties, applications, and prospects in the food industry. *Food Hydrocoll.* **2024**, *154*, 110089. [CrossRef]
21. Kaimal, A.; Singhal, R. Bigels for controlled gastric release of ascorbic acid: Impact on rheology, texture, thermal stability and antioxidant activity. *Food Hydrocoll. Health* **2023**, *4*, 100171. [CrossRef]
22. Yang, X.; Li, A.Q.; Li, D.; Guo, Y.R.; Sun, L.J. Applications of mixed polysaccharide-protein systems in fabricating multi-structures of binary food gels-A review. *Trends Food Sci. Technol.* **2021**, *109*, 197–210. [CrossRef]
23. Cornec, M.; Wilde, P.J.; Gunning, P.A.; Mackie, A.R.; Husband, F.A.; Parker, M.L.; Clark, D.C. Emulsion stability as affected by competitive adsorption between an oil-soluble emulsifier and milk proteins at the interface. *J. Food Sci.* **1998**, *63*, 39–43. [CrossRef]
24. Sakuno, M.M.; Matsumoto, S.; Kawai, S.; Taihei, K.; Matsumura, Y. Adsorption and structural change of β-lactoglobulin at the diacylglycerol-water interface. *Langmuir* **2008**, *24*, 11483–11488. [CrossRef] [PubMed]
25. McClements, D.J.; Jafari, S.M. Improving emulsion formation, stability and performance using mixed emulsifiers: A review. *Adv. Colloid Interface Sci.* **2018**, *251*, 55–79. [CrossRef]
26. Cheng, H.; Fan, Q.; Liu, T.C.; Wusigale; Liang, L. Co-encapsulation of alpha-tocopherol and resveratrol in oil-in-water emulsion stabilized by sodium caseinate: Impact of polysaccharide on the stability and bioaccessibility. *J. Food Eng.* **2020**, *264*, 109685. [CrossRef]
27. Tang, C.H. Emulsifying properties of soy proteins: A critical review with emphasis on the role of conformational flexibility. *Crit. Rev. Food Sci. Nutr.* **2017**, *57*, 2636–2679. [CrossRef] [PubMed]
28. Zhao, S.; Li, K.; Huang, H.; Zhao, P.; Su, S.; Mcclements, D.; Chen, S.; Ma, C.; Liu, X.; Liu, F. Regulation of goat whey protein complex interfacial structures by gum Arabic to improve emulsion performance for curcumin delivery and application. *Carbohydr. Polym.* **2025**, *366*, 123782. [CrossRef]
29. Li, T.; Wang, L. Improved curcumin bioaccessibility in Pickering emulsion fabricated by rice glutelin fibrils. *Food Biosci.* **2023**, *55*, 102988. [CrossRef]
30. Mao, L.K.; Roos, Y.H.; Miao, S. Volatile release from self-assembly structured emulsions: Effect of monoglyceride content, oil content, and oil type. *J. Agric. Food Chem.* **2013**, *61*, 1427–1434. [CrossRef] [PubMed]
31. Bao, H.Y.; Ni, Y.Z.; Wusigale; Dong, H.H.; Liang, L. alpha-Tocopherol and resveratrol in emulsion-filled whey protein gels: Co-encapsulation and in vitro digestion. *Int. Dairy J.* **2020**, *104*, 104649. [CrossRef]
32. Fei-Ping, C.; Bian-Sheng, L.; Chuan-He, T. Nanocomplexation between curcumin and soy protein isolate: Influence on curcumin stability/bioaccessibility and in vitro protein digestibility. *J. Agric. Food Chem.* **2015**, *63*, 3559–3569. [CrossRef]
33. Lin, D.Q.; Kelly, A.L.; Miao, S. Preparation, structure-property relationships and applications of different emulsion gels: Bulk emulsion gels, emulsion gel particles, and fluid emulsion gels. *Trends Food Sci. Technol.* **2020**, *102*, 123–137. [CrossRef]
34. Shen, X.X.; Zheng, H.; Han, M.H.; Xu, X.Y.; Li, B.Y.; Guo, Q. Intermolecular forces regulate in-vitro digestion of whey protein emulsion gels: Towards controlled lipid release. *J. Colloid Interface Sci.* **2023**, *649*, 245–254. [CrossRef]
35. Choi, M.; Choi, H.W.; Kim, H.; Hahn, J.; Choi, Y.J. Mimicking animal adipose tissue using a hybrid network-based solid-emulsion gel with soy protein isolate, agar, and alginate. *Food Hydrocoll.* **2023**, *145*, 109043. [CrossRef]
36. Qin, X.S.; Bo, Q.L.; Qin, P.Z.; Wang, S.F.; Liu, K.Y. Fabrication of WPI-EGCG covalent conjugates/gellan gum double network emulsion gels by duo-induction of GDL and $CaCl_2$ for colon-controlled Lactobacillus Plantarum delivery. *Food Chem.* **2023**, *404*, 134513. [CrossRef]
37. Zhang, H.; Huang, Z.; Guo, P.P.; Guo, Q.N.; Zhang, H.J.; Jiang, L.W.; Xia, N.; Xiao, B.W. Tuning egg yolk granules/sodium alginate emulsion gel structure to enhance β-carotene stability and in vitro digestion property. *Int. J. Biol. Macromol.* **2023**, *232*, 123444. [CrossRef]
38. Jo, Y.J.; Chen, L.Y. Gelation behavior of lentil protein aggregates induced by sequential combination of glucono-δ-lactone and transglutaminase. *Food Struct.* **2023**, *36*, 100312. [CrossRef]
39. Fang, H.C.; Li, J.Y.; Huo, T.Y.; Niu, Y.G.; Yu, L.L. Novel double cross-linked gels of soybean protein isolates and soluble dietary fiber from soybean coats with their functionalities. *Food Hydrocoll.* **2021**, *113*, 106474. [CrossRef]
40. Zheng, H.X.; Mao, L.K.; Cui, M.N.; Liu, J.F.; Gao, Y.X. Development of food-grade bigels based on κ-carrageenan hydrogel and monoglyceride oleogels as carriers for β-carotene: Roles of oleogel fraction. *Food Hydrocoll.* **2020**, *105*, 105855. [CrossRef]
41. Alnaief, M.; Obaidat, R.; Mashaqbeh, H. Effect of processing parameters on preparation of carrageenan aerogel microparticles. *Carbohydr. Polym.* **2018**, *180*, 264–275. [CrossRef]
42. Hu, K.; Huang, X.X.; Gao, Y.Q.; Huang, X.L.; Xiao, H.; McClements, D.J. Core-shell biopolymer nanoparticle delivery systems: Synthesis and characterization of curcumin fortified zein-pectin nanoparticles. *Food Chem.* **2015**, *182*, 275–281. [CrossRef]
43. Guo, Q.; Su, J.; Shu, X.; Yuan, F.; Mao, L.; Liu, J.; Gao, Y. Fabrication, structural characterization and functional attributes of polysaccharide-surfactant-protein ternary complexes for delivery of curcumin. *Food Chem.* **2021**, *337*, 128019. [CrossRef]
44. Sen, M.; Erboz, E.N. Determination of critical gelation conditions of kappa-carrageenan by viscosimetric and FT-IR analyses. *Food Res. Int.* **2010**, *43*, 1361–1364. [CrossRef]

45. Kharat, M.; Du, Z.Y.; Zhang, G.D.; McClements, D.J. Physical and chemical stability of curcumin in aqueous solutions and emulsions: Impact of pH, temperature, and molecular environment. *J. Agric. Food Chem.* **2017**, *65*, 1525–1532. [CrossRef]
46. Puligundla, P.; Mok, C.; Ko, S.; Liang, J.; Recharla, N. Nanotechnological approaches to enhance the bioavailability and therapeutic efficacy of green tea polyphenols. *J. Funct. Foods* **2017**, *34*, 139–151. [CrossRef]
47. Zhao, K.; Hao, Y.L.; Gan, J.L.; Ye, H.Q.; Shen, X. Development of quinoa protein emulsion gels to deliver curcumin: Influence of oil type. *J. Food Eng.* **2025**, *384*, 112260. [CrossRef]
48. Sun, Y.; Qin, R.; Gao, Y.; He, Q.; Sun, R. Influence of monoglyceride on curcumin-loaded nanostructured lipid carriers: Stability, antioxidant activity, digestion behavior, and intestinal absorption. *Food Biosci.* **2025**, *71*, 107150. [CrossRef]
49. O'Sullivan, C.; Davidovich-Pinhas, M.; Wright, A.; Barbut, S.; Marangoni, A. Ethylcellulose oleogels for lipophilic bioactive delivery—Effect of oleogelation on in vitro bioaccessibility and stability of beta-carotene. *Food Funct.* **2017**, *8*, 1438–1451. [CrossRef]
50. Witzleb, R.; Müllertz, A.; Kanikanti, V.; Hamann, H.; Kleinebudde, P. Dissolution of solid lipid extrudates in biorelevant media. *Int. J. Pharm.* **2012**, *422*, 116–124. [CrossRef]
51. Lv, P.; Wang, D.; Dai, L.; Wu, X.; Gao, Y.; Yuan, F. Pickering emulsion gels stabilized by high hydrostatic pressure-induced whey protein isolate gel particles: Characterization and encapsulation of curcumin. *Food Res. Int.* **2020**, *132*, 109032. [CrossRef]
52. Liu, J.; Yang, S.Q.; Liu, J.Y.; Liu, H.Z.; Wang, Z.Y. Preparation of transglutaminase-catalyzed rice bran protein emulsion gels as a curcumin vehicle. *Foods* **2024**, *13*, 2072. [CrossRef]
53. Li, X.M.; Meng, R.; Xu, B.C.; Zhang, B. Investigation of the fabrication, characterization, protective effect and digestive mechanism of a novel Pickering emulsion gels. *Food Hydrocoll.* **2021**, *117*, 106708. [CrossRef]
54. Xu, Q.Q.; Qi, B.K.; Han, L.; Wang, D.Q.; Zhang, S.; Jiang, L.Z.; Xie, F.Y.; Li, Y. Study on the gel properties, interactions, and pH stability of pea protein isolate emulsion gels as influenced by inulin. *LWT-Food Sci. Technol.* **2021**, *137*, 110421. [CrossRef]
55. Zhang, L.; Chen, D.L.; Wang, X.F.; Xu, L.; Qian, J.Y.; He, X.D. Enzymatically modified quinoa starch based pickering emulsion as carrier for curcumin: Rheological properties, protection effect and in vitro digestion study. *Food Biosci.* **2022**, *49*, 101933. [CrossRef]
56. Brodkorb, A.; Egger, L.; Alminger, M.; Alvito, P.; Assuncao, R.; Ballance, S.; Bohn, T.; Bourlieu-Lacanal, C.; Boutrou, R.; Carrière, F.; et al. INFOGEST static in vitro simulation of gastrointestinal food digestion. *Nat. Protoc.* **2019**, *14*, 991–1014. [CrossRef]
57. Zhang, R.J.; Zhang, Z.P.; Zhang, H.; Decker, E.A.; McClements, D.J. Influence of emulsifier type on gastrointestinal fate of oil-in-water emulsions containing anionic dietary fiber (pectin). *Food Hydrocoll.* **2015**, *45*, 175–185. [CrossRef]
58. Liu, F.G.; Liang, X.P.; Yan, J.; Zhao, S.L.; Li, S.Q.; Liu, X.B.; Ngai, T.; McClements, D.J. Tailoring the properties of double-crosslinked emulsion gels using structural design principles: Physical characteristics, stability, and delivery of lycopene. *Biomaterials* **2022**, *280*, 121265. [CrossRef] [PubMed]

MDPI AG
Grosspeteranlage 5
4052 Basel
Switzerland
Tel.: +41 61 683 77 34

Gels Editorial Office
E-mail: gels@mdpi.com
www.mdpi.com/journal/gels

www.ingramcontent.com/pod-product-compliance
Lightning Source LLC
Chambersburg PA
CBHW061444210926
51850CB00008B/1610
9783725858972